QH 366.2 .M397 1993
McKinney, Michael L.
Evolution of life

S0-ASM-046

DATE DUE

MAY 29 1996		
NOV 26 1995		
MAR 24 1996		
DEC - 4 1996		
APR 1 3 1998		
MAR 26 1998		
NOV 1 8 1998		
Dec 14		
DEC 30 1998		
AUG 22 2000		
9/12/00		
SEP - 8 2000		
GAYLORD		PRINTED IN U.S.A.

STOCKTON STATE COLLEGE LIBRARY
POMONA, NEW JERSEY 08240

EVOLUTION OF LIFE

Processes, Patterns, and Prospects

MICHAEL L. McKINNEY

The University of Tennessee

PRENTICE HALL, ENGLEWOOD CLIFFS, NEW JERSEY 07632

STOCKTON STATE COLLEGE LIBRARY
POMONA, NEW JERSEY 08240

Library of Congress Cataloging-in-Publication Data

McKinney, Michael L.
    Evolution of life : processes, patterns, and prospects / Michael
L. McKinney.
        p. cm.
    Includes index.
    ISBN 0-13-292939-2
    1. Evolution (Biology)   2. Human evolution.   3. Life—Origin.
I. Title.
QH366.2.M397   1993
575—dc20   92-40465
                                                                CIP

Acquisitions editor: Raymond Henderson
Editorial/production supervision and
   interior design: Alison D. Gnerre
Cover design: Bruce Kenselaar
Prepress buyer: Paula Massenaro
Manufacturing buyer: Lori Bulwin
Cover art: Mark Hallett, *Dawn of a New
   Day,* gouache, 1984, Collection Mark
Hallett.

 © 1993, by Prentice-Hall, Inc.
A Simon & Schuster Company
Englewood Cliffs, New Jersey 07632

All rights reserved. No part of this book may be
reproduced, in any form or by any means,
without permission in writing from the publisher.

Printed in the United States of America
10   9   8   7   6   5   4   3   2   1

ISBN 0-13-292939-2

Prentice-Hall International (UK) Limited, *London*
Prentice-Hall of Australia Pty. Limited, *Sydney*
Prentice-Hall Canada Inc., *Toronto*
Prentice-Hall Hispanoamericana, S.A., *Mexico*
Prentice-Hall of India Private Limited, *New Delhi*
Prentice-Hall of Japan, Inc., *Tokyo*
Simon & Schuster Asia Pte. Ltd., *Singapore*
Editora Prentice-Hall do Brasil, Ltda., *Rio de Janeiro*

STOKMAN/M.J.C. COLLEGE LIBRARY
POMONA, NEW JERSEY 08240

For my daughter, Jeannie V.,
for the prospects

Contents

PROLOGUE: ABOUT THIS BOOK ix

1   **HISTORY OF EVOLUTIONARY THOUGHT AND ITS RELEVANCE**   1

Overview: Darwin's Profound Impact, 1
History of the Idea of Evolution, 2
Social Impact of Evolutionary Thought, 10
Summary, 18
Key Terms, 19
Review Questions, 19
Suggested Readings, 20

2   **PROCESSES OF EVOLUTION: SELECTION, GENES, AND MICROEVOLUTION**   21

Overview: Evolution by Selecting Variation, 21

PROCESSES OF PHYSICAL EVOLUTION, 22
PROCESSES OF BIOLOGICAL EVOLUTION, 32

Part 1: Selection and Speciation, 32
Part 2: Inherited Variation, 43
Part 3: Evidence for Evolution by Selection of Variation, 53
Part 4: Misconceptions about Evolution and Natural Selection, 66

PROCESSES OF HUMAN EVOLUTION, 69

Summary, 81
Key Terms, 83
Review Questions, 84
Suggested Readings, 86

3   EVOLUTION PAST: HISTORY OF LIFE   87

Overview: From Big Bang to Big Brains, 87

PRODUCTS OF PHYSICAL EVOLUTION, 88

Part 1: History of the Universe, Stars, and Planets, 88
Part 2: History of the Earth, 101

PRODUCTS OF BIOLOGICAL EVOLUTION, 122

The Fossil Record, 122
Part 1: Evolution Toward Multicellular Life, 127
Part 2: Evolution of Life in the Oceans (Hydrosphere), 137
Part 3: Evolution of Life on Land, 152
Part 4: Evolution of Life in the Air, 188

PRODUCTS OF HUMAN EVOLUTION, 202

Human Biological Evolution, 202
Human Cultural Evolution, 213
Summary, 217
Key Terms, 220
Review Questions, 222
Suggested Readings, 224

4 **EVOLUTION PAST: PATTERNS IN THE HISTORY OF LIFE** 226

Overview: Rates and Directions of Evolution, 226

PATTERNS OF PHYSICAL EVOLUTION, 227

Directions and Rates of Physical Evolution, 227
The Sun and Universe, 230

PATTERNS OF BIOLOGICAL EVOLUTION, 230

Part 1: Origination Rates, 232
Part 2: Origination Directions, 248
Part 3: Extinction Rates and Directions, 275
Rates and Directions of Past Extinctions, 277

PATTERNS OF HUMAN EVOLUTION, 296

Human Biological Evolution, 296
Cultural Evolution, 299
Summary, 301
Key Terms, 302
Review Questions, 303
Suggested Readings, 304

5 **EVOLUTION PRESENT & FUTURE: EXTINCTIONS, BIOTECHNOLOGY, AND ?** 305

Overview: Extrapolating the Past, 305

PROSPECTS FOR PHYSICAL EVOLUTION, 308
PROSPECTS FOR BIOLOGICAL EVOLUTION, 318

Overview, 318

PROSPECTS FOR HUMAN EVOLUTION, 355

Overview: We Will Evolve; The Question is "Where"?, 355
Summary, 370

Key Terms, 372
Review Questions, 373
Suggested Readings, 374

6   **EPILOGUE: SOME PERSONAL AND SOCIAL IMPLICATIONS**   375

Basic Facts and Interpretations, 375
Social and Personal Implications, 377

GLOSSARY   381

ADDITIONAL CREDITS   395

INDEX   397

Prologue:
About this Book

RELEVANCE OF EVOLUTION

In a world increasingly concerned with short-term problems, learning about evolution would apparently rate as a low priority indeed. Dinosaurs, for example, may be entertaining, but their relevance may seem questionable given the pressing needs of today.

But let us look beyond the short-term view. Knowledge about evolution can lead to an enormously better understanding about both the present and the future. Regarding the present, nearly everyone wonders how they got here and how the world became the way it is. Without knowledge of evolution, you would be in the same position as our distant ancestors, driven to ponder such questions in an informational vacuum. Furthermore, there seems to be no end to the variety of enlightening answers that evolutionary knowledge can give us: Why are porpoises so smart? Why do we sleep about eight hours per day? Why are there millions of species on earth? Why do most of these live in the tropics? Why is sugar sweet? Why must we age and die? Any person struggling with the purpose and meaning of life will soon find himself or herself confronting such questions.

Knowledge of evolutionary processes can also tell us about the future. For example, by knowing that evolution occurs from selection on genes, you can see that human manipulation of genes in the laboratory (genetic engineering) can give us direct control over the evolution of many species, including ourselves. As another example, we know from the fossil record that widespread, "weedy" species tend to avoid extinction more often than localized species. This will give you some insight into what species may survive in the future if species extinctions continue to accelerate from human destruction of habitats.

GOAL OF THIS BOOK: A BROAD VIEW OF EVOLUTION

The goal of this book is to introduce the principles of evolution to the non-scientist. Therefore, I have tried to use an informal style that conveys some of the enthusiasm that many scientists feel toward the subject. Regarding content, this book attempts a much broader view than most books on evolution. It is broader in two ways. First, some discussion of physical and cultural evolution is included. Physical evolution is critical for understanding how life and its physical environment arose. Cultural evolution is important for a better understanding of ourselves. Furthermore, both physical and cultural evolution are continuous with biological evolution since life originated with physical processes and cultural evolution arose from biological processes. Therefore, attempts to understand the evolution of life while excluding these are incomplete and artificial.

PROCESSES, PATTERNS, AND PROSPECTS. The second way that this book presents a broader view is by examining the processes, patterns, and prospects of evolution. Most books about evolution focus on either patterns (history of life, emphasizing fossils) or the processes (on-going evolution, emphasizing genetics). As for the future, most evolution books give it a cursory treatment, if any at all.

However, such treatments reflect the biases of our educational system more than the reality of evolution. Past, present, and future evolution are arbitrary categories, determined solely by our own current position on what is really a continuous arrow of time. Evolution will occur just as surely in the future as it did in the past. And the future, especially to young people, is understandably of more interest. Besides, the most useful virtue of science is prediction and discussion of the future is a stimulating way to exercise one's understanding of concepts. Similarly, to focus only on the processes of the present is to ignore that those processes also operated in the past where their products can be directly observed to illustrate important principles. That over 99% of all species to have ever lived are now extinct succinctly shows just how much is omitted by the many evolution texts that give scant attention to the fossil record.

ACKNOWLEDGMENTS

I am indebted to the following reviewers for their thoughtful and useful remarks: Robert L. Anstey, Geology, Michigan State University; David G. Davis, Biology, University of Alabama; Bruce A. Fall, Biology, University of Minnesota; Douglas

S. Jones, Florida Museum of Natural History; Joel Kingsolver, Zoology, University of Washington; Arnold I. Miller, Geology, University of Cincinnati; Donald Prothero, Geology, Occidental College; Robert M. Schoch, Basic Studies, Boston University; Philip W. Signor, Geology, University of California, Davis; and A. Spencer Tomb, Biology, Kansas State University. I want to especially thank Arnie Miller for his exceptionally thorough, balanced, and clear-headed review.

The people at Prentice-Hall made this book possible: Mary Deluca, Holly Hodder, David Brake, and especially Ray Henderson and Alison Gnerre.

For help with the review questions, figure captions, glossary, and other tasks, I thank Stephanie Duncan and Mary Ruth Brewer for their industry and persistence.

Much of this book was written during support from the American Chemical Society, Grant #ACS-PRF 22635-AC8.

1

History of Evolutionary Thought and Its Relevance

So God created man in his own image. . . .
Genesis 1:27

To sum up:
1. *The cosmos is a gigantic flywheel making 10,000 revolutions a minute.*
2. *Man is a sick fly taking a dizzy ride on it.*
3. *Religion is the theory that the wheel was designed and set spinning to give him the ride.*

H. L. Mencken, Smart Set, *1920*

OVERVIEW: DARWIN'S PROFOUND IMPACT

Darwin's discovery that humans evolved from "lower" animals sent a shock wave through nineteenth-century society. Copernicus had already delivered the first humbling blow that earth was not at the center of the universe. Now Darwin showed that even our special place as inhabitants on earth was misconceived. We had not only originated from apes, but the process of natural selection that created us showed no evidence of conscious design. To many (as noted in the Mencken quote), we had suddenly gone from a divine being, with a special origin and purpose, to an insignificant "cosmic accident."

The impact of this blow to the human self-image can hardly be overstated. Darwin's book *On the Origin of Species* documented his discovery and is one of the most influential books in the history of humankind. Many long-cherished religious and philosophical tenets had to be discarded. Many basic values that supported the major moral and ethical rules governing society were called into question. Even practical affairs such as business and politics were affected. The "laws of evolution" were used to justify everything from imperial national aggression to unrestricted business practices. Literature, art, and virtually every sphere of the human condition was affected by the Darwinian Revolution.

Darwin's ideas continue to reverberate through society. Many of the long-held beliefs disproven by Darwin have not been easy to replace. Why are we here? Where are we going? These questions ultimately motivate every human endeavor. The facts of evolution, which we will outline in this book, certainly cannot provide simple answers to these questions. However, knowledge of evolution has provided far more insight than was gained in the thousands of years preceding its discovery. This is because scientific knowledge, by focusing on what can be proven, creates boundaries on the otherwise boundless human imagination.

HISTORY OF THE IDEA OF EVOLUTION

Evolution comes from the Latin word meaning "to unroll." In a general sense it refers to any change through time, but it is often restricted to biological change. (This lack of precise definition is one reason why Darwin himself preferred the phrase "descent with modification" for biological evolution.) In today's world, we take change for granted. However, aside from a brief period in Greek philosophy influenced by the teachings of Anaximander, Democritus, and Empedocles, for most of human history the dominant view was of an unchanging universe. This, in part, was due to the human condition of earlier times: up through the Middle Ages there was little change in society or technology from one generation to the next. It seemed reasonable therefore that the universe and the natural world had also always been the same, at least since the divine Creation taught by the Judeo-Christian

Man
Monkeys
Quadrupeds (Mammals)
Bats
Ostrich
Birds
Aquatic Birds
Flying Fish
Fish
Eels
Sea Serpents
Reptiles
Slugs
Shellfish
Insects
Worms
Polyps (Hydras)
Sensitive Plants
Trees
Shrubs
Herbs
Lichens
Mold
Minerals
Earth
Water
Air
Etherial Matter

Figure 1–1. A simplified version of The Great Chain of Being published in 1764. Modified from Peter Bowler, *Evolution: The History of an Idea* (Berkeley: University of California Press, 1989), p. 61.

tradition. It was acknowledged that there was a progression in nature, from simple nonliving things such as rocks, to simpler life forms such as worms, and up through "higher" animals to humankind. This **great chain of being** was thought to reflect the natural order as God created it (see Figure 1–1). Thus, the progression was not an evolutionary one but was static. The ranks on this chain represented steps toward "godliness" or some similarly immeasurable quality. Humans had more of this than insects, for example, but had less than the angels ranked above them. In its most extreme form even human races were ranked on the chain. (Not surprisingly, Europeans, who originated the chain, were usually placed as the highest race.)

Birth of Modern Evolutionary Thought

Beginning in the 1500s, the European Renaissance caused thinkers to question this view of an unchanging world. There was much social upheaval, as reflected by the French and American Revolutions. This was largely caused by rapid technological change which, in turn, created new social classes and social migrations. Such upheavals of the old order made the idea of change in general more apparent. Of critical importance was the rise of the **scientific method,** which relies on observed facts to generate hypotheses. These hypotheses are tested by gathering more facts. This replaced personal revelation and unquestioning reliance on dogma that typified previous views of the world.

DISCOVERY OF PHYSICAL EVOLUTION. The discovery of physical laws by Isaac Newton and others in the 1600s paved the way for the idea that the physical universe (such as the stars and earth) had evolved. They showed that a few simple forces, such as gravity, governed the motion of all large objects, both on earth and in the heavens. It was not long before scientists, such as Laplace, applied these laws to suggest that earth and the other planets condensed from a giant dust cloud and evolved into the solar system we know today.

The evolution of the earth was first accurately described in the late 1700s and early 1800s by the geologists James Hutton and Charles Lyell. Using detailed observations and measurements, such as how long it takes to erode a mountain and for a river to cut a deep gorge, they were able to show that the earth was probably many thousands, and perhaps even millions, of years old. (We now believe it to be over 4.5 billion years old.) This method of using present day processes to explain past events is called **uniformitarianism,** often summarized as "the present is the key to the past." This great age stood in stark contrast with religious beliefs at that time, which saw the earth as only about 6,000 years old. This was based on Archbishop Ussher's estimate, based on the number of generations ("begats") in the Bible, that the earth was created in 4004 B.C. Early geologists also cast doubt on biblical interpretations of past events. For instance, they showed that throughout history many great floods had covered entire continents on earth, not just the one Great Flood of Noah.

Discovery of Biological Evolution

Many intellectuals of the late 1700s and early 1800s believed that life had evolved along with the earth. A major reason was the emerging realization that fossils were the remains of long-dead species. Previously, fossils had usually been explained by any number of creative ideas. For instance, extinct mammoths were thought to be the remains of Hannibal's elephants from his march on ancient Rome. A fossil amphibian was said to be the remains of the one-eyed mythical god, Cyclops, because its skull cavity was mistakenly identified as an eye socket. However, the most common explanation for most fossils was that they were minerals that nature had "shaped" by some mystical process. Ultimately, detailed comparisons of fossils to living species, especially by the great French anatomist Georges Cuvier in the early 1800s, clearly showed that fossils were dead relatives of living species. This led to the evolutionary interpretation that ancestral fossil species had somehow been transformed into the living ones.

Even after fossils were recognized to be relatives of living species, many people refused to accept the reality of the death (extinction) and origin (speciation) of species. For example, it was suggested that fossil species still lived on in unexplored areas of the earth. Thomas Jefferson ventured that extinct North American elephants (known from fossils) would be found in the unexplored northwest territories of the United States. The main reason for this lack of consensus was that the evidence for evolution, such as the fossils, was strictly circumstantial. Until there was a proven mechanism explaining *how* evolutionary change occurred, alternative interpretations for any evidence were inevitable. Unfortunately, most early speculations about mechanisms were not testable and therefore could not be proven. For example, many early theories were based on religious tales of ancient catastrophes that caused new sets of species to be created. (In fact, most fossil deposits do superficially resemble this because erosion often removes overlying transitional fossils.) This idea of **catastrophism** had many proponents in the early 1800s (including Cuvier), but it was greatly hampered by its inability to explain the creation of new species. Many catastrophists simply resorted to supernatural creation, with Noah's flood being one of a number of catastrophes.

The Frenchman Jean Baptiste **Lamarck** is usually credited with popularizing the first *testable* mechanism for evolution in 1809. This is the theory of **inheritance of acquired traits.** According to this theory, giraffes evolved long necks because they stretched their neck muscles to reach leaves higher on the tree. The length gained during the parents' lifetime was then passed on to the offspring (Figure 1–2). However, while this idea is important as a first testable theory, it is easily shown to be false. Traits acquired during an individual's lifetime are not passed on to offspring. If a person works hard at weightlifting, it does not mean his or her offspring will be stronger. Or, as shown experimentally, if one cuts off the tails of mice, for even hundreds of generations, the offspring will still be born with tails. However, it was not until the 1880s that August Weisman showed why this is so: The germ cells (sperm and egg) are isolated from the cells comprising the rest of

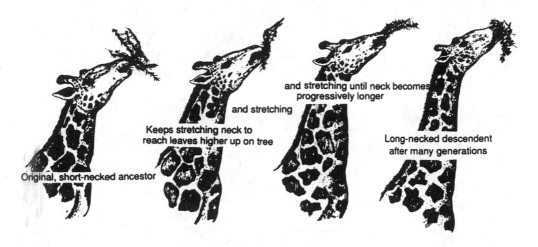

and stretching until neck becomes
progressively longer

and stretching

Keeps stretching neck to
reach leaves higher up on tree

Long-necked descendent
after many generations

Original, short-necked ancestor

Figure 1–2. The theory of inheritance of acquired traits stated that longer necks acquired through stretching were passed on to the offspring. From R. Wicander and J. Monroe, *Historical Geology*, West Publishing, 1989, p. 115.

the body. For instance, the genetic material in sperm cells in the testicles is normally unaffected by what happens elsewhere in the body.

DARWIN DISCOVERS NATURAL SELECTION. It was not until 1859 that **Charles Darwin** first publicly proposed the true cause of biological evolution in his book *On the Origin of Species by Means of Natural Selection: Or the Preservation of Favoured Races in the Struggle for Life* (usually referred to as *On the Origin of Species*). These ideas originated from his keen observations while on his famous voyage around the world (1831–1836) on the H. M. S. Beagle. However, Darwin's rather shy personality and extreme commitment to documenting all available facts led him to withhold formal publication of his ideas until 1859. Even then it seems he was only galvanized into publication by a letter from another biologist, A. R. Wallace, who had independently formulated the same ideas. Darwin's detailed observations of living species, fossils, embryos, and artificial breeding were essential in convincing people of the validity of his discovery.

Darwin's discovery is commonly called the **theory of natural selection.** We will discuss this theory in detail in Chapter 2, so just a summary is presented here. Three basic observations led Darwin to his conclusions:

1. Individuals compete for space, food, and other resources.
2. Individuals produce more offspring than can survive.
3. Individuals vary in their traits.

Darwin used these observations to reason as follows. Due to limited resources (1), only some offspring can survive (2). Therefore, those variations (3) that aid in

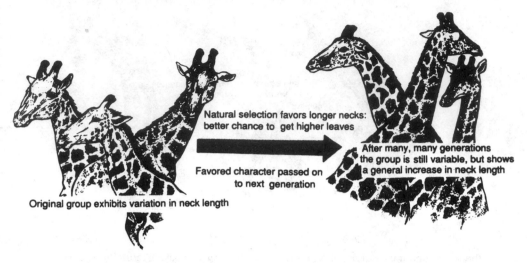

Natural selection favors longer necks: better chance to get higher leaves

Favored character passed on to next generation

After many, many generations the group is still variable, but shows a general increase in neck length

Original group exhibits variation in neck length

Figure 1–3. The theory of natural selection states that giraffe variants with longer necks tend to have more offspring because they were favored by natural selection. From R. Wicander and J. Monroe, *Historical Geology*, West Publishing, 1989, p. 118.

survival will tend to be passed on more often because individuals with them survive and reproduce. Thus, we say that the most advantageous traits are "selected by nature" (giving it the name natural selection). Evolution occurs from this process because the environment ("nature") is always changing. Therefore, traits selected (favored) by the new environment are changing too: if it becomes very warm, a heavy coat of fur now becomes a disadvantage. Or, to compare to Lamarck's example cited earlier, suppose environmental change killed off some of the tree species serving as food for giraffes. Increased competition for available leaves might then favor individuals with even slightly longer necks. This trait would tend to increase in the population through time (Figure 1–3).

A more compact way of expressing this process is:

variation + environmental change (selection) = evolution

If sufficient environmental change occurs, then many new variations will become favored in the population. As these new variations accumulate, the population as a whole will become altered. For instance, the change in climate in our example would lead to widespread fur loss, along with many other trait changes. Ultimately, enough changes would occur so that a new species would be created. This process is not more visible because it is so slow compared to the human lifespan.

Darwin's theory was widely read and debated, even by the general public. Because it was so well argued and documented, it was generally accepted as true, just as it is today (although, as we shall see, much has been added to it). Yet even though it has successfully withstood many attacks on its validity, Darwin's theory continues to be misunderstood by many people. Most of these misunderstandings are reviewed in Chapter 2, but let us briefly note two common ones related to the history and name of Darwin's theory. First, Darwin did not discover evolution—it was identified by many people before him. His contribution is that he was the first to discover and document the *processes* that cause evolution to occur.

Second, the label "theory" is misleading. While we call it the theory of natural selection, natural selection has been been proven so many times and in so many ways that it is more properly viewed as an *established fact*. In addition, the theory of natural selection is about much more than just natural selection. Referring to the "equation" above, recall that

$$\text{variation} + \text{selection} = \text{evolution}$$

In order for environmental change (selection) to cause evolution, there must be variation in the population to start with. For example, giraffes must vary in neck length if selection is to favor longer necks. Therefore, a more complete name would be the "theory of evolution by natural selection of variation." This unfortunate mislabeling of Darwin's work is no accident. It reflects Darwin's own lack of knowledge of how variation is produced and inherited. As we see next, it was not until after Darwin's death that the rules of inherited variation became widely known.

DECLINE OF DARWIN'S IDEAS: LATE 1800s–EARLY 1900s. While Darwin's theory gained wide acceptance shortly after its public presentation in 1859, the problem of the origin and inheritance of variation was seized upon by many critics. By the late 1800s, Darwin's ideas were not accepted in many circles, largely because of this. As if to overcompensate for Darwin's focus on the importance of selection, new theories began to emerge that emphasized variation as the more important force causing evolution. These are often called *internalist* theories—they suggest that variation arises within the species, while selection comes from the environment, which is external to the organism.

There were many internalist theories, but we shall only mention two of the most influential. One was the **orthogenetic school.** Orthogenesis roughly means "straight-line." This school of thought was especially popular among paleontologists because study of the fossil record often showed trends in many lineages. For example, horses have increased in body size and lost a number of toes over the last 40 million years. We now know that this trend was complex, involving many horse species, including some that actually decreased in size. However, early paleontologists, with incomplete data, produced simplistic diagrams that indicated these horse trends formed a straight line, as shown in Figure 1-4. This led to the internalist notion that such trends were somehow "programmed" into the organisms. At first, a vaguely described, somewhat mystical force was thought to be the cause. After genes were discovered in the early 1900s, it was thought that the genes themselves were somehow programmed to carry out such trends. For example, the early horses

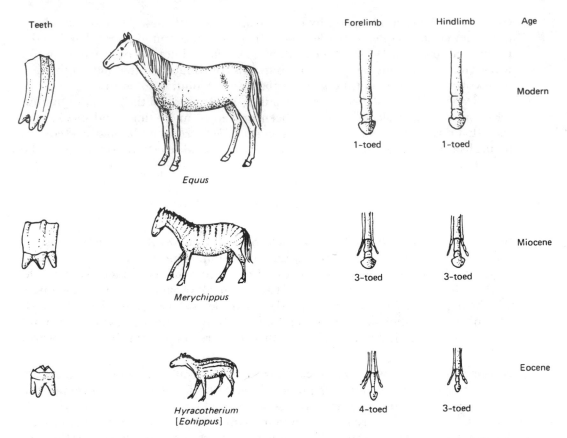

Figure 1–4. Orthogenetic ("straight-line") evolutionists visualized horse evolution as a directional sequence of changes. Modern evolutionists now know that such depictions as this are greatly oversimplified. In reality, such trends involved many branchings, deviations, and reversals on the evolutionary tree. From F. Racle, *Introduction to Evolution* (Englewood Cliffs, Prentice-Hall, 1979, p. 71).

would have had, in their genes, a program that predestined their descendants to grow larger and lose toes.

A second influential internalist school arose when the laws of variation were discovered in the early 1900s. Genes were found to be the source of variation and inheritance that Darwin had sought and the science of genetics grew rapidly. (Ironically, a monk, Gregor Mendel, had discovered the laws of genetics while Darwin was alive. Unfortunately, the laws were published in an obscure journal, an unread copy of which was in Darwin's library.) Geneticists soon found that mutations are the ultimate source of variation—they are spontaneous changes in the genetic "blueprint." This led to the **mutationist school,** popular among many geneticists. They held that evolution occurred whenever mutations caused new forms to arise. Because mutations could be drastic, a new species could be "instantly" created from them. For instance, some mutationists believed that the first bird hatched from a reptile egg.

RENEWAL AND REFINEMENT OF DARWIN'S IDEAS: 1930s–NOW. As data from genetics, paleontology, anatomy, and other areas of biology accumulated, it became clear that the various non-Darwinian schools were less accurate than Darwin's original ideas. We now know that drastic mutations are relatively rare. Instead, small mutations provide the raw material that selection acts upon. Thus, both mutations (variation) and selection are necessary for evolution, as the "equation" cited on page 6 indicated. The error of the mutationist school was in omitting the first half of the equation, asserting that mutations alone would drive evolution. Similarly, the orthogenetic fossil trends, such as in horses, are now known to be gross oversimplifications, as we will discuss in Chapter 4.

Therefore, by the 1930s the focus shifted back to Darwin's theory of natural selection as the centerpiece for explaining how evolution occurred. This renewal of Darwin's ideas is called the **modern synthesis** because new information gleaned from the various fields (genetics, paleontology, anatomy, and others) was synthesized under the theory of natural selection. Of course many refinements have been made as new information accumulates. For instance, Darwin's ignorance of variation has been greatly supplemented by the knowledge of population genetics and the biochemistry of genes (DNA). Similarly, increasing documentation of the fossil record has revealed many specific events and patterns in the history of life that "flesh out" Darwin's ideas. Indeed, we can now see that the various schools of thought that developed after Darwin's theory was presented were an important part of the testing process.

The refinements, beginning with the modern synthesis, have done much more than simply improve the technical correctness of Darwin's ideas. They have greatly modified our perception of evolution in two important ways. Darwin saw evolution as a generally *gradual* and *deterministic process*, whereas many modern evolutionists see a greater role for *rapid change* and *chance*. Darwin's "gradual" view arose because he saw natural selection as sorting small variations, such as minor differences in neck length. Evolution was characterized as a slow process, taking perhaps millions of years to create a new species. This is not wrong, because evolution undoubtedly does sometimes occur this way. However, new evidence indicates that evolution can also be relatively rapid, such as when small populations are subjected to intense environmental change. The greater role for chance has become popular because of recent evidence in paleontology. Global catastrophes have occurred a number of times in the history of life, extinguishing over 50% of all species. Such mass extinctions often do not operate in the deterministic manner that Darwin envisioned. Instead of favoring the individuals that are the most finely adapted to the local environment—the "fittest"—catastrophes are "random" because they annihilate virtually every living thing in the devastated area. Thus, survival becomes largely a matter of chance. Again, Darwin was right, for most of the time evolution *is* a generally deterministic process of adapting species to the local environment (by sorting variation). However, rare catastrophes are now thought to "reset" this background process by the introduction of chance. For example, humans would likely not exist today if the dinosaurs had not been eliminated by the impact of a huge meteorite with the earth (see Chapter 3).

Neither of these two refined views are new in concept. An expanded role for both rapid change and chance were suggested in Darwin's time. Darwin's advocate and friend, T. H. Huxley, criticized him for overemphasizing the gradual view. Similarly, recall how Cuvier had promoted catastrophism as a view of how evolution occurred. However, it is only since the 1970s that these views have been supported by substantial amounts of testable scientific evidence.

THE DISCOVERY OF CULTURAL EVOLUTION. The idea of cultural evolution soon followed behind that of biological evolution. This was a logical step because the same expeditions that brought back descriptions of primitive plants and animals also described "primitive" cultures, such as the aborigines of Australia and the American Indian. As a result, in the late 1800s a group of specialists, called anthropologists, began to study humans in an evolutionary framework. One of the most famous was Lewis Henry Morgan, who proposed that all cultures evolve through a three-stage sequence of **savagery, barbarism,** and **civilization.** Aborigines, American Indians, most Africans, and other indigenous societies were seen as cultures that had not yet evolved to the "higher" stage. We now know that Morgan was wrong in assuming that all cultures go through exactly the same stages. We also know that Morgan's ideas suffered from the same cultural biases found in the great chain of being. Evolution was often seen as a "ladder" on which nineteenth-century European society was at or near the top. Nevertheless, we should not let these excesses obscure the fact that human culture in general has tended to become more complex compared to our earliest tool-using ancestors. The details of how this occurred are still a topic of much debate in the social sciences, but there is no doubt that the basic cause was increasing brain size. Culture is transmitted by language, writing, and other types of learned information, as we will discuss in Chapter 2.

SOCIAL IMPACT OF EVOLUTIONARY THOUGHT

Darwin's ideas spread far beyond the limits of science, causing people to question many fundamental values. Efforts to revise, replace, or justify those values led to widespread debates and intellectual soul-searching that continue today. While many aspects of thought were affected, we focus on three of the most notable: (1) religion, (2) philosophy and ethics, and (3) business and politics (Figure 1-5).

Religion: Challenge to Four Traditional Views

In his masterful book, *Evolution: The History of an Idea* (1989), Peter Bowler discusses four major traditional religious views that were challenged by Darwin's ideas. These traditional beliefs and their challenges are listed as follows.

1. *Age of the Earth.* We have already seen how an age of about 6,000 years was estimated from a literal interpretation of the Bible. This was questioned with the rise of uniformitarianism in the early 1800s. The gradual pace of evolution (often millions of years) as predicted by Darwin tended to confirm this. Modern dating of radioactivity in rocks shows that the earth is over 4.5 billion years old.

2. *Static Universe.* Traditional thought held that nature's hierarchy had not changed since its creation by God, as shown by the great chain of being. Darwin's evidence for natural selection showed that evolutionary change accounted for the hierarchy because simple forms evolved into more complex ones.

3. *Conscious Design of the Universe.* A purposeful Creator was a crucial part of the traditional religious view. Darwin's view implied that life had evolved by the natural process of environmental selection acting on "random" mutations.

4. *Man Apart from Nature.* The Bible says that man was formed in the image of God and given dominion over nature. Moreover, man alone had a soul. Natural selection, along with fossil and anatomical evidence, showed that humans evolved from apes. This has been verified by many modern methods. Perhaps the most direct evidence is from DNA analysis: We share over 98% of our genes with chimpanzees.

HOW RELIGION REACTED TO EVOLUTION. Different religious groups have responded differently to these four challenges. Not surprisingly, the initial reaction was often outright denial, which resulted in many famous debates between evolutionists and traditionalists. T. H. Huxley, often called "Darwin's bulldog," was especially noted for his debating prowess against those who disputed Darwin's ideas. The evidence and logic presented by Darwin were so persuasive that within a matter of years, most "mainstream" religious groups modified their beliefs to allow for evolution via natural selection. They adopted the new views without changing their basic tenets. For instance, the majority of Judeo-Christian groups (Jews, Catholics, and many Protestants) now accept the great age of the earth, that life has changed via natural selection, and that humans evolved from apes (views 1, 2, and 4 above). They see evolution through natural selection as the mechanism by which God created humans. The great age of the earth is accepted because the Bible is not seen as

Figure 1–5. Summary of some of the major impacts of Darwin's ideas on society.

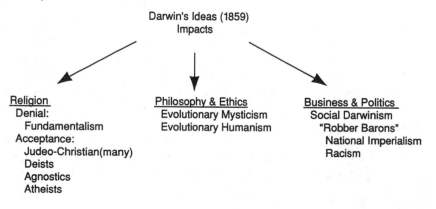

a document for literal historical interpretation. Rather, it is seen as an inspirational document that contains many allegories and metaphors.

The third traditional view is the one view that continues to be retained by Judeo-Christian groups. This is hardly surprising given that belief in a personalized God that created and cares for the universe is at the very heart of Judeo-Christian tradition. The best-known attempt to logically validate this belief in the face of natural selection was the *argument from design*. According to this argument, if you find a complex item, such as a watch, that has been clearly designed to function in a certain way, then such an item must have a designer. Therefore, living organisms, which are also designed to function as they do, must have a designer. The flaw in this argument is that the designer need not be a conscious, personalized God. Indeed, Darwin had already shown that the designer was natural selection. By favoring random variations (mutations) that improved an organism's way of doing things, it inevitably designed the organism to function in certain ways. Richard Dawkins, in his book *The Blind Watchmaker* (1986), details how natural selection is like a "watchmaker." It is "blind" because natural selection (unlike an all-seeing God) cannot predict the future, so that it favors only those traits that give immediate advantage. The result is that evolution, instead of following a predestined course, often shows winding pathways of change, with many reversals, and sometimes leads to evolutionary dead ends. Of course, just because natural selection is the visible process that causes evolution, it does not prove or disprove the existence of a conscious Creator ("God").

The reactions of the Judeo-Christian groups just discussed are only some of many possible reactions to Darwin's ideas. **Fundamentalism** generally refers to religious groups that prefer more traditional beliefs. For example, many Fundamentalist Christians believe that the Bible should be interpreted literally. Thus, they have refused to accept the loss of *all four* major traditional views, continuing to believe that the earth is only 6,000 years old, that humans did not evolve, and so on. Fundamentalism has persisted since Darwin, but in recent years has seen a revival in many countries. Nearly *half* of Americans hold this view (Table 1–1). Fundamentalists have had a widely discussed impact in their attempts to have **creationism** taught in public schools as an alternative to evolutionary theory. While creationism initially had some success in having the biblical interpretation of origins discussed in the texts and public schools of some states (such as Arkansas and Louisiana), recent court decisions have reversed most of these. This was done on the grounds that there is no acknowledged scientific basis for this literal interpretation. Creationist views are strongly correlated with lower educational levels (Table 1–1).

In addition to Judeo-Christian groups, many other religions also believe in a personalized God. Examples include Islamism and Hinduism (the latter being *polytheistic*, with many Gods). Along with Judeo-Christian groups, these are all classified as **theism,** the belief that a personalized God exists. Table 1–2 shows the most common categories of beliefs about God. **Deism** is the belief that there is an impersonal God who created the universe and its natural laws and then, like a clock, let it "run." An impersonal God would not guide or plan for everything that happens

TABLE 1–1. American Beliefs About Evolution, Determined by a Gallup Poll, November 21–24, 1991

STRICT CREATIONIST VIEW God created man pretty much in his present form at one time within the last 10,000 years.		CENTRIST VIEW Man has developed over millions of years from less advanced forms of life, but God guided this process, including man's creation.		NATURALIST VIEW Man has developed over millions of years from less advanced forms of life. God had no part in this process.	
All Americans	47%	All Americans	40%	All Americans	9%
Men	39%	Men	45%	Men	11.5%
Women	53%	Women	36%	Women	6.6%
College graduates	25%	College graduates	54%	College graduates	16.5%
No high-school diploma	65%	No high-school diploma	23%	No high-school diploma	4.6%
Income above $50,000	29%	Income above $50,000	50%	Income above $50,000	17%
Income below $20,000	59%	Income below $20,000	28%	Income below $20,000	6.5%
Whites	46%	Whites	40%	Whites	9%
Blacks	53%	Blacks	41%	Blacks	4%

Note: percentages total less than 100% because a number of people answered "I don't know".

at every moment. Deism developed in response to the discoveries of Newton, Darwin, and others, which described the "mechanisms" of natural events (such as planetary orbits and evolution). Thus, it was natural that a view of the universe as a great ("unattended") machine should arise, challenging the older beliefs in a personal God that formed when humans were unaware of what caused many natural processes. Deism today thrives in the form of numerous naturalistic religions—in particular, pantheism is common. Pantheism is essentially the belief that God is present in all things. Some prominent scientists have recently incorporated deistic ideas into the well-publicized theory of the **anthropic principle** ("anthropos" = human). According to this principle, if any of the most basic physical or chemical forces of the universe were altered even the slightest fraction, then life, and even planets or stars, would be impossible to form. For example, if the attraction between an electron and a proton were increased or decreased, stars would not condense from a cosmic dust cloud, nor would chemical interactions lead to cellular organisms. Thus, the anthropic principle states that, because all of these physical and chemical param-

TABLE 1–2. Some Basic Categories of Belief in God

RELIGIOUS CATEGORY	BELIEF IN GOD
Theism	Personalized God
Deism	Impersonal God
Agnosticism	Don't know
Atheism	No God

TABLE 1–3. Number of Agnostics (Nonreligious) and Atheists in the World

	NONRELIGIOUS	ATHEIST	TOTAL	%
East Asia	618,900,000	123,400,000	742,300,000	73.0%
USSR	83,100,000	60,600,000	143,700,000	14.1
Europe	49,400,000	17,400,000	66,800,000	6.6
South Asia	18,400,000	5,100,000	23,500,000	2.3
North America	19,000,000	1,000,000	20,000,000	2.0
Latin America	12,900,000	2,400,000	15,300,000	1.5
Oceania	2,900,000	500,000	3,400,000	0.3
Africa	1,300,000	100,000	1,400,000	0.1
Total	805,900,000	210,500,000	1,016,400,000	100.0%

Source: *Universal Almanac*, 1990.

eters *are* precisely attuned to forming a universe where life and planets apparently form quite easily, there must be some conscious design behind it all. Opponents of this argument point out that there is much circular reasoning here—if these parameters did not exist, then we would not be here to think about it. Thus, there may be (or have been) many other universes, with different properties, that exist without life.

Agnosticism (literally, "no knowledge") (Table 1–2) simply means that knowledge of God's existence is impossible given the available information. T. H. Huxley coined the word and was a major proponent of this view. Finally, **atheism** (literally, "no god") is the belief that God does not exist. However, neither of these seem very popular in most of the world, including the United States. As shown in Table 1–3, only 2% of the people in North America classified themselves as nonreligious (agnostic) or atheistic. Of the 98% that believe in God, only 60% claim affiliation with any particular denomination, with the vast majority of these being Christians (94%) and Judaists (4%). The largest number by far of agnostics and atheists live in the former USSR and China (East Asia).

Philosophy and Ethics: Search for Guidance from Nature

The reexamination of religious beliefs caused by Darwin led to uncertainty beyond the existence of God. Views on traditional values, rules governing human behavior, and even the purpose of life began to be questioned. In particular, the questioning of religious absolutes turned many thinkers to natural laws for guidance. The reasoning, which is still common today, was as follows: since humans evolved from nature by natural processes, we can look to natural laws for "lessons." In short, "Mother Nature knows best." Unfortunately, for those who want simple answers, this reasoning is rarely valid. Let us look at specific examples.

The most common natural "lesson" cited by observers is that of *progress* in evolution. According to this view, the fossil record indicates that natural selection has caused life to experience many kinds of progressions. Progressions in organization (from simple to complex) and size (from small to large size) can be cited, and the evolution from the first simple bacteria to the complex, large multicellular organisms of today provide evidence. This led many thinkers to create philosophical beliefs and ethical systems based on progress. Probably the two best-known attempts were by Teilhard de Chardin and Julian Huxley. De Chardin was a priest who wrote extensively on evolution as driven by a spiritual force toward progressively "higher" goals (most popularly discussed in his book, *Phenomenon of Man*, 1959). This basic philosophy is very old, but de Chardin's interpretation differed because he applied Darwinian concepts. However, most scientists have been highly critical because many of de Chardin's spiritual ideas are not based on any visible evidence. As a result, his philosophy has been called **evolutionary mysticism.** Huxley's philosophy is different because it is deeply rooted in facts, with no reference to spiritual qualities. It is often called **evolutionary humanism** because Huxley tried to derive values and ethical codes for human behavior that were not spiritually based. (*Evolution in Action*, 1953, is probably Huxley's most widely-read discussion of this.) One of his basic premises was that evolution revealed a progressive increase in the ability to transcend limits imposed by the environment. Humans represented the highest expression of this quality. This logic illustrated the "specialness" of humans and was used to justify a number of humanistic principles.

While writings such as de Chardin's and Huxley's have been widely discussed, the general consensus of evolutionary scientists and philosophers is that none of these attempts to use nature as a guide for human values and activities has been successful. To cite Peter Bowler's conclusion about such attempts: "The principles they establish are either so vague or so tenuously linked to the logic of scientific evolution that their arguments could justify almost any ethical position." There are two basic reasons why these attempts have failed. First, *nature simply does not show clear-cut patterns* that provide easy lessons to be inferred. For example, while biological evolution does show various kinds of general progressions, such as size and complexity, these trends show many reversals, branchings, and dead ends. These occur because evolution, as we have said, is "blind" to the future, adapting organisms only to the immediate environment. Thus, the common notion of evolution as a linear and predestined (goal-seeking) progression is quite wrong. The veteran philosopher of science, David Hull, has stated in the book *Evolutionary Progress* (1988) that evolution does not show one "big" direction. Rather it shows many "little" directions (such as evolutionary trends in horse body size or human brain size), and even these are never "straight-line."

The second reason for the failure of nature to provide a philosophy or system of ethics is that even if there were simple patterns that could be interpreted as "lessons" for humankind, there is no reason why humans must adopt them. Thus, thinkers who seek to apply nature to humans have fallen into the **naturalistic fallacy:** Whatever is natural must therefore be good. There are many cases where this is obviously not so. For instance, many of us are alive today because of advanced "artificial" medical technology. Should we let people die from all diseases so that natural selection can run its course? Consider, too, that many natural substances (such as natural toxins) are just as lethal as manmade pollutants.

A PHILOSOPHY OF SELF-RELIANCE. The apparent conclusion is that we must look *internally* for guidance. According to Bowler (1989): "If Darwinism comes anywhere near the truth, the universe has not been designed to show us, its products, where we should go in the future. If there is an ethical message in the theory, it tells us simply that we cannot look outside ourselves for guidance. With or without the knowledge of evolution, each of us must look into his or her own conscience for a source of moral values."

This interpretation indicates that the long human search for absolute rules outside ourselves (such as from God or in nature) is often an attempt to escape making our own painful decisions. Having created an organ (the mind) so distinct from the rest of nature, evolution has itself freed us from many of the constraints of other species. Yet, as many people recently freed from the yoke of totalitarian regimes have stated, freedom is often more unpleasant than having decisions made for us. This same motive is why people are constantly projecting more into evolutionary patterns than is really there. They are, in a sense, searching for advice.

Business and Politics: Survival of the Fittest?

The effects of Darwinian thought were not limited to the "higher" human pursuits of religion and philosophy. Darwin also had, and continues to have, a direct impact on daily business and politics. That nature selects only certain individuals to survive and have offspring has conjured many colorful images, such as "survival of the fittest." The "fittest" are often thought of as the mightiest, fastest, meanest—qualities associated with aggression. Such images have had a strong appeal to many people in the business and political arenas where it has been used to justify aggressive behavior. **Social Darwinism** refers to this application of "survival of the fittest" to social interactions, especially in business and politics. The logic of Social Darwinism is roughly this:

1. Evolution is driven by the differential survival of competing individuals.
2. Evolution has resulted in "progress," such as more efficient and complex organisms.
3. Therefore, progress in human affairs is also driven by competition and differential survival of human beings, such as competition among nations (war) or corporations (uncontrolled free market).

This logic has been applied since the late 1800s to justify many types of aggression, from cutthroat business practices, to national imperialism, to racial and ethnic repressions. That the logic is flawed on a number of counts has proven less important than the intuitive appeal of this idea to many people. Thus, Social Darwinism continues to be a widely held idea, at least in modified form. Let us see how this idea has been misapplied to business and politics before specifying what is wrong with it.

BUSINESS AND SOCIAL DARWINISM. Social Darwinism as a business practice peaked in the late 1800s with the "robber barons." These were the industrialists who made huge fortunes in railroads, petroleum, and other areas during the massive industrialization of the United States. Many of these industrialists, such as Rockefeller, literally preached Social Darwinism as a policy to be widely used. This "no holds barred" approach (often called laissez-faire capitalism) led to extensive bribery of officials and competitor's employees, infiltration of competing companies with spies, and even acts of violence such as industrial sabotage. Even aside from moral considerations, it was soon discovered that society as a whole was not served by unrestrained competition. Often the winner was not the entrepreneur who produced the best product more cheaply, but was simply the one who was most resourceful in eliminating competitors through whatever means. Furthermore, the progressive domination of the industry by one or a few of these "barons" (monopolies and oligopolies, respectively) allowed them to raise prices and control the markets, much to the detriment of the consumer. Therefore, government regulations (such as antitrust laws) were introduced in the early 1900s to limit the unrestrained practices.

POLITICS AND SOCIAL DARWINISM. The same "survival of the fittest" logic has also been used to justify nationalistic aggression. The late 1800s was a time of imperialism for many countries, including England, and many politicians espoused that domination of other countries by the "fittest" was nature's way of making progress. Adolf Hitler took this to the extreme and combined political imperialism with racial imperialism, preaching that the Aryan race (of northern Europe) was genetically superior and was simply carrying out the laws of nature by subjugating other nations and races. The same logic has also been applied to exploitation of minorities within national boundaries.

Proponents of Social Darwinism have tended to embrace another biological argument that reinforces it: that most human behavior is genetically determined (see Chapter 5). For instance, it is easier to justify the underclass's plight in terms of Social Darwinism (they are intrinsically, or genetically, "less fit") than to attribute their position to lack of education.

FLAWS IN SOCIAL DARWINISM. Most people reject extreme versions of Social Darwinism in both business and politics because they find it morally repugnant. Few modern people would want to live in a society where "might makes right" in every situation. However, many people often resort to moderate kinds of Social Darwinian logic to rationalize various social injustices. While every individual has a right to an opinion on this, it is crucial to realize that, ethics aside, Social Darwinism is based on invalid reasoning. Therefore, let us try to objectively critique Social Darwinism, specifying two basic flaws in: (1) understanding the scientific facts of biological evolution, and (2) their application to humans.

First, success in natural selection as a biological process usually involves much more than simply being the biggest, fastest, or meanest. As we will see in Chapter 2, the fittest individuals are often the smaller ones (who can conceal themselves), those most attractive to mates, and many other nonaggressive traits, depending on the particular environment. Most important, cooperative behaviors are often advantageous, such as in bees or among other colonial organisms. Thus, aggressive behaviors in nature often render individuals *less fit*. Another misconception about natural selection is that because it causes evolution, it also causes "progress." We have already discussed that while biological evolution shows statistical progressions in many kinds of traits, most evolutionary theorists agree that there is no overall theme of "progress" visible.

The second flaw involves the application of natural selection to society. Here we return again to the naturalistic fallacy: even if natural selection provided valid analogies for society, there is no necessary reason why we should embrace it. Indeed, the very fact that governments often step in to regulate unrestricted competition is a tacit acknowledgement that humans can operate by whatever rules we find to be most beneficial to society.

SUMMARY

Evolution refers to any change through time. There are three types of evolution: physical, biological, and cultural. The rise of **rationalism** during the European Renaissance led to advances in the understanding of evolution as reliance on religious ideas as the basis of understanding was replaced by reliance on observation and logic.

The discovery of **physical evolution** came about when geologists were able to show that the earth was much older than 6,000 years, as determined by the religious beliefs of the time. This discovery was made based on principles of **uniformitarianism,** which is often summarized as "the present is the key to the past."

The discovery of **biological evolution** occurred with the realization that fossils were remains of past species and as global exploration brought about the discovery of exotic plants and animals.

The first widely discussed mechanism for evolution, the **inheritance of acquired traits,** was proposed by Lamarck. This theory suggested that traits acquired during an individual's lifetime would be passed on to the offspring. This idea was disproven and the actual mechanism for evolution, **natural selection,** was later proposed by Charles Darwin. His findings were published in *On the Origin of Species*, one of the most significant books in the history of western civilization.

The discovery of **cultural evolution** came about as global expeditions described non-European cultures. Lewis Henry Morgan, a famous anthropologist, proposed a three-stage sequence of cultural evolution: **savagery, barbarism,** and **civilization.** This is now known to be a gross oversimplification.

The ideas of evolution and natural selection had a great impact on society far beyond science. For instance, **Social Darwinism** applied the ideas of evolution to justify social and economic exploitation.

Evolution had a profound effect on religion and philosophy. Judeo-Christian beliefs such as **theism**—the belief in an involved, personal God—were questioned. **Deism**—the belief in an impersonal, uninvolved God—arose in response to the discoveries of Newton, Darwin, and others, and described mechanisms for natural events. **Scientific creationism** was created as religion's rebuttal to science by those who interpret the Bible literally, even when it contradicts science. The **anthropic principle** presents a scientific argument for the existence of God: because physical and chemical parameters are precisely right for forming a universe and life forms, this suggests to some that it was by conscious design. **Agnosticism** pleads ignorance about the existence of God, while **atheism** denies the existence of God.

Humanism states that human affairs are not predestined; we have control over our own fates, making it unnecessary to rely on absolute, external principles such as religion to guide us. This view was often tied to technological **progress.** However, the recent rise of **nihilism** has led to the worry that **self-fulfilling prophecy** may itself cause decline.

KEY TERMS

evolution	civilization
great chain of being	fundamentalism
scientific method	creationism
uniformitarianism	theism
catastrophism	deism
inheritance of acquired traits	anthropic principle
theory of natural selection	agnosticism
orthogenetic school	atheism
mutationist school	evolutionary mysticism
modern synthesis	evolutionary humanism
DNA	naturalistic fallacy
savagery	Social Darwinism
barbarism	

REVIEW QUESTIONS

Objective Questions

1. What three observations led Darwin to his theory of natural selection?
2. Define: theism, deism, atheism, agnosticism.

3. What is Social Darwinism?
4. What did the orthogenetic school of thought propose?
5. What is uniformitarianism?

Discussion Questions

1. What is the naturalistic fallacy? Cite examples.
2. List and discuss the four major challenges to traditional religious views initiated by Darwin's ideas.
3. What are your religious beliefs and how do you interpret evolution by natural selection in light of them?
4. Do you think that business today is dominated by "survival of the fittest"? Explain.
5. Look in the dictionary for a definition of "progress" that you think best applies to the history of life. Discuss why you think it applies. Include how it compares with the meaning of progress in this statement: "The forest was destroyed in the name of Progress."

SUGGESTED READINGS

BOWLER, P. J. 1989. *Evolution: the History of an Idea*. University of California Press, Berkeley.
DAWKINS, R. 1986. *The Blind Watchmaker*. W. W. Norton, NY.
NITECKI, M. (ed.). 1988. *Evolutionary Progress*. University of Chicago Press, Chicago.

A recent thorough treatment of evolution and creationism

BERRA, T. M. 1990. *Evolution and the Myth of Creationism*, Stanford University Press, Stanford, CA.

2

Processes of Evolution: Selection, Genes, and Microevolution

OVERVIEW: EVOLUTION BY SELECTING VARIATION

The processes of physical evolution were a prerequisite for the evolution of life for two reasons: (1) they produced a habitat for life, by producing the earth (life's abode) and the sun (life's energy source), and (2) they produced life itself. Life is thought to have originated from the ongoing flux of physical processes, wherein many complex molecules are constantly forming, breaking up, and re-forming. The transition to natural selection occurred when one of these molecules began to copy itself. Copying is the key to evolution because the copying process (reproduction) gave the molecule a "memory" so that the molecular organization could be passed on instead of breaking up. Furthermore, errors in the copying process would create variation, which in some cases resulted in improvements and increased organization of the molecules. Such improvements and increased organization would be passed on and accumulate quickly because of the copying process. Thus, any such system with imperfect reproduction will become *intrinsically additive*, with a "built-in" tendency to accumulate complexity.

Ultimately, the increasingly complex molecules gave rise to cells, the basic unit of life today. Cells use genes to store the information that allows the cells to function and produce new cells. This genetic information is passed on during reproduction. As with the ancestral molecules, errors during reproduction can result in variation. These can sometimes be beneficial and are favored by natural selection. One evolutionary product of this process is the formation of organisms with enhanced learning abilities leading to cultural evolution. Cultural evolution occurs much faster because learned information is passed on very quickly. Whereas natural selection is dependent on errors in copying to generate and transmit change, changes in learned information are passed on immediately through speech or other symbols.

PROCESSES OF PHYSICAL EVOLUTION

Physical processes are those that directly involve energy and nonliving matter. In order to produce life and planetary abodes for life to inhabit, physical processes must involve attractive forces between units of matter, causing atoms to "clump" together.

FOUR BASIC FORCES. All physical processes are based on just four fundamental kinds of forces. These cause smaller units of matter to combine into progressively larger and more complex units. As shown in Figure 2–1, **nuclear forces** cause the nuclei of atoms to form from smaller (subatomic) particles, such as protons and neutrons. There are two types of nuclear forces—the **strong** and **weak**—so that nuclear forces comprise two of the four basic forces. The third force, the **electromagnetic force,** attracts negatively charged electrons to the positively charged protons in the nucleus. This causes atoms to bond, forming molecules. Chemical bonding of atoms to form molecules occurs when atoms share or exchange electrons. Thus, the myriad chemical reactions that led to the molecules forming the minerals and rocks of the planets are based on this force. The fourth and final force is the **gravitational force.** This causes the mineral and rock molecules to come together to form planets. Once planets are formed, more chemical bonding causes complex molecules of life to form on the planets.

Figure 2–1. Forces that lead to increasing complexity of matter, and the scientific disciplines which study them. (p = proton, n = neutron, e = electron)

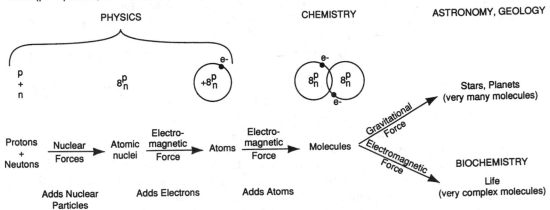

Whatever the ultimate cause, this "clumping" into more complex units first began after the **Big Bang** created the basic subatomic particles composing the universe some 15 to 18 billion years ago. Our own solar system formed about 5 billion years ago. The process continues today according to astronomers, who have recorded a number of stars and solar systems in the process of forming.

The Chemistry That Is Life

As noted, the force most directly involved with the origin (and continued functioning) of life is the electromagnetic force, which causes chemical bonding among atoms. This produces chemical reactions and creates molecules from atoms. Of course, such bonding is not limited to life. Virtually all matter on earth, living and nonliving, is composed of atoms that can undergo bonding.

An atom consists of just three kinds of particles: **electrons,** which "orbit" much larger **protons** and **neutrons** in the nucleus (Figure 2–2). There are 103 different kinds of atoms, each making up one of the 103 elements that comprise everything in our universe. The simplest element, hydrogen, has atoms with only one proton. The most complex element has 103 protons. Notice how the number of protons designates the element. This is because the chemical properties of the element are determined by the number (and location) of electrons, and it is the number of protons that help determine the electron number and arrangement. Atoms with more than 103 protons do not occur because they break up and are too unstable to form elements. The reason is that all protons have a positive charge and tend to repel each other as they accumulate in the nucleus. Even the largest atoms are very small to us—this page is about 1 million atoms thick.

Molecules can range in complexity from two simple hydrogen atoms bonded

Figure 2–2. A methane molecule forms when one carbon atom combines with four hydrogen atoms. From R. Buffaloe and J. Throneberry, *Concepts of Biology* (Englewood Cliffs, NJ: Prentice-Hall, 1973), p. 84.

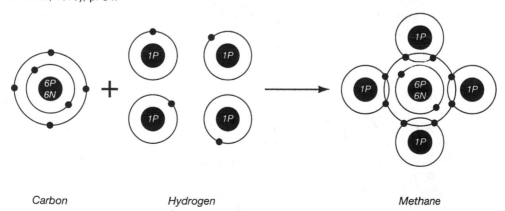

Carbon   Hydrogen   Methane

together to biochemical molecules with thousands of different kinds of atoms arranged in intricate, complex ways. The simpler elements, with few protons (such as hydrogen), are the most common in the universe because they generally form more easily. It is perhaps no surprise then that the six most important elements of life have relatively simple atoms. *These six comprise over 99% of all living tissue*: carbon, hydrogen, oxygen, nitrogen, phosphorus, and sulfur. These atoms are used in molecules important in energy storage (fats), reproduction (DNA), muscle tissue and enzymes (proteins), and other uses. Carbon, in particular, can form many different kinds of molecules because it bonds readily to many kinds of atoms. It is therefore often said that life is carbon-based. Other elements besides the six just listed are also important for life, but they are generally needed in much smaller (trace) amounts.

Origin of Life Through Chemical Processes

EXPERIMENTAL AND THEORETICAL EVIDENCE. As long ago as the last century, there was speculation (by Darwin, among others) that chemical reactions on the earth led to the origin of life. However, it was not until 1953 that direct experimental evidence for this was gathered. In one of the most famous experiments in the history of modern science, Stanley Miller and Harold Urey recreated the physical and chemical conditions approximating those on the early earth. We will see in Chapter 3 that the earth is thought to have condensed from a huge cloud of interstellar gas and dust. The cloud, rich in hydrogen, led to an atmosphere of hydrogen and such hydrogen-based gases as ammonia and methane. Other gases, such as water vapor and carbon dioxide, were also present. Miller and Urey put all these gases in an experimental device, as shown in Figure 2–3, and subjected them to an electrical energy source, which simulated sunlight and lightning on the early earth.

The dramatic result of Miller's and Urey's experiment was that this combination of gases and energy produced **amino acids.** These are molecules that serve as the building blocks of proteins. Proteins are among the most essential materials in a living organism, making up much of the muscle tissue, enzymes, organs, and other parts. Since 1953, the experiment has been performed many times and under varying conditions (such as changing the proportions and even types of gases). Amino acids are produced in many of these. This shows how easily these building blocks originate under natural conditions resembling early earth. Further evidence for the ready production of amino acids is that they have been found in meteorites. (Indeed, some scientists speculate that the massive meteorite bombardment of earth during its early formation contributed to the origin of life.)

However, to produce complex molecules is a far cry from producing life. A diagnostic feature of life is not what it is composed of, but how it is organized. The basic organizational unit is the cell: All self-sufficient life forms are composed of at least one cell. Therefore, another important experiment on early life was done by Sidney Fox, who showed that the amino acids, when occurring in hot waters (as were common in the early earth) formed cell-like structures, called **microspheres.**

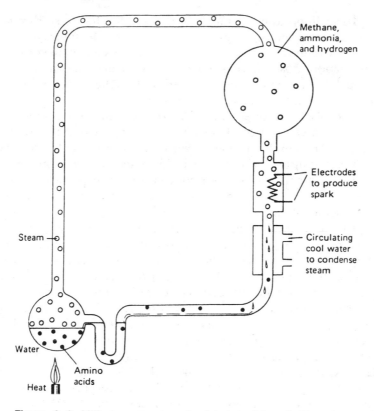

Figure 2–3. Miller's apparatus simulated earth's primitive atmosphere (prior to the introduction of oxygen), that consisted of methane, ammonia, hydrogen and water, and electricity. Amino acids formed from this reaction. From D. Eicher and A. McAlister, *History of the Earth* (Englewood Cliffs, NJ: Prentice-Hall, 1980), p. 115.

These microspheres occur because many of the "proteinoids" formed by the amino acids have extremely strong chemical bonds. While far simpler than true cells, microspheres possess membranelike properties that serve to create an internal environment for the amino acids. This was a crucial step in creating the right concentration of chemicals to produce chemical reactions that would not otherwise occur outside the microsphere's internal environment.

This experimental evidence on how organized structures may originate has recently been supplemented by theoretical evidence from chemistry and physics that many chemical and physical systems are **self-organizing.** That is, when a chemical system is subjected to an outside energy source (such as sunlight), it will sometimes be driven far out of equilibrium and into an organized state. There is a growing body of mathematical theory that seeks to understand why such behavior occurs. Apparently it is quite common in many dynamic systems.

In addition to being highly organized, life has one other key feature: reproduction. Each organism is capable of recreating at least part of itself, thereby perpetuating its organization in the face of the parent's inevitable death. As to exactly how reproduction arose, there is much less experimental evidence. One reason is that the reproductive process today (based on DNA) is so complicated that it is difficult to reconstruct intermediate steps. Nevertheless, lab experiments show that nucleic acids, the basic components of DNA, are like amino acids—they are easily produced under primitive earth conditions. However, many scientists think that even nucleic acids are too complex to have been used by early molecules. Thus, one popular theory is that instead of using the DNA molecule as a template (that is, a surface to line up on), early molecules reproduced using clay particles as the template. Clay is very common on earth and it is a flat particle with a strong electric charge at each end to attract atoms.

In summary, laboratory experiments indicate that many of the major steps toward life could have occurred under the natural conditions of the early earth. However, such experiments do not come close to actually creating life in the lab. The molecules and microspheres produced are extremely primitive compared to even the simplest true life form. For example, the simplest of all known cell-based life forms still contains about 2,000 protein molecules and carries out about 600 chemical reactions to sustain itself. On the other hand, the early earth was a much vaster and more complex place than any human laboratory. The early oceans have been called a **primordial soup,** teeming with a myriad of "trial and error" chemical reactions. It thus seems very unlikely that we could ever reproduce the exact sequence of steps that occurred. Also, there is the problem of the long time scale. The earth was not only a much larger "laboratory," but there were millions of years of "experimentation" involved.

FOSSIL EVIDENCE. Aside from the experimental and theoretical evidence just discussed, the fossil record provides yet a third line of evidence that life quickly arose from natural physical processes. As discussed in detail in Chapter 3, fossils of primitive single-celled organisms appear in rocks about 3.5 billion years of age. This seems astonishingly soon after the earth became habitable: It was not until sometime after 4 billion years ago that the earth's molten crust had cooled enough to support life. Furthermore, given the rarity of fossil preservation, it seems likely that the first life arose even sooner than 3.5 billion years ago, but was not fossilized. Traces of certain carbon compounds in rocks dating to around 3.8 billion years ago imply this.

DID LIFE ORIGINATE ONLY ONCE ON EARTH? Given the ease and speed with which life arose from natural conditions, you might suspect that life originated more than once on earth. However, two important lines of evidence strongly indicate that all living organisms are ultimately related to the same ancestor. One, all life forms on earth, from microbes to elephants to trees, use the *same DNA code* to reproduce. (Because this genetic language is so similar, humans can "splice" together genes of different kinds of organisms, with huge practical benefits, as discussed in Chapter 5.) Second, all living organisms have the *same basic biochemistry*. They use similar biochemical reactions to process oxygen, metabolize food, and so on.

Given that all life today appears to have the same ancestor, we can infer the following. Either (1) life arose more than once, but only one life form survived long enough to leave living descendants, or (2) life arose only once. In the first scenario, some scientists have suggested that life originated a number of times in the early earth, but the massive meteorite bombardment then killed off the fragile life forms. (Consider that many of these meteors were the size of California moving at 40,000 miles per hour.) In this scenario, it was only after the bombardment diminished, about 4 billion years ago, that one of these life forms was able to persist. If the second scenario is true, one reason might be that the precise chemical conditions are so unlikely that it occurred only one time. Another, perhaps more likely, reason is that once life did originate, it would soon start to modify its environment, thereby *disturbing the original conditions* that gave rise to life in the first place. There are at least two major ways that the first life form must have disturbed the original conditions. First, in the early stages, life forms must have depleted the environment of readily available nutrients. Most scientists think that life began in the ocean, which was a vast sea of "primordial soup" containing amino acids and other complex molecules that would serve as food. Once life began to consume such molecules, which form the precursors to life, there would be less chance for the precursor molecules to evolve. Given how rapidly life can reproduce, most estimates indicate that the oceans would soon be depleted of nutrients. A second major way that life modified the environment was to alter the atmosphere. Photosynthesis by early plants produced oxygen as a byproduct. This was gradually added to the atmosphere until it composed about 21% of the atmosphere, as it does today. Oxygen is an extremely reactive gas so that, once common, it would likely have broken down any complex molecules in the process of evolving into life. Only life forms that originated before oxygen was common, and evolved in the presence of it (and thus developed ways of coping with it) could exist.

DID LIFE EVOLVE ELSEWHERE IN THE UNIVERSE? Assuming that life can originate through natural processes, it seems almost inevitable that it would also arise elsewhere in the universe. There are billions of galaxies in the universe, each with billions of stars. Many millions of these stars are thought to have habitable planets. Because the same 103 elements and the same chemical and physical laws occur in all parts of the universe as occur here, there is no scientific reason to suppose that, under similar conditions, similar events would not occur. This is the so-called **principle of mediocrity,** which says that earth is not particularly unique in its conditions for the origin of life.

In addition, life is not restricted to "life as we know it." The properties of being highly organized and able to reproduce do not restrict life to being chemically or anatomically similar to ours. Many speculations have been suggested about other kinds of life: different biochemistries (such as copper-based blood instead of the oxygen-based blood of humans), different genetic codes, and even life not based on carbon. Physical considerations also apply. For example, life that evolved on planets with higher gravity would be built much more massively and low to the ground.

This is not to say that life will occur everywhere, or even on many planets. Two key criteria for a habitable planet are: (1) *temperature* and (2) *size*. Temperature is crucial because if it is too low all liquids will freeze, essentially stopping all major biochemical reactions. If temperature is too high, the liquid will boil, breaking up all of the large molecules needed for life. Therefore, most scientists believe that life will arise only on planets that are at a moderate distance from their sun. This distance will vary depending on the star's temperature and size, but it must not be so close that liquid boils or so far that it freezes. This is called the **Goldilocks's paradox:** not too hot, not too cold, but "just right," as the porridge in the children's tale. Thus, in our solar system, Venus is too close and Mars is barely too far, as shown in Figure 2–4. Planetary size is important because if the planet is too small, it will not have enough gravity to retain its atmosphere. It will drift off into space, as has happened on our moon. Internal heat will also be lost too quickly so that small planets soon become geologically dead. If the planet is too large, it will retain the highly volatile lighter gases (such as methane and ammonia), unlike the earth, where these have trickled into space. This would cloak the planet in a thick atmosphere and almost certainly not lead to life as we know it. Figure 2–4 shows that Venus and Mars are adequate for life in terms of size, so it is mainly their distance from the sun that renders them unfit for life (at least as we know it).

Figure 2–4. Life's balancing act, set by gravity and distance from the sun. Modified from *Ad Astra* magazine, April 1990, p. 27.

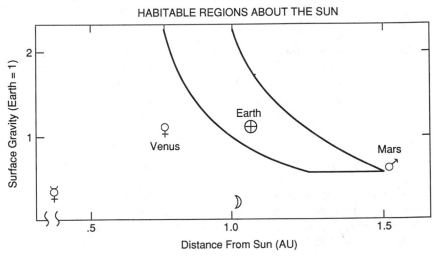

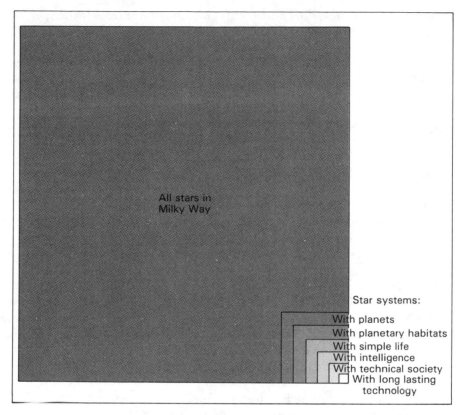

Figure 2–5. Of all the star systems in our Milky Way (represented by the largest box), progressively fewer have the qualities typical for a longlasting technological society (represented by the smallest box at the lower right corner). From E. Chaisson, *Universe* (Englewood Cliffs, NJ: Prentice-Hall, 1988), p. 540.

How many planets in other star systems fit the temperature and size criteria? Estimates by astronomers indicate that there are about 300 billion stars in our galaxy alone. However, as shown in Figure 2–5, only a fraction of these have planets. Of these, only a fraction are suitable for life, and of these, only a fraction actually develop life. Yet a smaller fraction of these would develop intelligent life. In his popular book, *Cosmos*, Carl Sagan describes how astronomers have estimated these fractions, based on knowledge of how planets form and other information. He shows that even though only a fraction of planets may be habitable, this could still mean that many millions of stars have habitable planets in our galaxy alone; even a tiny fraction of 300 billion stars is still a large number.

Where are these life forms? Unfortunately, we may never know if our speculations are correct because the stars are so far apart. Even the closest star to earth is so far that light takes over four years to reach us. Even the fastest spacecraft (probably even for advanced technologies) can move nowhere near the speed of light and would take many tens of years to reach even the closest stars. The most likely means of communication then is to use energy (such as radio waves). There are projects (such as SETI, Search for Extraterrestrial Intelligence) that are now searching the heavens for possible signals from other worlds, using computers to monitor hundreds of frequencies at once. Life elsewhere in our own solar system does not seem likely since there is no place nearly as hospitable as earth. Chemical tests for microbes in the soil of Mars by the Viking spacecraft (in the 1970s) were not entirely conclusive, but generally indicate the absence of life.

Does Life Violate Physical Laws?

Life is clearly "different" when compared to the physical world. A rabbit moves, reacts, reproduces, and has many qualities that are absent from a rock. This has led many people, over the years, to argue that life has some special quality that sets it apart from the physical universe. These **vitalists** have tried time and again to show that life has a "soul" or some other substance that is not possessed by nonlife. However, no such claims have ever been proven. The closest vitalists have come to scientific proof was in the 1800s when chemists thought they had found that only living tissues contained the element carbon. (Thus, today "organic" chemistry is the study of carbon-based reactions.) But, it was soon found that nonliving matter (such as the mineral graphite, used in pencil leads) also contained carbon, so this logic had to be abandoned. Indeed, for all its profound diversity and complexity, life seems amazingly unremarkable in composition. As previously noted, over 99% of living tissue is composed of just six common elements and the remaining 1% is composed of just a few more elements in trace amounts. Thus, it is often stated that all of the chemicals that make up a human could be bought for a few dollars.

Then what makes life so obviously different from the physical world? The answer is that it is not what composes life but how those components are arranged. It is the *organization* of their atoms (and molecules) that makes living things so distinctive, that allows them to move, grow, and reproduce. Our bodies are far more than a loose collection of atoms. It is the incredibly intricate manner in which these atoms are arranged that makes us able to behave in such complex ways compared to a rock. This illustrates the concept of **emergent properties.** These are properties that "emerge" as the degree of organization increases: muscle tissue, nervous systems, and so on are the result of highly organized cells, which, in turn, are composed of highly organized molecules. A simpler example would be a car, which could be called a "collection of metallic atoms." Yet a car is obviously very different from the same metallic atoms found in an ore deposit. In the car, they are fashioned into a carburator, precisely made cylinders, and so on, while in the rock they are disorganized and distributed at random. Because of this organization into cylinders, tires, and so on, cars have such emergent properties as being capable of movement and even more qualitative ones such as being "sleek" in appearance.

By now you may wonder just what is meant by "organization." In general, something that is highly organized contains much more information than something that is not. For example, information in computers is often in "bits." Thus, the organization (amount of information, in bits) in even the simplest single-celled organism, with all of its structural molecules and programmed chemical reactions, is many *thousands of times greater* than that of any nonliving thing, even a highly "ordered" crystal: a bacterium has about a million bits of information in its genetic repertoire. In this sense, humans are not just being egocentric when we think of ourselves as highly complex life forms because our brains do indeed hold much more information than that of any genetic system, or the brains of any other organism. Human genes contain about 5 billion bits of information, while our brains contain up to about 100 trillion bits.

LIFE AS AN OPEN SYSTEM.   Having found no compositional differences between life and nonlife, many vitalists have tried to argue that life's great organizational complexity must somehow break the laws of physics. They point to one of the most basic laws of physics that life seems to break, the second law of thermodynamics. This law states that, in any isolated system, there will be an inevitable tendency toward disorder. Another, perhaps more familiar, name for this law is the **law of entropy,** where entropy essentially means disorder. This observation fits in with everyday experience. Mountains tend to erode over time, rooms become cluttered, roads need to be constantly repaired, and so on. How, then, does life grow (from a fertilized egg) and maintain its complexity in the face of all this disorder in the physical world?

The answer (unfortunately for those seeking vitalistic answers) is simply that living systems are *not isolated*. They take energy from their environment: plants use and store energy from sunlight and animals eat the chemical energy stored in plants. In short, organisms are **open systems,** meaning that they take energy from the environment and use it to construct complexity. This same principle also applies to our car analogy. What makes a car cost so much more than a pile of raw minerals is the huge amount of work (energy expended) that goes into arranging those atoms into something that functions as a unit. This is the fallacy in saying that the human body is worth only a few dollars. It ignores the vast amount of energy ("work") that goes into organizing those components into a functioning unit.

Because an organism (or a car, or any highly organized object) is an open system, it can temporarily circumvent the law of entropy by taking energy from its environment to maintain its order. However, the law of entropy is not broken because the net disorder of the entire system is still increased. In taking energy from the environment, animals increase the order within their bodies but they *decrease the order in their environment*. In absorbing and using energy, a great deal of less usable waste products (body heat and digested matter) are produced. This increases the total amount of disorder in the environment more than if the organism were not there. In addition, all organisms eventually die (as cars and everything else eventually degenerate), so this "fight against entropy" is only a temporary victory anyway. In fact, this second point proves that life is bound by the laws of physics. Ultimately, we must all pay homage to physical reality.

It may still seem that life is "different" in even temporarily circumventing (not breaking) the law of entropy. Yet even this is true only to a limited extent. Many nonliving systems also create temporary pockets of local organization. Many machines, such as cars or refrigerators, do this. Even natural phenomena can spontaneously do this: a growing crystal will align atoms into an ordered lattice, taking energy from its environment to do so. Obviously, life is much more effective at doing this, taking more energy from the environment to create greater organization. Yet the difference is one of *degree and not kind*.

So far we have dealt with the origin of life from nonliving, physical processes. However, this approach does not explain why the complex systems we call life have evolved ever greater organization by taking ever greater amounts of energy from the environment. To answer that, we turn to a discussion of the processes of biological evolution.

PROCESSES OF BIOLOGICAL EVOLUTION

Overview

Darwin's major contribution was not the discovery of evolution but the discovery of the *process* that causes biological evolution to occur: natural selection of inherited variation. We saw in Chapter 1 how this process can be summarized into two steps:

$$\text{variation} + \text{environmental change (selection)} = \text{evolution}$$

In this section, we examine this "equation" in detail, starting first with selection (the second step of the evolutionary "equation"), followed by a discussion of variation (the first step). This is followed by two more parts, one concerning evidence and the other misconceptions about evolution and natural selection.

In summary, this section is divided into four parts:

Part 1: Selection and Speciation
Part 2: Inherited Variation
Part 3: Evidence for Evolution by Selection of Variation
Part 4: Misconceptions about Evolution and Natural Selection

PART 1: SELECTION AND SPECIATION

Selection

Selection as an evolutionary force is not difficult to understand: It is simply the favoring of some individuals over others because of certain traits (or genes, as we see later) that the individuals possess. By "favoring" we mean not only that certain individuals will preferentially survive, but that certain individuals will have *more offspring*.

The main roadblocks to understanding selection are the many subtle implications and fine points of its workings. For example, there are actually a number of kinds of selection, with natural selection being the most familiar to the nonscientist. For convenience, we will consider the two most basic ways of classifying selection: (1) by the *agent* of selection, and (2) by the *direction* of selection.

THREE AGENTS: NATURAL, SEXUAL, AND ARTIFICIAL SELECTION. Table 2–1 shows the three basic categories that Darwin used to classify selection by the agents involved. By "agent" we mean who or what is doing the selecting. Let us discuss each of the three categories, beginning with natural selection.

Natural selection is selection by the natural environment. This is a large category because the natural environment includes many different elements. For instance, both physical and biotic elements of the environment can be selective agents. As shown in Table 2–1, biotic natural selection include predation, parasites, and competition. Our discussion of competition among giraffes for food in Chapter 1 (see Figure 1–3) would go in this category. Examples of physical natural selection include temperature and humidity. A change in climate would affect individuals in a species differently because they usually vary in tolerance to temperature changes. For instance, although polar bears in general cannot stand warm (tropical) temperatures, some individual bears are relatively more tolerant than others. Therefore, if the polar climate were to gradually warm, the more warm-tolerant individuals would be favored.

Sexual selection is selection by mates such that some individuals have traits that allow them to have more offspring than others. Many scientists now include

TABLE 2–1. Three Basic Types of Selection

SELECTIVE FACTOR	MEANS OF ACTION	TYPE OF SELECTION
Man	Breeding of domesticated plants and animals to suit man's desires	**Artificial selection**
Physical environment	Cold, heat, humidity, etc. select for types that can survive	
Biotic environment	Predators, parasites, etc. select for speed, strength, resistance, etc.	**Natural selection**
a. Other species		
b. Same species, with no sexual discrimination	Parental care and the response of the young to it	
c. Same species, with sexual discrimination	Male-male competition for mates and resources	**Sexual selection**
	Female selection of mate	
	Male selection of mate	

sexual selection as a kind of natural selection, because it is "natural" and mates are part of an individual's biotic environment. As shown in Table 2–1, sexual selection may include traits that are advantageous in male-male competition for females, or traits that make an individual more attractive to the opposite sex. An example of selection in male-male competition would be the favoring of males with larger body size in combat. There are many examples of selection on males to attract females through the use of displays, one being the ornate tail feathers of male peacocks.

The third basic type is **artificial selection,** which is selection caused by the direct involvement of humans. Animal or plant domestication is a good example, where humans allow only those individuals to breed that have traits that we favor: fattest cattle, whitest rabbits, sweetest apples, and so on. A kind of unintended artificial selection occurs when pests develop immunities to poisons. As shown in Figure 2–6, the few individuals with resistance will survive treatment and will produce offspring that are also resistant.

KEY CONCEPTS: FITNESS AND ADAPTATION. Evolution by selection of variation is often referred to as "survival of the fittest." Two major points need to be made about this widely-used and misleading phrase. First, much more than just survival is involved. As we have noted, "reproduction" of the fittest is more accurate in that selection is favoring not those that survive, but those that leave more offspring. Two individuals may survive for equal periods of time, but the "most fit" is the one leaving more offspring. Thus, biologists define **fitness** as the measure of relative reproductive success of an individual or trait. This leads us to the second major point: the fittest individuals are often not the biggest, meanest, or fastest. Instead, being smart, inconspicuous, most attractive to mates, or most cooperative can lead to having more offspring than brute force or being aggressive may afford. (We saw in Chapter 1 how this scientific fact undermines arguments for Social Darwinism.)

By favoring the fittest individuals, natural selection acts to constantly "adjust" a species to its environment. Because those individuals that perform most effectively in the environment are favored, the whole species tends to accumulate those traits that make the individuals successful. **Adaptation** describes this process wherein the whole species becomes better suited to perform in its environment. It is also used to describe those traits involved in the process: organs or behaviors that enhance performance are "adaptations." The exquisite adaptedness of species to their environments was a source of wonder to early naturalists (and still is to anyone who appreciates nature). They noted that virtually every feature of an organism is precisely molded to provide maximum effectiveness, as illustrated in Figure 2–7 (see page 36). We see here that this occurs because natural selection molds traits to keep pace with the environment by favoring individuals with the most advantageous variations.

DIRECTIONAL VERSUS STABILIZING SELECTION. Many traits in a species vary according to a so-called "normal," or bell-shaped, distribution. For instance, Figure 2–8 (page 37) shows how tan coloration in a mouse species may be much more abundant than

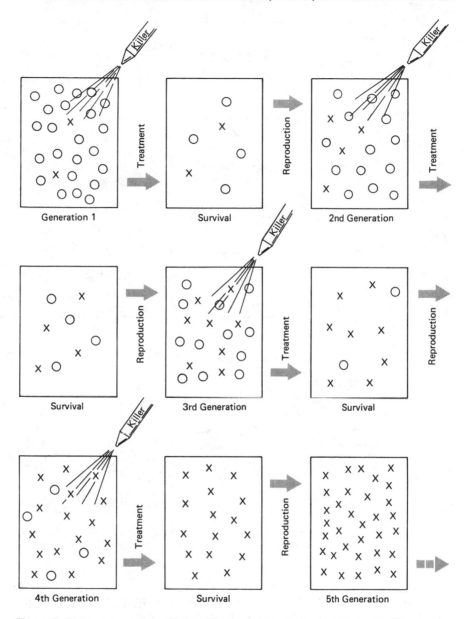

Figure 2-6. An example of artificial selection which eliminates nearly all of the sensitive organisms (○). The few individuals which are resistant (×) survive to reproduce. As treatment continues, the population eventually consists entirely of resistant individuals. From B. Nebel, *Environmental Science* (Englewood Cliffs, NJ: Prentice-Hall, 1981), p. 68.

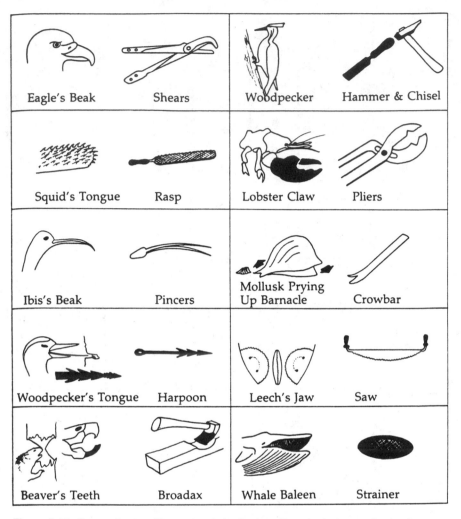

Figure 2–7. Comparisons with manmade tools show the adaptedness of animal organs. From R. Augros and G. Stanciu, *The New Biology* (Boston: New Science Library, 1987), p. 205.

white or brown mice in a particular environment. Now suppose that the environment changes and begins to favor darker individuals. Perhaps less snowfall occurs (physical natural selection) or females begin to prefer darker males (sexual selection). Or maybe humans find it easier to find and kill the lighter mice (artificial selection). Figure 2–9 shows that the curve will move to the right as darker individuals begin to leave more offspring. This is called **directional selection,** defined as selection on one extreme ("tail") of any trait distribution, leading to a shift in the distribution average.

Note that a prerequisite for directional selection is the very existence of variation. Without brown mice, for instance, the whole species might have become ex-

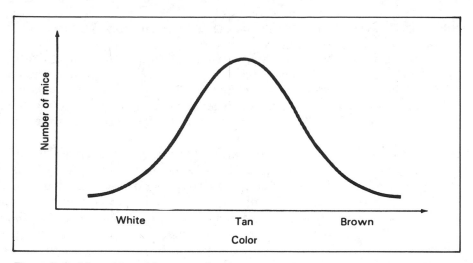

Figure 2–8. All members of any species show some variation of traits, often in a "bell-shaped" frequency distribution. In this example, the number of mice with different fur colors have been plotted. Plots of such traits as height or weight would also resemble this frequency distribution. From E. Chaisson, *Universe* (Englewood Cliffs, NJ: Prentice-Hall, 1988), p. 479.

Figure 2–9. Directional selection enhances one extreme (a) population and eventually leads to a shift in the distribution of that trait in the species (b). From E. Chaisson, *Universe*, p. 479.

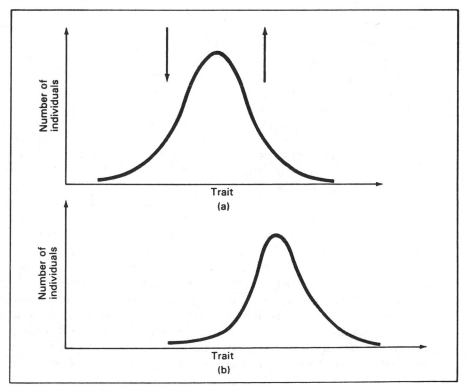

tinct. But why was such variation maintained before the change, when there was selection against brown mice? There are many reasons why such variation is maintained in nature. One of the most important is that virtually no natural environment is totally uniform. There are usually local areas ("microhabitats") within any environment where different traits are favored. For instance, perhaps the brown mice originally inhabited darker, more shadowy areas.

Directional selection will continue until the trait average for the species is once again at an optimum (Figure 2–9). When this optimum is reached, **stabilizing selection** will occur. This is selection that operates on both extremes of the distribution simultaneously. It will continue as long as the environment is relatively unchanged. Notice that even where there is no change in a species (bottom of Figure 2–9), *selection is occurring*. Stabilizing selection acts to maintain the species at its current optimum state by selecting against individuals that have traits too extreme in either direction from the central point of distribution.

Speciation

If it acts over a sufficient period of time, directional selection will eventually cause significant change in a number of traits. If the entire species is affected, it will eventually become transformed into a new species. Shown in Figure 2–10 (uppermost portion), this is called **nonbranching evolution.** (More technical terms for this are phyletic evolution and anagenesis.) Notice that the graduational nature of nonbranching evolution makes the distinction between the old and new species somewhat arbitrary. Thus, we can say that a new species has evolved from an older one but we cannot point to the exact time that it occurred, much as one cannot say exactly when daybreak begins and ends, even though we can be sure it happened.

The second major kind of evolution, also seen in Figure 2–10 (lowermost portion) is **branching evolution** (also called cladogenesis). In this case the whole species is not transformed. Instead, separate populations of the old ("parent") species undergo different directional selection, causing them to branch off in different directions. This has been very important in creating the diversity of life: If only nonbranching evolution occurred, there would still be only one species of life on earth, changing over and over again. A major reason that branching evolution is so common is that many species have geographic ranges that are spread over many areas. Therefore, it is very unlikely that a single environmental change will affect the entire species in the same way. Instead, local populations of the species are subjected to differing local environmental pressures (Figure 2–11, p. 40). Note that differences between the different populations will at first be gradational: populations form a **cline,** defined as a continuum of traits. For instance, in Figure 2–11, a cline would occur if the artic fox and gray fox populations blended together in the middle to form a population that had traits intermediate to both, such as medium-sized ears and legs. However, in many cases, a population becomes cut off from the rest and no such gradient forms. Thus, the formation of a mountain range or river channel may isolate a population. Such isolation will hasten the process of

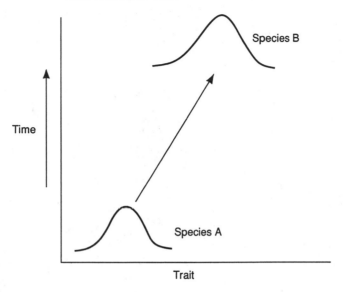

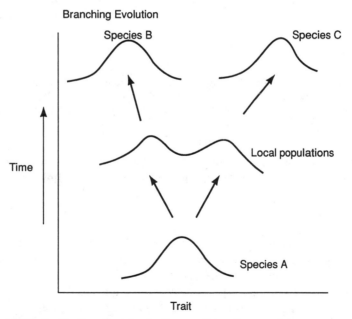

Figure 2–10. Directional selection on the entire species causes nonbranching evolution. Different directional selection on separate populations causes branching evolution.

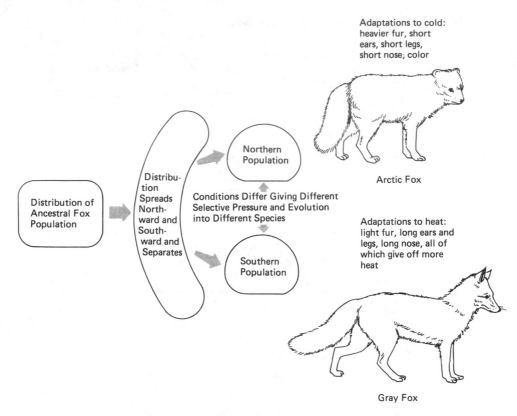

Figure 2–11. Populations spread over broad distances experience different selective pressures which may cause them to eventually evolve into separate species. This shows a hypothetical distribution of the ancestral fox population that speciated into the modern fox and gray fox. From B. Nebel, *Environmental Science*, p. 72.

splitting because it prevents mixing with members of the other populations, allowing selection to produce change more rapidly, without "dilution" of traits from other populations.

Sometimes the isolated population is initially very small (say, 20 to 30 individuals, or even a single pregnant female). When this happens, it is likely to evolve (branch off) very rapidly—with few individuals, selection can affect overall offspring production very quickly. In addition, such small populations are often subject to random sampling effects. That is, whenever small samples are taken from a large group, it is highly probable that some traits will be disproportionately represented. (Would you rather base a marketing strategy on a poll that questioned 10 people or 1,000 people?) For example, consider a large parent population of mice in which 10% exhibit white coloration; calculations show that there is about a 50% chance that among a random sample of seven individuals, *none* would be that color. Thus, a randomly isolated population might lack white coloration even though as much as 10% of the parent species has it. This nonrepresentative sampling of a parent species due to the small size of the founding population is called the **founder effect.** It has been especially important in the evolution of island organisms. For example, many of the native species of the Hawaiian Islands originated from a few individual insects, plant seeds, and birds that were carried by the storm winds or oceanic currents.

WHAT IS A SPECIES? We have referred to "species" (Latin for "type") rather loosely, as groups that are visibly distinct because directional selection causes accumulating changes in traits through time. However, there is a much more objective definition of species based on the *criterion of interbreeding*. To a biologist, a **species** is a group of individuals that can interbreed to produce fertile offspring. A new species has evolved not when it "looks" sufficiently different from its ancestors or neighboring populations but when it can no longer successfully interbreed with them. This definition of reproductive isolation is important because many closely related species look quite similar yet cannot interbreed.

There are many ways for reproductive isolation to occur. As the groups continue to change, one or more *reproductive isolating mechanisms* will evolve that will prevent mating. As shown in Table 2–2, such mechanism can be divided into two basic types: (1) premating, and (2) postmating. Premating isolating mechanisms are those that prevent mating from ever occurring. For example, changes in behavior such as mating dances in birds or songs in crickets will prevent copulation. Physical changes in reproductive organs have the same effect. Postmating isolating mechanisms allow mating to occur but prevent production of healthy, fertile offspring. For instance, the sperm and egg will not properly unite or, if fertilization does occur, the developing embryo will be unhealthy or sterile. A good example of postmating isolation is the mule, which is the offspring of a horse and a donkey. The mule is healthy but reproductively sterile.

SPECIES ORIGINATION. You can now see that branching evolution via geographic separation does not actually create new species until the separated groups can no longer

TABLE 2–2. Types of Reproductive Isolating Mechanisms

PREMATING	EXAMPLE
1. Ecological isolation	Different habitats
2. Behavioral isolation	Different mating seasons
3. Mechanical isolation	Different genital shapes

POSTMATING	
1. Gamete death	Sperm die before reaching egg
2. Embryo death	Spontaneous abortion
3. Offspring subnormal	Offspring sterile or crippled

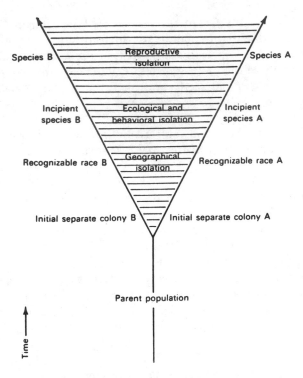

Figure 2–12. Steps in the evolutionary divergence of a single parental population. This divergence can lead to the formation of two distinct species and possibly a third if the parental population survives. From C. Swanson, *The Natural History of Man* (Englewood Cliffs, NJ: Prentice-Hall, 1973), p. 155.

interbreed. This is shown in Figure 2–12; at first the separated populations are recognizably different but are still able to interbreed. In such cases, they are called **races** or **subspecies** because they still belong to the same species. However, as separation continues, one or more isolating mechanisms will develop, preventing successful mating. At this point, two species exist where there was only one.

How can we recognize this process in the fossil record, which is the only record we have of the origination of nearly all species? There are many cases of fossil lineages where groups clearly undergo branching evolution. However, interbreeding ability and reproductive isolating mechanisms do not fossilize, so that we have no idea when reproductive isolation occurred between the branches. This is called the **paleospecies problem:** how to distinguish among fossil species because we cannot use the interbreeding criterion. The problem is partially solved by classifying fossil species based on detailed anatomical description. This is probably usually valid because different species generally do differ anatomically. However, without knowledge of interbreeding ability, there will always be some uncertainty, especially among closely related, anatomically similar groups.

WHY IS LIFE PACKAGED AS SPECIES? This question seems nonsensical at first because we take for granted, for example, that cats do not interbreed with dogs or rabbits. However, natural selection works to isolate even closely related populations if they occupy different environments. This is not just some byproduct of anatomical change. It occurs because each species is finely tuned to its environment and way of life. If interbreeding were to occur with individuals adapted to a different environment, it would be mixing traits that were not as well adapted to each individual's way of life. To take an extreme example, consider what would happen if a cat could breed with a rabbit. The mixture of traits would make the offspring less effective a meat-eater than a cat and less effective a plant-eater than a rabbit. Therefore, natural selection acts to eliminate such mixing.

This leads to a very fundamental point: why life on earth has produced millions of species, instead of just millions of individuals that can all interbreed. Recall that all life has the same DNA code, so that the genes, say, of a plant can be "read" by those of a dog. Yet they clearly cannot interbreed because the ancestors of the dog took up a life so different from the plant's ancestors that they became reproductively isolated. This is why the species is the fundamental unit of life. When adaptations become too different between groups, mixing their traits only "dilutes" the adaptations of both. Thus, natural selection acts to isolate those groups.

PART 2: INHERITED VARIATION

Genes: What They Are, What They Do

So far, we have focused on selection, the second half of our "equation":

$$\text{variation} + \text{selection} = \text{evolution}$$

However, without variation, evolution would cease because selection would have nothing to act upon. Let us now turn to how variation is produced.

The basic "laws" of inherited variation were discovered by Gregor Mendel in 1866 while Darwin was wrestling with the problem of how variation arose. Unfortunately, Mendel's work lay unnoticed until the early 1900s when other scientists independently rediscovered the **gene** as the basic unit of heredity. Genes are now known to be molecular "blueprints" which are repeatedly copied within each cell. They contain instructions on how to: (1) build the organism (create cells during development), and (2) how to maintain the organism (tell cells how to function).

GENES AND DEVELOPMENT. Genes are passed on to the offspring when the gametes (sperm and egg) unite, as shown in Figure 2–13. The resulting fertilized egg, called

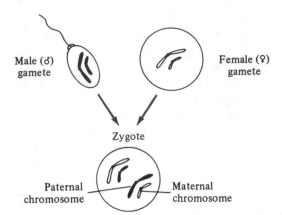

Figure 2–13. Fertilized egg (zygote) is a combination of chromosomes from the mother and the father. Only four chromosomes are shown for simplicity. Modified from J. Barrett et al., *Biology* (Englewood Cliffs, NJ: Prentice-Hall, 1986), p. 607.

the **zygote,** consists of one cell. Within the zygote cell nucleus, the genes occur on strands called **chromosomes.** The chromosomes occur in pairs, with one member of each pair from the father and one from the mother (Figure 2–13). Humans have 23 pairs of chromosomes containing a total of about 100,000 genes. As the embryo grows, this original cell will multiply into trillions of new cells, but the genes in *each new cell* will generally be exact copies of those in the original 23 pairs of chromosomes. This cell multiplication, with duplication of all chromosomes, is called **mitosis.** As growth continues, certain genes in each cell will be "read" and will give instructions on what happens next. For example, some cells will migrate and change into liver cells, others will become brain cells, and so on. This reading and performance of instructions are accomplished by a series of very complex biochemical reactions that are poorly understood. Just how an embryo develops from a single fertilized egg is the study of developmental biology and it is a major area of current research.

GENES AND THE ADULT. Once the embryo has developed into an adult, the genes in the cells still play a key role in the individual by governing maintenance. Maintenance is a huge task, covering millions of chemical reactions that occur in our cells to keep us alive. Examples include directing how cells metabolize food, produce enzymes, and fight disease.

In addition, genes must be copied for reproduction. This occurs in the production of gametes, the sperm and egg cells. In these cells only one of each chromosome pair is found. This is called the **haploid** state, in contrast to the **diploid** state of nongamete (body) cells, which have both chromosome pairs. The reason that gametes exist in the haploid state is obvious: If one person's egg or sperm is to unite with the egg or sperm from another person, only one of each pair is appropriate (see Figure 2–13). The process of producing haploid gamete cells is called **meiosis.** Figure 2–14 shows how meiosis differs from mitosis. In addition to producing half the normal complement of chromosomes, meiosis also differs in that genes are shuffled about during the copying process. This shuffling is called **recombination.** For example, even though brothers and sisters have the same parents, they clearly differ (except for identical twins, formed when the zygote splits into two zygotes). This is because of recombination among the parent's sperm and eggs so that each offspring gets a different subset of each parent's genes. (On average, siblings have 50% of the same genes.) Recombination is crucial in evolution because it greatly increases the amount of individual variation for selection to act upon.

How Genes Work. Genes are composed of the molecule **DNA** (deoxyribonucleic acid). DNA is shaped like a helix, often called a twisted ladder. As shown in Figure 2–15, the rungs of the ladder are made of "base pairs." The four kinds of base pairs are adenine (A), cytosine (C), guanine (G), and thymine (T). These four bases form the "genetic alphabet" in that *gene expression is completely determined by the sequence of the bases.* Genes are expressed when the DNA helix unzips down the middle and the bases undergo **transcription,** whereby messenger **RNA** (ribonucleic acid) molecules line up in a sequence determined by the DNA. This transcribed sequence

Figure 2–14. (Top) A cell undergoing mitosis to produce two cells identical to the original cell. (Bottom) A cell undergoing meiosis to produce two germ cells with half the normal complement of chromosomes.

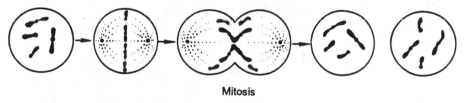

Mitosis

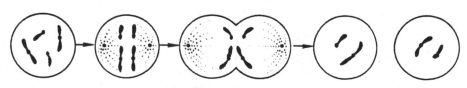

Meiosis

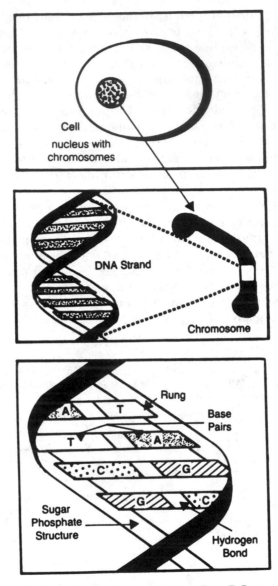

Figure 2–15. The structure of DNA. From R. Brennan, *Levitating Trains & Kamikaze Genes* (New York: Wiley, 1990), p. 46.

of messenger RNA then travels outside the cell nucleus to the ribosome, which is the biochemical "factory" of the cell. At the ribosome, the messenger RNA sequence determines which proteins are manufactured by the cell. As shown in Figure 2–16, this is because amino acids are linked according to the RNA sequence (and proteins are composed of amino acids chains). **Translation** is the name of the process whereby the messenger RNA sequence is converted into proteins. Because proteins make up most enzymes, tissues, organs, and many other parts of the body, genetic determination of these proteins therefore controls how the body is assembled (development) and how it is maintained. Interestingly, most DNA in humans, and indeed in all mammals, is not translated. This "silent DNA" does no apparent harm and seems to consist largely of copying errors that have accumulated through time. The sheer amount of DNA is awe-inspiring: If you could uncoil all the DNA helixes in just one human cell, it would be 6 to 8 feet long.

If all of the original 23 chromosomes are copied into later cells, then why does the human body have over 100 kinds of cells? In other words, if all cells have the same genes, why do cells behave differently? For instance, how does a liver cell "know" it is supposed to become a liver cell and not a brain or muscle cell? The answer is that only a *limited part* of the total DNA (genes) in any cell are actively being copied and used. As the embryo grows, biochemical stimuli from adjacent cells "tell" each cell where it is in the growing body. Thus, if a cell is in the location where the liver is to form, only the genes that cause liver cells to develop and function are "switched on." Obviously, this is an incredibly complex process, with

Figure 2–16. This four-stage process shows the assembly of a polypeptide chain. (1) The DNA double helix is contained in the nucleus as part of the chromosomes. (2) Messenger RNA (ribonucleic acid) transcribes the DNA code and carries it to the site of protein synthesis on the ribosomes. (3) Small chains of transfer RNA bring the correct amino acids into position by temporarily attaching to the messenger RNA. (4) The completed protein chain detaches and becomes a part of the cell's cytoplasm.

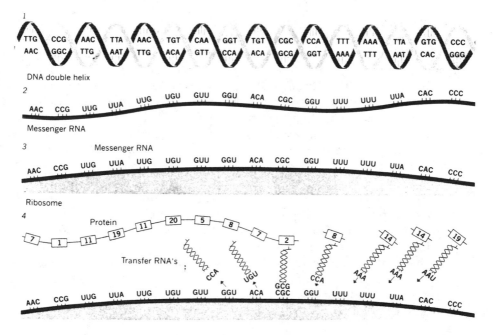

much interplay and feedback among the cells and genes in the developing embryo. Once the adult has developed, only those genes appropriate to cell function continue to be expressed. For instance, genes for the production of digestive enzymes will be expressed in the liver cells, but not in skin cells (even though skin cells contain those genes).

Genes in Evolution

With this added knowledge of genes as the units of heredity, we see that natural selecti · acting on individuals selects not only traits but also selects the genes that serve to determine those traits. Thus, as well as being a change in species' traits through time, evolution is also often defined as a *change in the gene pool of a species* through time. The **gene pool** is the total of all genes contained in a species. When the environment changes, newly advantageous genes will become more common in the pool, while disadvantageous genes will tend to be lost. For example, any existing genes that produce enzymes that promote resistance to insecticides will become common in insect species when poison becomes a new part of their environment.

There are two basic factors that increase the amount of variation in a gene pool:

1. *Shuffling* of genes within the gene pool—increased by sex
2. *Size* of the gene pool—increased by mutation and gene flow

Let us discuss each of these.

SHUFFLING THE GENE POOL: SEX. Selection does not act directly on genes or as though genes were isolated units. Genes are only sorted on the basis of how they come together to create and maintain an individual organism. In other words, it is the success of the individual (the set of that individual's genes as an *integrated whole*) that determines whether a gene will be passed on. This has led some biologists to point out that organisms may therefore be profitably viewed as "survival machines" constructed by genes for their own perpetuation. For example, in his widely-read book *The Selfish Gene* (1976), Richard Dawkins notes that genes, and not organisms, are the basic units that persist through evolutionary time. Thus, many of the genes in our bodies are millions of years old, having been passed along from our ancestors.

Because selection acts on the individuals that genes produce and not the genes directly, variation among these individuals is important. This is the key role of sexual reproduction: it *produces variation* among individuals. It does this in two ways. One, each offspring is different from both parents because the offspring is a combination of 50% of the genes from each parent. Two, because of recombination in the creation of gametes (meiosis), no two offspring are alike (except when identical twins are produced).

This production of variation is illustrated in Figure 2–17. In this case, both father and mother possess the combination Ww at the gene site for fur color. This

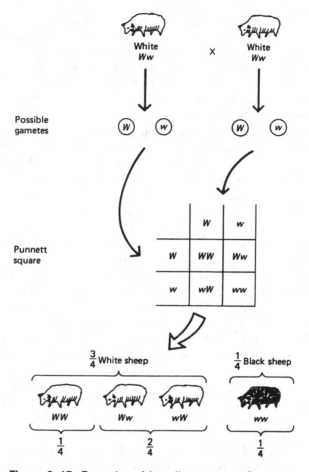

Figure 2–17. Examples of breeding crosses. Cross between heterozygous white sheep produces different genotypes and phenotypes. Modified from F. Racle, *Introduction to Evolution* (Englewood Cliffs, NJ: Prentice-Hall), p. 44.

means that on a certain chromosome pair, one chromosome codes for W and the other w, on both parents. The alternate forms of genes at this site are called **alleles.** W is the allele that codes for white fur color and w is the allele for black. Both parents have white fur because white (W) is **dominant** and black (w) is **recessive,** meaning that only W is expressed when it is paired with w. (For some traits, such as some flower colors, alleles are codominant and the traits mix; for example, red and white alleles would yield pink flowers.) The main point is that each parent has two possibilities to donate to each gamete. In this example (Figure 2–17), the gamete will receive either a W or w; the male's sperm has a 50% chance of carrying a W and a 50% chance of a w. This is also true for the female's egg.

A good way to visualize this is with a **Punnett square,** shown in Figure 2–17. Each block represents the possible combination of alleles that an offspring might receive. In this case, there is a 25% chance of WW, a 25% chance of ww, and a 50% chance of Ww. Therefore, regardless of how many offspring are produced, about half would have the Ww combination. This is called the **heterozygous** combination because the alleles are different. In contrast, the other half of the offspring would be **homozygous,** having the same alleles at the site: about 25% would be homozygous WW and 25% would be homozygous ww. The **phenotype** describes how these **genotypes** (allele combinations) are expressed: about 75% of the offspring would have a white phenotype, because W is dominant. Only where the genotype ww occurred would the offspring have the black phenotype. This means that heterozygous individuals can be carriers of an allele (such as genetic diseases), but the allele is masked if it is recessive. This allows disease genes to persist much longer because individuals with the allele are not impaired by it.

The fur color example has illustrated how variation in a gene can be passed on to produce offspring that are different from the parents and each other. For example, the black offspring differs from the white parents and offspring. When you consider that we looked at only *one gene* and that complex organisms have many tens of thousands of genes, the vast potential for variation in offspring is obvious. This potential is increased even more in that sometimes even a single gene can control many traits. This is called **pleiotropy.** In other cases, many genes are involved in a single trait; height is such a **polygenetic** trait.

The importance of the variation produced by sexual reproduction is made clear when we look at those organisms (such as some single-celled forms and some of the other simpler animals and plants) that are **asexual:** they do not reproduce sexually. Instead of shuffling alleles during reproduction, they simply clone themselves, meaning that they reproduce exact copies of themselves. From the point of view of an individual, this may seem like a good thing; it is a chance to live longer, in a sense, by being reborn. However, from an evolutionary viewpoint this greatly reduces the variation available for selection to act on, crippling the ability of the species as a whole to change if the environment changes. Indeed, we will see in Chapter 3 that the evolution of life was greatly accelerated when sex was "invented," for just this reason.

How, then, do these nonsexual organisms evolve? One reason is that even in the absence of sex, there is one way of producing individual variation. This is by mutation, which we discuss next.

EXPANDING THE GENE POOL: GENE FLOW AND MUTATION. While sexual reproduction produces individual variation by shuffling and combining genes in the gene pool, this does not explain where the different genes and their varieties (alleles) originated. For example, where did the W and w alleles originate? There are two answers: gene flow and mutation. **Gene flow** is the movement of genes from one gene pool to another, usually by migration of individuals. Thus, gene flow into a gene pool will increase the size of the gene pool, thereby expanding the total amount of variation available for sexual reshuffling. If gene flow is very small and the size of the gene pool is also small, **genetic drift** will occur, meaning that chance events play a large role in determining the genes in the gene pool. The classic example of this is the "founder effect," discussed above, where just a few individuals may initially make up the whole gene pool on an island. The genes of these few individuals will supply the entire gene pool for generations to come, unless incoming gene flow increases the variety.

Although gene flow is one way to increase variation in a gene pool, we have still not answered the question of where genetic variation initially comes from—why is there variation in the incoming genes? The answer is that *mutation is the ultimate source of variation* in a gene pool: genes and their alleles originate as mutations. A **mutation** is defined as a heritable change in a gene, such as a change in the DNA. For example, a single change can lead to a different amino acid in a protein, greatly changing the properties of that protein (such as an enzyme or a pigment).

Any given gene has a very low chance of mutation. On average, a mutation occurs about once every million times a gene is copied. But since there are so many genes, the overall chance of a mutation somewhere is high. In humans, for example, there are often one or two mutations in each sperm or egg cell produced. However, these mutation probabilities are not evenly distributed: Some genes tend to mutate much more often than others. Mutation frequency varies not only among organisms, but also among genes in the same organism, as shown in Table 2–3. The cause of mutations can also vary: chemicals and intense radiation can cause them. Other mutations appear "spontaneously" and have no clear cause.

Many mutations are not expressed. For example, they contribute to the "silent DNA" noted above. This is fortunate because those mutations that are expressed often kill or handicap the offspring. The reason is that any organism is a highly integrated, complex system and a major mutation is therefore likely to disrupt it. (Consider a finely tuned watch: A random disturbance is more likely to be detrimental than beneficial.) However, rare improvements do occur and it is these mutations that are passed on and increase in the gene pool. It is also possible for mutations to be expressed and be adaptively "neutral"; in other words, they are neither improvements nor detriments. For example, blue eye color in humans is a relatively recent

TABLE 2–3. Approximate Mutation Rates of Selected Traits

ORGANISM	TRAIT	MUTATION RATE
Bacterium (*Escherichia coli*)	Streptomycin resistance	1 in 1,000,000,000
Fruit fly (*Drosophila melanogaster*)	White eye	4 in 100,000
Humans	Achondroplasia	2.8 in 100,000
	Albinism	2.8 in 100,000
	Hemophilia	3.2 in 100,000
	Total color blindness	2.0 in 100,000

mutation that probably has no significant (if any) selective advantage, but has managed to become part of the human gene pool. Where such neutral mutations manage to spread through the gene pool, it is by chance sampling processes (such as genetic drift) rather than by selection.

POPULATION GENETICS: SUMMING UP. Population genetics provides a good summary of genes in evolution because it studies how shuffling of genes occurs in the gene pool. For example, suppose we know the relative proportion of two alleles in a gene pool. If p = proportion of dominant (A) alleles and q = proportion of recessive (a) alleles in the gene pool, then:

$$p = \text{proportion of AA individuals (homozygous dominant)}$$
$$2pq = \text{proportion of Aa individuals (heterozygous)}$$
$$q = \text{proportion of aa individuals (homozygous recessive)}$$

To illustrate with the example of W and w alleles above, suppose that W alleles compose 0.80 (or 80%) and w alleles compose 0.20 (or 20%) of the alleles in a given gene pool. Then, we would expect individuals with the WW genotype to make up about 64% of the population ($0.8 \times 0.8 = 0.64$). Heterozygotes would make up about 32% of the population ($2 \times 0.8 \times 0.2 = 0.32$), while ww individuals would make up only 4% ($0.2 \times 0.2 = 0.04$), even though ww alleles are 20% of all alleles in the pool. (Thus, in terms of phenotype, only 4% of the population would be black.) Notice that $64\% + 32\% + 4\% = 100\%$. This will always be the case because:

$$p^2 + 2pq + q^2 = 1.0.$$

This is called the **Hardy-Weinberg equation** in honor of the two men who simultaneously discovered it in 1908.

Figure 2–18 displays the outcomes of the Hardy-Weinberg equation for all proportions of A and a. The following major points should be noted:

1. Heterozygotes (Aa) result most often when both alleles are at roughly equal proportions.
2. The number of heterozygotes decreases rapidly as the proportion of a increases and A decreases (and vice-versa).
3. Conversely, as alleles of a increase from 0, the number of aa genotypes is very low at first (note how a = 0.2 predicts aa = 0.04, as we calculated). Then aa shows an exponential rise as a increases. (The same is true for A and AA genotypes.)

These predictions are based on the assumption that the allele frequencies in the gene pool are not undergoing change from mutation, gene flow, genetic drift, selection, or some other process. Thus, the predictions are valuable as "null hypotheses" because deviations from them indicate that the alleles in the gene pool may be undergoing evolutionary change from one or more of these processes.

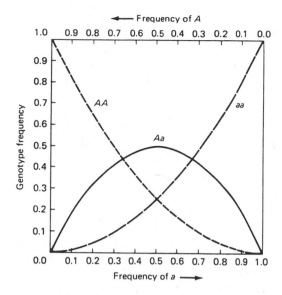

Figure 2–18. Relationship between gene frequencies and genotype frequencies. From F. Racle, *Introduction to Evolution* (Englewood Cliffs, NJ: Prentice-Hall, 1979), p. 55.

PART 3: EVIDENCE FOR EVOLUTION BY SELECTION OF VARIATION

Evidence for evolution by selection of variation is shown by indications that life has changed through descent with modification (to use Darwin's phrase for evolution). In other words, species would show patterns of divergence from some common ancestor because evolution occurs by *tinkering with existing species*. There are three major lines of evidence that show this:

1. Similarities among living organisms—these reveal hierarchical patterns of common ancestry.
2. Fossil record—it provides "snapshots" of past evolutionary events.
3. Ongoing evolution—direct evidence of evolutionary changes from selection in living organisms.

Let us discuss each of these, in order.

Similarities Among Living Organisms

If descent with modification has occurred, then we would expect that, going back in time, all life ultimately originated with a single ancestor whose descendants have since progressively diversified and branched off. Even though life has become highly varied through branching, we would still expect some very basic similarities that unite all life on earth, as an imprint from that common origin. Since life begins with chemical evolution, we would expect these most basic similarities to be chemical in nature. In addition, we can expect similarities in anatomy and embryological development. Indeed, this is what we find.

BIOCHEMICAL SIMILARITIES. As mentioned in our origin of life discussion, all life on earth today, from bacteria to whales to plants, shows two kinds of biochemical similarities. One, in the chemistry of heredity, the same DNA code is used by all organisms. Thus, a gene in a rose can, for example, "read" that from a human. Two, in the chemistry of physiology, many molecules of cellular function are the same, using the same biochemical pathways to respire, use energy, and so on. For instance, adenosine triphosphate (ATP) and cytochrome proteins are involved in respiration in every organism. Furthermore, biochemical traits are more similar in groups that have a more recent common ancestry. For example, blood proteins are similar in all mammals, but those of humans are more similar to the blood proteins of other primates than to those of dogs.

ANATOMICAL SIMILARITIES. Similarity of organs also reveals patterns of common ancestry. For instance, as shown in Figure 2–19, the forelimbs of humans, whales, dogs, and birds are all composed of the same bones. This is in spite of the forelimb's vastly different functions in each animal: flight in the bird, swimming (steering) in the whale, running in the dog, and grasping in the human. This indicates that all of these four creatures shared a common ancestor with a forelimb that was modified in different ways. Selection has variously "tinkered" with the ancestral forelimb as

Figure 2–19. Homologous organs can serve different functions. For example, the forelimbs of humans, dogs, whales, and birds are different in function, but are composed of the same elements and have similar embryological origins. From R. Wicander and S. Monroe, *Historical Geology* (St. Paul, MN: West Publishing, 1989), p. 138.

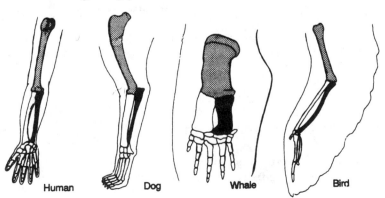

Human   Dog   Whale   Bird

the groups separated and took up different ways of life. These are called **homologous** organs because they are derived from the same ancestral organ. As you might expect, homologous organs often differ most between groups that separated in the more distant past, especially where they have taken up very different ways of life. Thus, the bird forelimb is more different from the human forelimb than the dog's is (Figure 2–19). This is because, in addition to a flying lifestyle, the birds separated from the line leading to dogs and humans over 100 million years ago, whereas dogs and the human lineage separated many millions of years after that.

An interesting kind of homologous organ is the **vestigial organ.** Such organs are no longer useful, but have not yet been fully eliminated by natural selection. For instance, horse evolution involved the reduction of all digits ("toes"), except for the middle (third) one on which horses now run (see Chapter 3). However, the remnants of the digits next to this middle toe are still present, being called "splint bones." These splint bones are homologous to the second and fourth toes on humans, modified from a common mammalian ancestor with five-digit limbs. Another example is the python snake, which has small, useless hind limbs, retained from its lizard ancestor. The human appendix is yet another of many possible examples, being the remnant of a once useful digestive organ.

Sometimes an individual has a vestigial organ that is well developed (often called an atavism or throwback). For instance, some horses are born with the splint bones grown into well-developed toes. This indicates that while natural selection may ultimately remove the vestigial organ entirely, the genes coding for such traits are still present in the species. Under rare genetic combinations, they become fully expressed. Vestigial traits are especially good evidence for evolution because they show that not all aspects of organisms are perfectly "designed." If life did not evolve, why do these largely useless (and sometimes harmful, such as the appendix) traits occur?

DEVELOPMENTAL SIMILARITIES. You may be wondering how we can recognize organs as being homologous, when they differ between two groups (such as the bird's wing and the human limb). The answer is that each bone and organ can be traced as it originates and develops during the growth of the embryo. Homologous organs come from the *same embryonic precursors*. This is a very crucial point because it shows that evolution occurs by modifying development. People usually visualize evolutionary change by noting species' differences among adults. But in reality, the only way to create different adults is to *alter how they develop* (are "assembled") into adults.

In general, only the later part of development is modified—early changes tend to alter too many later events. For instance, if a small group of cells in an embryo eventually forms the entire digestive system, even a slight change of those few cells could interfere with the development of the entire system. An entire organ could be lost if the few precursor cells that multiply into that organ are removed. On the other hand, a change in a few cells after the digestive system has formed (and grown to consist of billions of cells) would hardly be noticed. Thus, most evolution has occurred by altering only the later parts of development. Mutations that alter earlier parts are much more likely to be fatal and not passed on. There-fore, when we compare the entire developmental sequence of various groups, we see many developmental similarities among them. For example, a comparison of development in the major vertebrate groups, shown in Figure 2–20, reveals the most similarity at earlier stages. At eight weeks old, humans have gill structures, a primitive tail, a circulatory pattern, and the general shape of a fish. As shown in the fossil record, this is because humans are descended from certain fish that be-came the first land vertebrates. This is true of all the other vertebrates, from am-phibians to birds, which also have these fish as ancestors. As humans (and other

Figure 2–20. A comparison of development in major vertebrate groups. Similarities in appearance tend to occur in the earliest stages. From J. Barrett et al., *Biology,* p. 755.

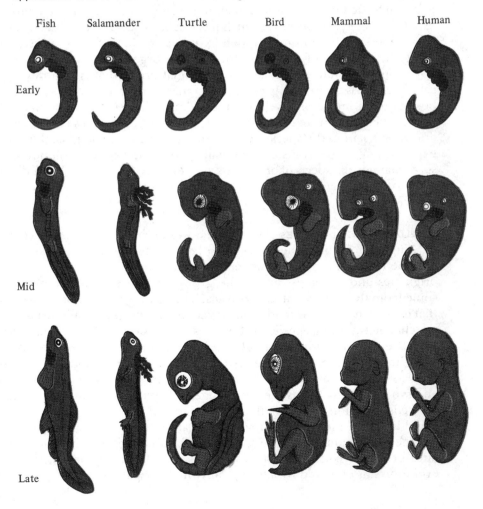

vertebrates) evolved, the newer traits (such as lungs, body hair, arms, larger brain, and so on) tended to be added on at the end of development.

This process of evolution through modification later in development leads to the following two basic observations. One, more closely related groups will tend to have more similar developmental sequences, usually not diverging until near the end of development. Compare, for instance, the human versus mammal and human versus salamander sequences (Figure 2–20). Human and mammal development diverge later than that of human and salamander. Two, in a very general way, the developmental sequence will repeat the evolutionary changes of the past. Because modifications are made at the end of the sequence, old evolutionary stages are retained. For instance, the human retention of gill structure and a tail give evidence that fish are ancestors of humans. When this was first observed by embryologists, this led to the **theory of recapitulation,** which said that developmental sequences always "recapitulated" (in other words, "repeated") the organism's evolutionary past. However, we now know that this is an oversimplification for a number of reasons. One of the most important is that early changes in development can sometimes (albeit rarely) successfully occur so that not all evolutionary changes occur as a simple "adding on" at the end of development. Nevertheless, the basic pattern of evolutionary modification can often be seen in the developing embryo, even though development does not exactly repeat each evolutionary change of the species' ancestry.

SIMILARITIES AND THE CLASSIFICATION OF LIFE. The chemical, anatomical, and developmental similarities discussed above are inevitably going to affect how we perceive and organize living things. It is obvious that some groups, such as snakes and lizards, are more closely related than others, such as snakes and dogs. Generally speaking, descent with modification means that *more similar groups have more recently become separated.* Conversely, the longer that groups have been separated, the more they anatomically (and behaviorally) diverge, as they continue to be modified in different ways. (For example, races tend to be more anatomically different than species—see Figure 2–12.) This progressive separation, much like a family tree, inevitably leads to a hierarchical classification, or nesting, whereby groups can be classified into more closely related subgroups and less closely related supergroups. In short, descent with modification causes a hierarchical pattern of similarity.

Such a pattern is shown in Figure 2–21, the formal biological system of classification, using the coyote as an example. So apparent is the hierarchy of life that this formal scheme—species, genus, and so on—was proposed by Carolus Linnaeus in the 1700s, well before evolution through natural selection was even heard of. Similar genera are grouped into families, which are grouped into orders, which are grouped into classes. The process of classifying organisms into this scheme is called **taxonomy.** In the case of the coyote (Figure 2–21), it is a mammal (specified by the class) and a member of the order Carnivora (meat-eaters). Within this order, it belongs to the subset that includes only dog relatives, the family Canidae. (Thus,

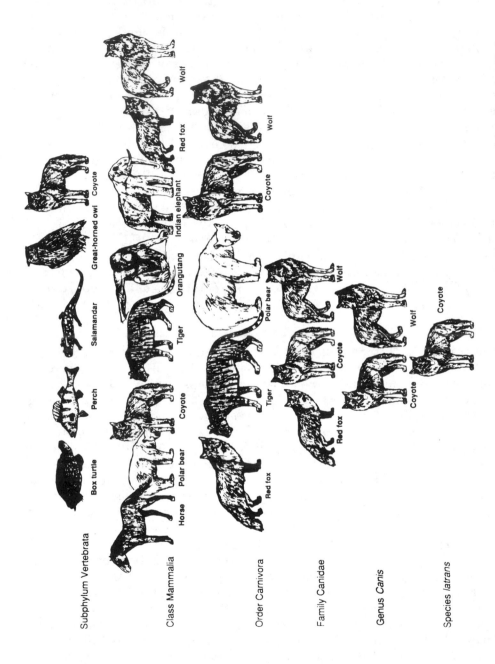

Figure 2–21. A classification of organisms according to their shared characteristics. Members of the subphylum **Vertebrata**, including fish, amphibians, reptiles, birds, and mammals, have a segmented vertebral column. Among these, however, only warm-blooded animals with hair or fur and mammary glands are considered mammals. Eighteen orders of mammals are recognized, including the order **Carnivora**, which is distinguished by specialized teeth necessary for an all-meat diet. The family **Canidae** and genus **Canis** include only doglike carnivores and closely related species. From R. Wicander and S. Monroe, *Historical Geology* (St. Paul, MN: West Publishing, 1989), p. 137.

cats fall within a different family, the Felidae, although they are also in the order Carnivora).

Taxonomy may seem straightforward enough, but there are many practical problems in classifying life forms. For instance, "similarity" is a very subjective perception—to some taxonomists, two groups may seem much more closely related than to other taxonomists. The main problem is deciding which traits are most important as criteria for classifying. **Cladistics** is one of the new methods that attempts to objectively classify traits. For example, **primitive traits** are those that are shared by many organisms (such as hair in mammals). In contrast, **derived traits** are traits that are more recently evolved, usually being therefore limited to fewer organisms (such as grasping hands in primate mammals).

The problems of classification are greatly magnified in fossils, where most traits and many species are not preserved. Yet since over 99% of all life is now extinct, taxonomic classification of fossils must be attempted if we are to understand evolutionary relationships. Also, the sheer number of species makes that job very difficult. Even ignoring the fossil record, over a million living species have been described and classified; yet there may be 30 times that number undescribed (many of these are insects).

Fossil Record

In addition to similarities among living organisms, similarities among fossils also provide evidence for evolution. Unfortunately, these similarities are often not as visible to us because only the more durable parts of organisms are preserved. (Indeed, many groups leave virtually no fossils at all.) Thus, chemical similarities and most developmental similarities are not visible in fossils. Nevertheless, we can recognize many anatomical similarities among hard parts, such as bones, teeth, and shells, that provide excellent evidence for descent with modification. In fact, where available, they are even better than similarities among living organisms because fossil similarities directly capture evolutionary modification, as a "snapshot" in time.

The most informative "snapshots" are fossils that record the transition between groups. That is, they are *anatomical intermediates* between groups. A more common (but oversimplified) name for such intermediates is "missing link." A good example of a fossil intermediate is one of the best known of all fossils, the "bird-reptile" *Archaeopteryx*, illustrated in Figure 2–22. This is one of the earliest birds to appear, dating to roughly 150 million years ago. More importantly, *Archaeopteryx* has anatomical features that are both reptilian and birdlike. In fact, when it was first discovered, it languished for many years in a museum drawer because it was thought to be a fairly typical small reptile, on the basis of its skeleton: clawlike hands, teeth in the jaw, sternum and other breast bones too poorly developed for flying, and many other features. However, closer inspection revealed the imprints of feathers. Nor is this specimen a fluke, as a number of specimens have now been classified.

Figure 2–22. *Archaeopteryx* is a fossil considered to be a transitional "missing link" between reptiles and birds. From F. Racle, *Introduction to Evolution*, p. 72.

Many other transitional fossils also exist, as we will see in Chapters 3 and 4. For instance, there is an intermediate fossil organism between fish and the first land vertebrates, the amphibians. Human evolution also shows many transitional forms, especially in the later ones leading to modern humans.

Ongoing Evolution

Ongoing selection of populations and gene pools is the most direct evidence for descent with modification. There are many cases of both artificial and natural selection that we can observe.

ARTIFICIAL SELECTION. One of the most common kinds of artificial selection occurs when humans purposely breed certain individuals for scientific, aesthetic, or practical reasons. Thus, a scientist can cause evolution in the lab, often with fruit flies. (Fruit flies have very large, visible chromosomes and short generation times.) Animal breeders have been doing such selection for much longer—dogs have been domesticated for over 10,000 years. Breeding of food plants and animals is also many thousands of years old. By selecting individuals for size, productivity, beauty, or whatever traits we desire, humans are causing evolution. In essence, we, and not nature, set out to define what is "fittest."

A second common kind of artificial selection occurs when humans try to eradicate pests such as disease microorganisms or insects. This second kind is sometimes called "negative" artificial selection because humans actively try to *eliminate* types of individuals. In contrast, positive selection, discussed above, is where we *encourage* production of certain individuals. Because resistant individuals almost always exist and therefore become the "fittest," negative artificial selection leads to the evolution of populations that are resistant to the poisons. Figure 2–23 shows how directional selection favoring insects resistant to DDT causes the whole population to generally become more resistant. (Compare this to Figure 2–6.) As a result, new poisons must be constantly invented (see Chapter 5). Such resistance evolves quickly in small organisms because they have very short generation times and produce many offspring (and therefore more chances for mutations).

Figure 2–23. DDT resistance and the directional effect of selection. (A) Resistance to DDT varies within the population. Upon exposure to DDT, only the resistant members of the population survive. (B) Because survivors from A represent 100% of the breeding population, a greater percentage of B will be resistant to DDT.

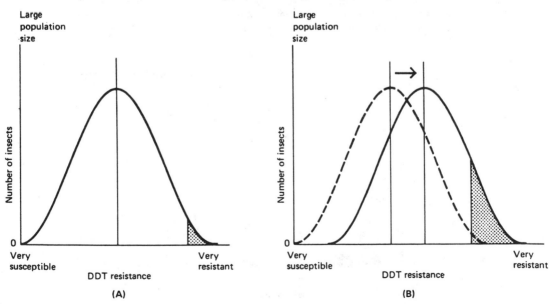

NATURAL SELECTION. In addition to ongoing artificial selection, we also observe ongoing selection in nature. Widespread alteration of the landscape has led to many examples whereby natural selection adapts populations to their new environments. For example, pollution of the soil around some mines has led to the origination of plant populations that are highly tolerant of toxic soils. Figure 2–24 shows how directional selection of resistant individual grasses causes the whole population to eventually become resistant.

Probably the best-known case of such natural selection is that of the peppered moths. These moths have two basic colors, the light form and the melanic (dark) form. As shown in Figure 2–25, the light form is less visible to predators in light-colored trees. Such trees were common in England before 1850 so that the dark form made up less than 1% of the population in most areas. After 1850, a byproduct of the industrial revolution, soot, covered many trees, making them darker. Because the darker form was now less visible to predators, by 1900 the dark form of peppered moths made up over 90% of the population in highly industrialized areas.

Note that the above examples differ from artificial selection because humans are not actively targeting individuals to reproduce or die. Instead, natural selection is adapting the species for a new environment, just as it would if climate or some other natural agent were causing the environmental change. As human alteration of the global environment rapidly increases, this is becoming increasingly common (see Chapter 5). However, we can still observe natural selection occurring in areas without human interference. We have already discussed races (subspecies), which

Figure 2–24. Frequency distribution of height in seedlings for nontolerant grass, compared with tolerant, after growth on copper waste for a period of four months: seedlings more than 30 mm are survivors.

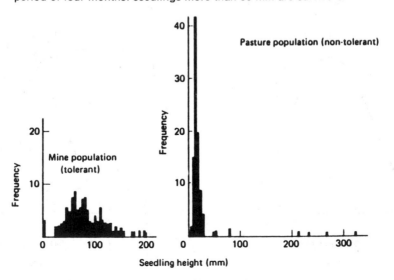

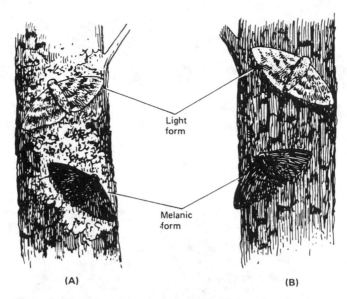

(A)   **(B)**

Figure 2–25. Inductrial melanism. (A) The light-colored and melanic (dark) forms of the peppered moth on an unpolluted, lichen-covered trunk. (B) The light-colored and melanic forms of the peppered moth on a soot-darkened tree trunk. From F. Racle, *Introduction to Evolution,* p. 80.

are essentially species in the process of formation. Races are populations of a species that are geographically differentiated. This is because the local areas they inhabit subject them to differing selective pressures. Given more time and isolation from gene flow, races would eventually become separate species. However, until then, they are still able to interbreed. An example is shown in Figure 2–26 (the rat snake has many races, varying drastically in color and marking pattern).

Finally, even though our technology has greatly diminished many types of natural selection on humans (a topic discussed in Chapter 5), it is worth noting that *natural selection is still important in human evolution.* For example, Figure 2–27 on page 65 shows that only about 12% of all human zygotes conceived today live to reproduce. This figure also reemphasizes another key point made before, that natural selection *acts at every stage in development.* An individual must be "fit" from the moment he or she is conceived as a fertilized egg and then on through adulthood. Thus, 50% of all human zygotes die before birth, often because they cannot survive in the embryonic environment.

MICROEVOLUTION: BUT WHERE ARE THE NEW SPECIES? Given the evidence just cited, it is surprising to many people that no one has ever actually seen a new species evolve. The standard answer to this is that natural selection takes many thousands, even

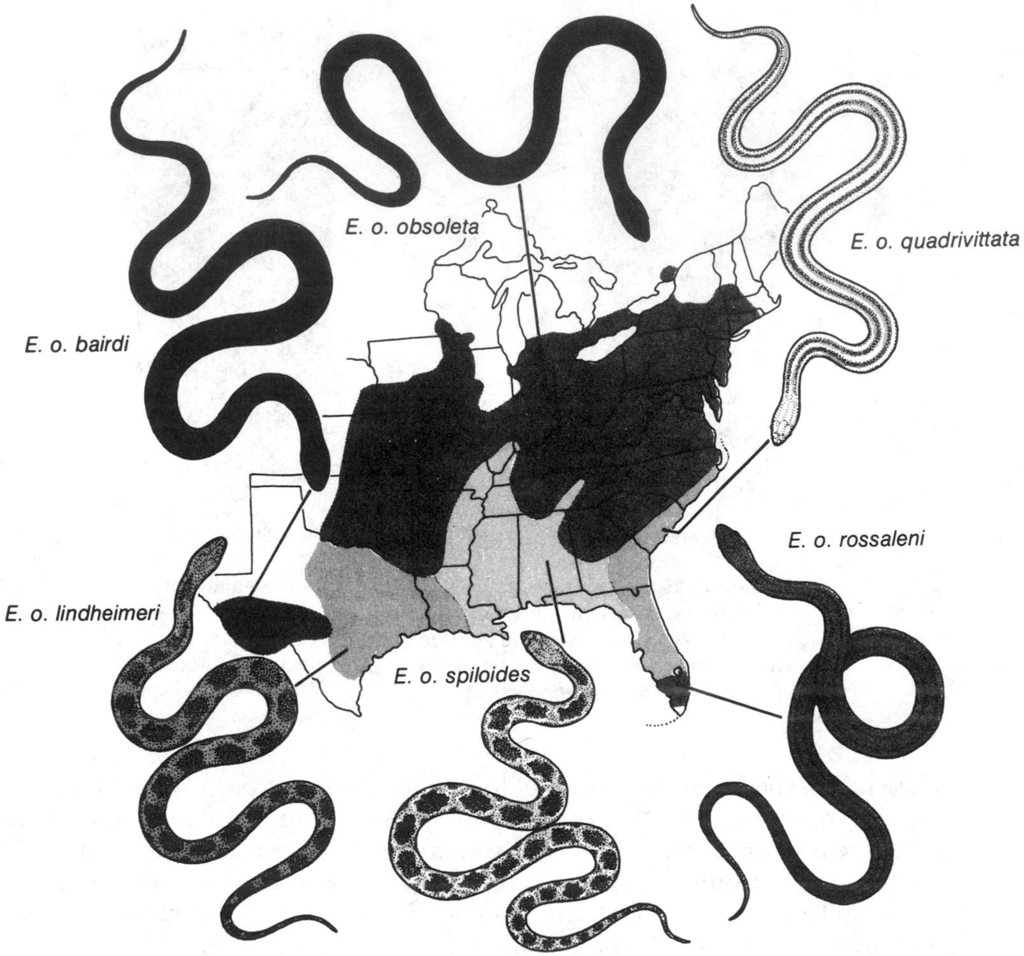

Figure 2–26. Classical subspecies in the rat snake. These geographic races interbreed where their habitats overlap. From D. Futuyma, *Evolutionary Biology* (Sunderland, MA: Sinauer, 1986), p. 106.

millions, of years to produce a new species because it takes that long for the gradual pace of most environmental change to alter the anatomy, behavior and produce reproductive isolation.

While this seems plausible, skeptics may argue that human interference has radically accelerated this process in two important ways so that evolution ought to be observable. One is by rapidly increasing the rate of change in the natural environment—consider the effects of mining and industry, to name just some of ways society has worked to alter the environment. The second way is even faster: artificial selection, either positive or negative, is thousands of times faster in removing individuals than the vast majority of natural processes. This is because human

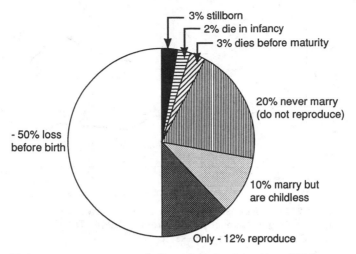

Figure 2–27. Natural selection today. Only about 12% of all human zygotes conceived ever reproduce. From H. Nelson and R. Jurmain, *Introduction to Physical Anthropology* (St. Paul, MN: West Publishing, 1991), p. 145.

intelligence can be much more precise in targeting individuals (and therefore genes) for directional selection. This is especially effective in organisms such as insects or microbes that can reproduce in weeks or even days. What happens under such conditions is that a species (such as a fruit fly species) will indeed show an initial rapid change in the selected traits. However, there is almost never the creation of a totally new species, with truly new traits accompanied by reproductive isolation. This is called **microevolution,** referring to selection over a few weeks or years that has caused changes in the frequency of traits and genes in a population. However, no new species originates.

Very few biologists interpret this to mean that evolution is not caused by natural selection. Instead, it emphasizes the *importance of variation* in the Darwinian "equation" of variation + selection = evolution. This is demonstrated in Figure 2–28: After many generations (and years) of selecting two populations of corn plants for high and low protein content (respectively), each of the two lines begins to level off. This reflects the limits of genetic variation. There is only so high or so low that the protein content can go, given the genes in the original parent population. Here, then, is the reason that natural selection takes so long to produce new species. No matter how intense selection is, it is limited by the amount of variation available in the gene pool. The long periods are needed to allow mutations to accumulate. Thus, even though mutations are random with respect to selection, some may eventually prove favorable and change the species in substantial ways. If time permits favorable mutations to accumulate, a new species will emerge. (If not, the species will simply become extinct, as the vast majority of species have done.)

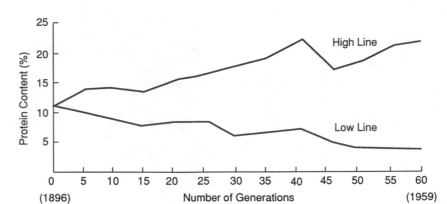

Figure 2–28. Response to selection for high or low protein content of grains in corn during a 60-generational period. From V. Grant, *The Evolutionary Process* (New York: Columbia University Press), p. 104.

Macroevolution refers to the long-term process of mutation and selection, occurring over thousands of generations, that result in new species, genera, families, and even higher taxonomic categories.

PART 4: MISCONCEPTIONS ABOUT EVOLUTION AND NATURAL SELECTION

ALL SPECIES' CHANGES ARE EVOLUTION. Some long-term changes in a species or population are not evolution because they do not occur from changes in the gene pool. For example, the height of the average Japanese citizen has increased substantially since 1945. This has occurred not because of any significant change in the Japanese gene pool, but because of an improved diet, which includes more protein and other nutrients. Similarly, most readers will be familiar with the concept of "stunted" growth in animals and plants because of poor environmental conditions such as poor diet. This is called **phenotypic plasticity,** defined as changes (in an individual or group) resulting strictly from environmental conditions alone, and not the gene pool. In time, such environmental changes would probably alter the gene pool, causing evolution. However, until that occurs, over a number of generations, any species changes, such as size change, are not due to changes in genes.

EVOLUTION IS RANDOM. Many people have interpreted Darwin's "equation" of the interaction between variation and selection as meaning that evolution, and hence human existence, are due to random events. This is because they think that mutations and environmental changes are products of chance. This reasoning is wrong for a number of reasons.

Most important is that the term "random" is widely misunderstood and misused. Random does *not* mean that something has no cause. Rather it means that any human observer, because of ignorance, cannot fully predict the outcome. That is, there are *too many unknown variables*. For example, the "random" flip of a coin would be fully predictable if one knows the exact velocity of the thumb, air pressure, and all the other variables of the process. Similarly, the process of evolution is "random" only in the sense that there are so many variables that the exact outcome at any time is unpredictable. Specifically, no one can predict just which mutations will occur (variation) and what environmental changes will occur (selection) over the next, say, 10,000 years. Since evolution is the outcome of the interplay of these two unpredictable variables, evolution would seem random to us over those next 10,000 years.

Thus, evolution is not random in the sense of having "no cause." Furthermore, it is not even completely random in the sense of human unpredictability. For example, mutations occur with considerable regularity, and the probability of different genes mutating is well known (Table 2–3). Similarly, the selection process is clearly nonrandom in how it sorts individuals based on specific traits and genes.

NATURAL SELECTION AND GENES ANTICIPATE CHANGE. We discussed in Chapter 1 how genes do not predetermine evolution. We note here that not only do genes lack a "plan" for the future, they cannot even anticipate (predict) future needs of the species, even in the short run. For instance, genes will not mutate in response to environmental changes that have not yet occurred, nor even respond to changes that have just occurred. Instead, genes mutate *without reference* to a species' immediate or future needs. If the mutations are helpful, then they are incorporated into the gene pool. However, many species have become extinct because the right mutations did not occur at the right place or right time.

Similarly, in modifying a species by sorting the gene pool, natural selection cannot anticipate future needs. Natural selection acts only to adapt the species to an *immediate situation*. But as with everything in life, what is best for the present, may prove detrimental later on. (Recall Richard Dawkin's review of this topic in his book, *The Blind Watchmaker*.) For example, through natural selection the koala of Australia has become highly specialized for eating only one food source, the eucalyptus tree. This has been a highly successful adaptation because the eucalyptus is a naturally abundant tree and there are few competitors for the food. However, should something happen to this single food source, the fate of the koala is sealed. Likewise, the panda feeds mainly on the bamboo plant and is therefore in danger of extinction, as that food source is rapidly being destroyed by humans.

That genes and selection have no foresight may seem obvious when explained this way. Yet, this misconception arises intuitively in many people. This is because humans *do* have foresight: we can anticipate problems by considering the consequences of our actions before we carry them out.

SPECIES DIE OF OLD AGE. **Racial senescence,** meaning that species die of "old age," was once advanced as a cause of extinction. However, we now know that individual organisms die because they have an internal metabolic clock that clearly limits the time that they can stay alive. In contrast, species are unlike individual organisms in that there is no such internal "clock" mechanism. Because the individuals that compose any species reproduce themselves, there is no preset limit on how long the species can persist. Instead, species are terminated when some "external" event in the environment overwhelms the species' ability to adapt.

Because a species' ability to adapt is determined by its pool of variation, a more sophisticated argument for racial senescence might be that species lose their variation through time. Indeed, we have just noted that some species, such as the panda, have become more specialized (and extinction-prone) through time. However, increasing specialization is not the rule for all older species. The oppossum, for instance, is often called a "living fossil" because it is one of the oldest mammals. Yet it is just the opposite of the koala and panda: it has a wide range of foods and habitats, and is flourishing as a result. The difference is that in the oppossum's case, the "blind" series of "choices" made by natural selection produced an animal that was much more broadly adapted.

Theory aside, there is also direct evidence from the fossil record that the age of a species has little to do with its likelihood of extinction. Statistical studies of species' ranges indicate that species of any age are about equally likely to die out at any given point in time.

COMPLEX ORGANS CANNOT ARISE THROUGH SELECTION. A major criticism of natural selection has been the "half-organ problem": How can something as complex as an eye evolve when the early stages of its evolution are of no use? In other words, what good is half an eye? If it is of no use, then natural selection cannot account for the evolution of the eye because individuals with half an eye would not be favored.

Actually, this assumption is invalidated two ways. One, an organ (or trait) may be of use in some *other function* in the partial state. Thus, selection favors its growth in that function, until it reaches a threshold, at which point it becomes useful in the new function. For example, some paleontologists think that the ancestors of birds used their incipient (partially developed) wings for catching insects. While such a wing was of no use in flying, it offered a selective advantage in catching food. At some point, the wing became large enough to be of use in gliding, and selection began to mold the wing to that function. This process, whereby an organ used for one function later becomes used in another function, is called **preadaptation.** In this case, wings were preadapted for flying because of their previous design for catching insects. There are many other cases of this, discussed later in the book. One prominent example is the human hand. We use it to write, drive cars, and so on, but it originally evolved as an adaptation for climbing trees.

The second argument that disproves the "half organ" premise is that many organs are *useful in early stages*, even without preadaptation. To return to the eye, many simple life forms have light-sensitive nerve outlets that are far simpler than a human eye but serve the organism very well in detecting environmental conditions. Furthermore, the animal kingdom exhibits a gradient of organs, wherein the simple organs are progressively modified to permit ever greater resolution of vision.

EVERY TRAIT IS ADAPTIVE. There is a common tendency toward **adaptationism.** This is the tendency to attribute virtually every feature of an organism to some adaptive function. There are two basic problems with this tendency. One is that not all traits are adaptive. We have already discussed vestigial traits, such as the human appendix, which is not only essentially useless but dangerous. Further, even nonvestigial traits may have no real function. For instance, the color of human eyes has little if any provable adaptive value. Such traits arise by mutations and can become entrenched in the gene pool, not because they are useful but because they are adaptively neutral and therefore increase by chance.

Note the phrase "provable adaptive value" just used. This brings us to the second problem: It is almost impossible to prove the adaptive value of many traits, even if they are useful. This has led to all sorts of unprovable rationalizations, especially about the adaptive origins of human behavioral traits. For instance, one often hears that territoriality, or deviousness, is a human trait that once had adaptive value. Yet trying to accurately measure such traits is nearly impossible. Even trying to prove that longer ears on a rabbit population are useful would require observing many generations of rabbits. This provability problem is common because the human imagination is often more creative than nature and can attribute function where natural selection had no role. For instance, the philosopher Voltaire once joked in his writings that the reason humans had noses was to support eyeglasses. The difficulty of measuring adaptiveness is compounded because many traits have many uses. Consider the human mouth: it is used for talking, eating, kissing, breathing, and many other activities. But which ones were most important to natural selection in designing the human mouth? It was not necessarily all of them and surely not all were of equal importance.

In short, it is true that organisms are impressively designed with features finely attuned to what they do. However, we must use caution in inferring function and how traits originated through natural selection.

PROCESSES OF HUMAN EVOLUTION

The rapid rate of cultural evolution is best illustrated by comparing humans today with our ancestors of 50,000 years ago. There has been no significant change in our biology: An individual from that time would pass unnoticed in a city street when properly attired. Yet our technology has changed dramatically, from stone tools to spacecraft. Technology is just one aspect of **culture,** which can be roughly defined as elaborate learned behavior. Besides technology (tools), other aspects of culture would include social customs, values, and many other kinds of learned information that we assimilate in our lives as social animals. Thus, whereas biological evolution is based on the transmission and natural selection of genes (chemically coded information), cultural evolution has been based on the different process of transmission and selection of learned information. As an analogy with genes, the basic units of learned information are called **memes.** These are simply the ideas that are stored and passed on. Thus, like a gene pool, each population has a meme pool. Also like genes, the amount and variety of this information is important for survival of the group. Should a radical environmental change occur, the variation is a reservoir for coping with it. This is an important reason why diversity of cultures as well as species should be preserved as much as possible.

Comparing Cultural with Biological Processes

A comparison of cultural versus biological information is shown in Table 2–4. This comparison reveals five main reasons why cultural evolution is so much more rapid and efficient than biological evolution. One, learned information is transmitted directly between individuals by speech, writing, or other forms of communication. This is obviously much faster than information passed on solely by reproduction of new individuals. Often a new gene will take thousands of years to diffuse throughout a population. In contrast, a single idea can diffuse throughout a human population, even without modern technology, in just a few days. In this sense, cultural inheritance can be called Lamarckian because, unlike biological inheritance, information acquired in an individual's lifetime can be passed on.

Two, mutations are terribly inefficient because of their randomness with respect to need. For instance, if an environmental change leads to a need for white fur, the species may well become extinct if no genes for that trait exist in the gene pool. If mutations do not occur in sufficient time, the species will die out. In contrast, memes can be consciously created to meet an immediate situation. Under stress of cold, for example, new kinds of clothes can be designed and produced. Such "cultural mutations" then become part of that society's meme pool.

Three, genes cannot be blended because they consist of discrete DNA molecules. On the other hand, memes are often blended with other memes. For instance, the Japanese have assimilated many Western cultural ideas (such as those related to technology), but they have blended them with many of their own original ones (such as social customs). This ability to combine useful ideas obviously leads to a great deal of flexibility and adaptability.

Four, the kind of information encoded in genes is restricted to instructions on how to build and maintain bodies and limited types of "programmed" behavior (such as instinct, to be discussed shortly). On the other hand, learned information

TABLE 2–4. A Comparison of Cultural Versus Biological Information

	BIOLOGICAL	CULTURAL
Basic unit of information	Gene	Meme
Transmission of units	Sexual reproduction	Communication
Alterations of units	Random mutations	Inventions
Blending of units	No	Yes
Kind of information	Restricted	Much greater
Amount of information	Restricted	Much greater

contains a much greater variety of information. Abstract literary themes, architectural blueprints, or instructions for social interactions are but a tiny fraction of the kinds of information encoded.

Finally, there is a much greater amount of information transmitted in human learning. As measured in bits (the yes/no or binary data units of computers), the total amount of biological information in a human's 100,000 genes is about equal to the amount of information in about 1,000 library books. In contrast, the amount of information in the average adult human brain has been estimated to be the equivalent of about 20 million library books.

Given these huge advantages, it is no wonder that cultural evolution is so much faster and efficient at adapting us to our environment. In fact, the analogy of the hare and the tortoise is sometimes made. Cultural evolution (the "hare") is so much faster that biological evolution in humans has virtually ceased relative to it. Reports indicate that the amount of cultural information produced has been increasing exponentially, and continues to do so. This is because our accumulated knowledge is providing ways to augment the advantages of learned information just discussed. For example, we can now transmit information even faster with electronic devices (telecommunications) and store evermore information outside our brains (in computers). This has allowed us to accumulate and transmit growing amounts of information. This is a classic case of **positive feedback,** which means that a previous change causes an even greater amount of the same kind of change (a "snowball" effect). Many social scientists now worry (justifiably) that the pace of change has become so great that even humans cannot properly adapt to it.

Not only is cultural evolution intrinsically faster for the above reasons, but culture is *actively slowing down natural selection*, making the rate difference between them even greater. Because our technology (knowledge and mastery of nature) has become so great, humans are basically free from natural selection. We no longer need to have our genes change to adapt to our environment because we alter the environment to suit ourselves. For instance, we can construct artificial ecosystems or at least wear warm clothes in cold weather. This is also true for our domesticated plants and animals in that we have modified their natural environment, when necessary, to make it less adverse.

However, this is not to say that there is no evolutionary change in humans or other species. Indeed, there will likely be great changes in store for the future for both. It is just that natural selection is no longer the main determinant of much evolutionary change. Instead, artificial selection is becoming increasingly important and refined through such methods as genetic engineering (see Chapter 5).

The Brain: The Central Organ of Culture

Since culture is based on learned information, the brain is clearly the basis by which cultural evolution proceeds. Thus, to understand the processes of cultural evolution, we must understand the brain and its evolution. As the brain is the product of biological processes, this will also tell us important things about how biological processes gave rise to the qualitatively different processes of cultural change. The human brain is specialized for learned behavior in two ways: (1) its relatively *large size*, and (2) its *complex organization*.

BRAIN SIZE. The average adult human brain weighs about 3 to 4 pounds and takes up a volume of about 1,300 cubic centimeters (a quart is about 1,000 cubic centimeters). This is not large compared to the brain of an elephant or whale. However, these animals have much larger bodies so that much of the brain's additional size is devoted to simple mechanical muscle movement and sensory reception. Thus, what best measures the amount of brain mass available for information processing or "intelligence" is *relative brain size*. This is defined as the size of the brain relative to body mass. As shown in Figure 2–29, humans have relatively large brains. Mammals (and birds) in general have brains some 10 to 100 times greater than reptiles and other older life forms of a similar body size. This is because, as discussed in Chapter 4, there has been an evolutionary "arms race" among animals so that maximum brain size has been generally increasing through evolutionary time (with some groups advancing much more than others). Primates are generally even larger-brained than other mammals. Humans have extended the trend of mammals and primates. Note also the unusually large relative brain size of dolphins. This is the only group to rival us in relative brain size and is a major reason that many

Figure 2–29. A plot of brain mass versus body mass for a variety of animals. Open circles represent reptiles (including some dinosaurs) and fish; the filled circles represent mammals; and the ×s represent primates. From E. Chaisson, *Universe*, p. 527.

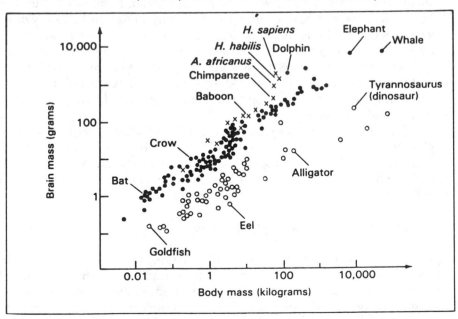

people are concerned over the often needless slaughter of these animals. This is not to say that porpoises think like us—their brains are organized in a very different way. But they do seem very capable of rapid learning, as shown by their ease of training and advanced communication skills using sound waves.

Why does a larger brain support learned behavior? The brain is the center for information brought in by the five senses of smell, sight, touch, taste, and hearing. As such, one of its primary functions is *information storage*. This is accomplished by nerve cells called **neurons,** which are connected to the senses to help form the nervous system. Nerve cells in the brain and the rest of the nervous system have many branches that connect with one another. Along these branches, electrochemical impulses travel at about 225 miles per hour, carrying sensory information to the brain and motor information ("commands") away from it. Upon reaching the brain, some stimuli are chemically stored in the memory in the form of molecular arrangements in the neurons. Thus, our large human brain has more neurons on which to store information as chemical "data bits." When we forget something, these chemical bits have been lost, sometimes to be rearranged to store new information. Thus, there is a constant turnover of information. Other times the information is lost because neurons die as we age. These are not replaced: All the neurons we will ever have are formed before we are born.

What we have said so far does *not* mean that people with larger brains are generally more intelligent. First, as we have said, body size influences brain size. Smaller-bodied people tend to have smaller brains simply because fewer neurons are needed to control body movement and receive sensations. Second, even after accounting for body size, there is clearly much more to intelligence than relative brain size alone. Allocating ever-greater numbers of neurons simply increases the space available to store information. Yet, a key aspect of intelligence is what we do with that information. Specifically, what kinds of connections and associations are abstracted from it. Brain size alone is not going to tell us anything about organization. There is much direct evidence for this; many "smart" people (such as Einstein), whose brains have been removed and measured after death, do not have particularly large brains. Conversely, many "average" people have very large brains for their body size. However, there are limits: Serious defects, which result in exceptionally small brains (below about 900 cubic centimeters), do lead to learning disabilities.

BRAIN ORGANIZATION. To understand culture and learning more fully, we must turn to other features of the brain besides size alone, particularly those features that relate to brain organization—interconnectivity, surface area, and localization. Perhaps the key anatomical feature that allows us to think (make associations among stored information bits) is the *interconnectivity* of the brain's neurons. Each neuron can have up to 200,000 connections so that an adult human brain with its billions of neurons has an impressive 100 *trillion* total connections. This is one of the major reasons that humans are still greatly superior to computers. Computers can theoretically have a much greater storage capacity, but they are far inferior in their ability to interrelate the information that they store. They therefore cannot "reason." Current attempts to create artificial intelligence are therefore focusing on such ideas as neural networking, which aims to create more interconnections in the computer's "brain."

This is also why it is said that we use so little of our brains. In our daily lives we use many of the same connections over and over, but we do not make many of the associations that we are capable of, given the vast amount of information stored in our brains. However, when we do make exceptional connections, we often have "insight" by tying together ideas that previously had not been interrelated. Evidence shows that while Einstein had a fairly average-sized brain, he may have had an exceptional number of connections among certain key types of brain cells.

Aside from interconnectivity, another important aspect of the brain's organization is that only a small part of the brain actually participates in the "higher functions" of thought. These are the neurons in the outermost layer, called the **cerebral cortex.** This layer is less than a half inch deep. All of what we do, think, and learn is accomplished not with the whole brain, but only in this even smaller outermost layer. To increase this crucial part, the brain has become convoluted with many ridges and valleys. This allows for increased surface area with little, if any, increase in the absolute volume. This is important because our heads are already so big that females have difficulty giving birth. In fact the head grows so big that human babies must be born about one year before they "should be" (that is, compared to birth times in our relatives, the apes). This is why human babies are so helpless for the first year while all other ape babies are able to function largely on their own. So effective are these convolutions in increasing surface area without increasing volume that if the cortex of an average adult brain (which is about 1.3 quarts in volume) were to be removed and spread flat, it would cover an area of nearly 10 square feet.

A third aspect of the brain's organization, after interconnectivity and surface area (cortex), is that the brain shows much **localization,** meaning that the cortex is subdivided into areas of local control and function. For example, an area at the rear of the brain takes in and synthesizes visual information (Figure 2–30). Another example is Broca's area, which is important for speech development and production. We know about such localizations because when brain damage occurs, people continue to function in some ways but not in others. Two areas that are particularly important for learning and therefore for cultural evolution are the **uncommitted cortex** and **frontal lobes.** The uncommitted cortices are areas of the brain that are not specialized for any one function. Instead they contain neurons, which store and synthesize information from other areas. The frontal lobes are where long-range planning and analytical abilities occur. At one time surgeons often performed lobotomies on mentally disturbed patients, a procedure in which a portion of the frontal lobes is removed. The surgery left patients passive and docile. (This practice has since been greatly restricted.)

There are a number of major differences in the size of various localized areas of the human cortex as compared to the other primates. Human frontal lobes and the uncommitted cortex are much larger. At the same time, areas devoted to other

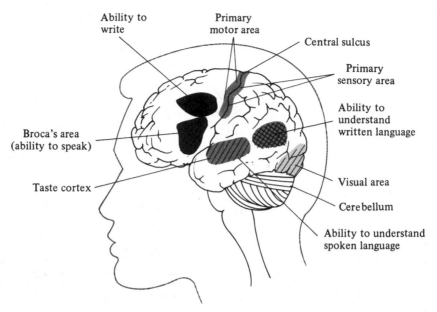

Figure 2–30. Labeled parts represent some of the motor and sensory areas of the human cortex. From Barrett et al., *Biology*, p. 414.

functions, such as smell, are smaller in humans. However, there have been *no new major structures* added to human brains: The difference between human brains and those of apes is one of degree and not kind.

Evolution of the Human Brain

The human brain is clearly our most distinctive organ, but just how distinctive is it? To answer that, let us compare human behavior with the behavior of other animals. As shown in Figure 2–31, biologists discern four general levels of behavior. The simplest is **reflex.** This is a basic response to some stimulus such as recoiling from light. Reflex is seen in the simplest of animals, even single-celled life. **Instinct** represents greater complexity in that there is a repertory of responses to a stimulus. This is the "programmed" behavior characteristic of insects, for example. Animals showing mainly instinctive behavior have nervous systems that have begun to centralize neurons into an actual brain. In simpler "reflex" animals, the nervous system is quite diffuse, with much less centralization. **Learning** is a still more complex characteristic, wherein the animal can modify its responses to stimuli from experience. Here the centralized brain has grown to a much larger size as uncommitted cortex is added to make associations and to process incoming information. This kind of behavior is most pronounced in mammals, although some less familiar groups show this as well. For instance, an octopus is quite capable of much learned behavior, equivalent (in some estimates) to that of a dog. Finally, **reasoning** is the

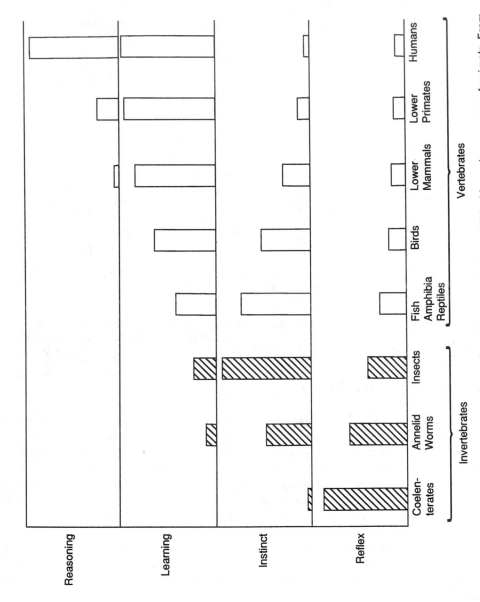

Figure 2–31. Some of the major differences in the types of behavior exhibited by various groups of animals. From S. Luria, S. Gould, and S. Singer, *A View of Life* (Redwood City, CA: Benjamin/Cummings, 1981), p. 558.

most complex form of behavior, wherein associations and synthesis of learned information occur as already discussed. This is made possible by the addition of still more uncommitted cortex and an increase in the interconnections of the cortex in general.

Thus, we see a general evolutionary sequence (summarized in Figure 2–31) of progressive increases in behavioral complexity, with progressively more learning involved. There is a trend toward organisms that assimilate more environmental information and depend less on genetic information passed on from the ancestors. This obviously allows a great deal more flexibility for changing conditions. With the vast store of information in the cortex, humans are able to call up and associate enough information to make abstractions and generalizations about the environment, and behave accordingly. We learn not only that fire is hot, but go beyond that to more complex kinds of reasoning that allows foresight and planning about how to build and sustain a fire. We are able to build a complex series of "if-then" statements that allow the consideration of many contingencies (alternative outcomes) of any given course of action.

ADDITION, CONTINUITY, AND EMERGENCE. The above shows that our evolution to the level of reasoning has been a largely additive process: We retain the general types of behavior of our ancestors while adding "higher" abilities. We still have reflexes, instincts, and, of course, learning. This is reflected in our brain's anatomy, where more primitive parts still exist. Indeed, in a very general way, we may think of our brain as a kind of onion—the more primitive parts exist in the deeper areas, while the outermost cortex is the most advanced part. For instance, the limbic system, rather deep in the brain, stimulates states of arousal and strong emotions like fear and rage. These instinctive behaviors dominate many animals but become progressively under the control of "higher" thought in animals with a well developed cortex, such as humans. While such basic emotions were once highly adaptive, and still can be if personal danger arises, they can do a great deal of harm in today's highly formalized and densely populated society.

The additive nature of our brains also has profound implications for understanding human nature. Because we retain many "levels," from deep emotions to pure logic, most of what we think and do is a mixture of these levels. This gives a great deal of range to our behavior and sensations. For instance, reading or thinking ("higher functions") can also be closely connected with strong emotional impressions (consider the reaction to exciting passages in a novel). This has led to much debate over how much of human nature is instinctive (the genetically programmed, older behaviors) versus how much is learned. We will discuss this so-called *nature-nurture debate* in Chapter 5 because it has so many implications for humanity's future. However, for now let us say that human behavior is clearly a mixture of both evolutionarily older, genetic ("nature") components and younger, learned ("nurture") components.

Does this additive aspect of brain growth mean that animals—especially our closest relatives, the apes—have mental processes that approach human capabilities? The answer seems to be a qualified "yes," although just how much we share is a matter of much debate. Chimpanzees use tools such as twigs to extract termites from logs. Even more impressive, they have a vocabulary of up to 400 symbols when taught sign language (their vocal chords are unsuited for speech). Further, they show some ability to combine words and concepts, indicating some ability of association. Such rudimentary behavior has been called **protoculture.** However, these abilities are not generally regarded as anywhere near the complexity of thought and reasoning shown by humans.

WHY HUMANS? PRIMATE PREADAPTATIONS FOR CULTURE. In addition to having a large brain capable of storing and associating large amounts of information, other biological traits are also necessary for the kind of cultural development that humans have. Along with a larger brain, these other traits are all possessed by our living primate relatives (such as monkeys and apes), usually in less developed forms. Such traits are preadaptations for the evolution of culture. Thus, apes and monkeys have dexterous *hands* that are easily modified for toolmaking and other manipulations. A factor that contributes to the development and use of this trait is that apes can sometimes walk on *two legs.* Modifications of the pelvis, legs, and feet have allowed us to become permanently two-legged. Apes have a tendency toward *stereovision.* That is, their eyes face forward allowing overlapping fields of vision for depth perception. This is important for toolmaking and other cultural activities as well. *Social behavior* is important in apes and there is much visual and auditory communication involved. Obviously, social interaction is crucial to culture.

While these characteristics point to the primate origins of our cultural abilities, they do not explain why we extended them as far as we have. In other words, it may seem obvious that advanced learned behavior is adaptive, but millions of species have survived without it, including our close relatives, the apes. So why did our species alone experience natural selection for advanced culture? Many anthropologists believe the need arose when our ancestors ventured to the grasslands of Africa. This made advanced, learned social behavior important for two reasons. One, they were now more exposed to predators, whereas in the forests, where apes originated, escape into the trees was possible. Two, there was a greater emphasis on meat-eating, which led to hunting. Now consider these two factors in relation to the ape's physical makeup. Apes are not physically adapted for either defense against predators or aggressive killing for food. They are slow, their teeth are not useful for fighting, and they have no sharp claws. But they did have those traits that predisposed them for cultural adaptations. The ability to cooperate socially, make use of primitive tools, and in general use their advantageous learning powers, allowed them to survive.

In sum, it was not the grassland lifestyle alone (many animals live there), or the ape ancestry alone (there have been many apes who did not evolve advance culture that led to their survival and our evolution. It was the *combination* of the grassland environment and ape ancestry that led to human culture. This is really just the Darwinian "equation" restated: variation (determined by our ape ancestry) + selection (grassland environment) = evolution.

Would advanced culture evolve again on earth if we disappeared? No one can say for sure, although one thing is clear: it would not necessarily take human form. Our particular kind of thought processes, behavior, and culture is strongly affected by our primate heritage. Certainly the same sequence of events that led to humans would never be repeated. Whatever intelligent creatures did evolve would not necessarily be limited to, for instance, having hands, the same senses, feeling the same kinds of emotions, and so on, especially if they were not primates. (For example, some people have speculated that intelligent raccoons or bears would supersede us.) Another likelihood is that, as in our case, advanced culture would occur only in one species. The reason is that, as happened with humans, this one species would likely be able to modify the environment so drastically that any other capable species would likely be exterminated as a competitor.

TOO SMART FOR NATURAL SELECTION? The above scenario of evolution of advanced culture by natural selection has often been questioned by what might be called the "overintelligence problem." This problem states that our brains, capable of complex mathematical calculations producing intricate novels, and many other advanced cultural activities, could not possibly have evolved from the selective pressures found in the primitive situation of hunting and gathering.

This reasoning underestimates the highly complex mental demands faced by our ancestors. As known from the few remaining "primitive" hunters and gatherers today, people in such societies are very knowledgeable—they know hundreds of folk tales, remember family lineages for many generations, and can identify (for hunting or collecting purposes) thousands of species of plants and animals.

In addition, other humans form the most important part of our environment. Most of the information in our brains is social data, concerning our daily habits, customs, and our communications with others. Dealing with others, either as enemies or as part of a cooperative group, has played a major part in the rise of intelligence. Making deals, outwitting enemies, and so on make this "social intelligence" extremely important for survival so that selection would favor it.

In conclusion, information is information, whether it deals with game trails, kinship data, social relationships, or mathematical symbols, or whether it is found in textbooks or assimilated through listening. The brain has evolved to store and process large amounts of information, without being limited to a specific kind of data.

Developmental Mechanisms: Growing a Large, Complex Brain

We have noted that natural selection acts on embryos, infants, and juveniles, as well as adults. Therefore, we must observe the entire developmental sequence in viewing cultural adaptations, not just human adult characteristics. When we do so, we find that humans show a pronounced *delayed development* compared to our relatives. As shown in Figure 2–32, each phase of our growth is prolonged so that we spend

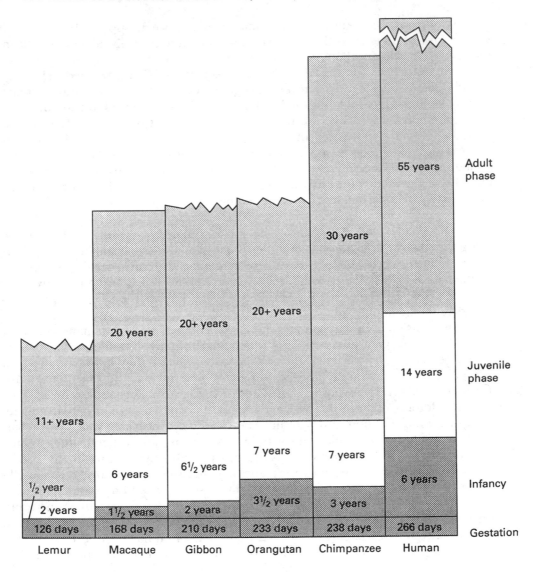

Figure 2–32. Life stages of various primates. Note the progressive increase in the durations of stages.

more time in infancy, childhood, adolescence, adulthood, and even old age than other primates.

Such developmental delays are caused by **regulatory genes** that control the timing of developmental events. The power of regulatory genes is seen in that about 99% our DNA sequence is the same as the chimpanzees' sequence: Our differences from chimps are largely the result of mutations in a small number of regulatory genes, representing only about 1% of our genes. This has recently been discussed in a stimulating book called *The One Percent Advantage* by John and Mary Gribbin (1988). Regulatory genes often act by controlling the timing of hormone production or response. For instance, humans become sexually mature (through sex hormone production) at about 12 to 13 years of age, while chimpanzees and gorillas mature through hormone onset at about 7 or 8 years.

This delay is the developmental mechanism producing most of our cultural abilities for three reasons. One, delayed embryonic growth is the direct cause of our *large brains*. All of the brain neurons originate in the first few months after conception. By prolonging this embryonic phase, more brain neurons are deposited. The brain continues to grow after this, but it is caused by increase in the size of existing neurons and supporting cells. Two, the juvenile and adolescent stages are the major learning stages in all organisms. By *prolonging the learning stages*, humans are able to learn more. Three, in delaying each stage, death itself is delayed (for instance, a chimpanzee dies of old age at about 40 to 50 years). This allows older individuals more time to *pass on acquired knowledge*, skills, and so on, to the juveniles.

SUMMARY

The processes of physical evolution were prerequisites for life because they produced habitats for life, an energy source to sustain life, and life itself. Four fundamental forces make up these physical processes: both strong and weak nuclear forces, electromagnetic forces, and gravitational forces. These forces caused atoms, molecules, and planets to form. The **Big Bang** created the three basic subatomic particles (the electron, proton, and neutron), which make up atoms.

Chemical reactions occurring on the early earth led to the origin of life. **Miller** and **Urey** conducted a landmark experiment in which the early atmospheric and climatic conditions were simulated. As a result they were able to synthesize **amino acids,** the building blocks of proteins. This supports the idea that steps toward the origin of life could have occurred under natural conditions in the primordial soup of the early earth. **Fossils** also provide evidence that life arose quickly from natural physical processes.

Life is thought to have only evolved once on earth. Evidence for this is that all life uses the same DNA code to reproduce and all organisms have the same biochemistry. This strongly indicates that all living organisms ultimately have the same ancestor.

The **principle of mediocrity** states that earth is not unique in its conditions for the origin of life. In fact it is likely that life exists on other planets. **Goldilocks's paradox** can be used to estimate planets that would be habitable.

Vitalists tried to prove that life has some substance not possessed by nonlife. However, life and nonlife cannot be distinguished by composition but by organization. As organization increases, **emergent properties** can be seen to develop which distinguish living organisms from the physical environment. Life appears to violate the law of entropy by maintaining order. However, the law of entropy applies only to closed systems and life is an **open system** because it takes energy from the environment.

Darwin discovered the process that causes biological evolution: **natural selection** of inherited variation. Selection is the differential success of an organism because it possesses certain traits. There are three types of selection: natural selection, sexual selection, and artificial selection. Evolution by selection of variation is often called "survival of the fittest," where fitness is a measure of the relative reproductive success of an individual and can be influenced by many different traits. **Adaptation** is the process by which species become better suited to an environment. There are two main types of selection: directional selection and stabilizing selection.

There are two major kinds of evolution: **nonbranching evolution,** in which an entire species evolves into a new species, and **branching evolution,** in which different populations of a species undergo different types of selection and branch off in different directions. A geographic continuum of traits among populations is called a **cline.**

A **species** is a group of individuals that can interbreed to produce fertile offspring. Species origination by **branching evolution** can occur when two populations of a species can no longer interbreed. This can occur due to premating or postmating **reproductive isolation mechanisms.** Populations that are recognizably different but are still capable of producing fertile offspring are called **races** or subspecies. The **paleospecies problem** occurs because interbreeding cannot be used as a criterion for recognizing species in the fossil record.

The most important advance in evolution was the "invention" of sexual reproduction because this greatly increased the amount of variation among individuals for selection to act upon. Asexual organisms, which reproduce by cloning themselves, have substantially less variation.

The basic unit of heredity is the **gene.** Genes are made up of **DNA** and occur on **chromosomes.** They encode instructions for building and maintaining the organism. Gene expression is determined by the sequence of the four bases that comprise DNA. **Regulatory genes** control the rate and timing of development. **Transcription** and **translation** are processes by which the DNA sequence is converted into proteins, which make up all of the cells of the body.

Mitosis is when the cell divides to produce two identical cells. The chromosomes are duplicated to produce a **diploid** cell. **Meiosis** is where reproductive cells are produced with half the original chromosomes. These are called **haploid** cells.

The **gene pool** is the total of all the genes of a species. Two factors affecting variation in the gene pool are (1) shuffling of genes, which is increased by sexual reproduction, and (2) size of the gene pool, which is increased by **mutations** (spontaneous changes) in the genes. **Gene flow** is the movement of genes between gene pools.

Alternate forms of a gene are called **alleles** and are either **dominant** or **recessive.** A **Punnett square** allows different combinations of alleles, or **genotypes,** to be calculated, and the probability of expression of a particular trait (known as a **phenotype**), estimated. **Homozygous** genes have identical alleles at a particular location, while **heterozygous** genes have different alleles at a particular location.

There are three lines of evidence for evolution by selection of variation: similarities of living organisms (such as biochemistry, anatomy, and development); the fossil record, which shows intermediates or "missing links"; and ongoing evolution, which allows direct observation of both artificial and natural selection. These similarities among organisms allow them to be classified into biological systems by a process called **taxonomy. Cladistics** is the classification of organisms on the basis of primitive or derived traits.

The short-term changes in genes of a population not resulting in the origination of a new species is known as **microevolution. Macroevolution** results from long-term processes acting on the gene pool over thousands or millions of years. These produce new species and evolutionary innovations.

Culture is elaborate, learned behavior. The basic unit of learned information is the **meme.** Cultural evolution is much more rapid and effective than biological evolution because information is transmitted faster, memes can be created as needed and blended with other memes, and memes can contain a wide variety of information. Information is passed on at an accelerated pace by **positive feedback** mechanisms.

The brain is the basis by which cultural evolution proceeds. It is equipped for learned behavior by large brain size and complexity. **Neurons** are nerve cells that allow information storage. The **cerebral cortex** participates in thought processes. Localization is the subdivision of the cortex into areas of local control and function. The uncommitted lobe is not specialized for a specific function, while the frontal lobes are used for analytical thinking.

There are four levels of behavior: reflex, instinct, learning, and reasoning.

Primate traits, which are **preadaptations** for evolution of culture, include dextrous hands, two legs, stereovision, and social behavior. A combination of ape ancestry and grassland environment led to human culture.

KEY TERMS

nuclear forces	amino acids
electromagnetic force	microspheres
gravitational force	self-organizing system
Big Bang	primordial soup
electron	principle of mediocrity
proton	Goldilocks's paradox
neutron	vitalists

emergent properties
law of entropy
open systems
natural selection
sexual selection
artificial selection
fitness
adaptation
directional selection
stabilizing selection
nonbranching evolution
branching evolution
cline
founder effect
races
subspecies
paleospecies problem
gene
zygote
chromosome
mitosis
haploid
diploid
meiosis
recombination
DNA
transcription
RNA
translation
gene pool
alleles
dominant alleles
recessive alleles
Punnett square
heterozygous
homozygous

phenotype
genotype
pleiotropy
polygenetic trait
asexual
gene flow
genetic drift
mutation
Hardy-Weinberg equation
homologous organ
vestigial organ
theory of recapitulation
taxonomy
cladistics
primitive trait
derived trait
microevolution
macroevolution
phenotypic plasticity
racial senescence
preadaptation
adaptationism
culture
memes
positive feedback
neurons
cerebral cortex
localization
uncommitted cortex
frontal lobes
reflex
instinct
learning
reasoning
protoculture
regulatory gene

REVIEW QUESTIONS

Objective Questions

1. What four basic forces are responsible for the attraction between units of matter? What formed as a result of each type of force?

2. Briefly describe the Miller-Urey experiment. What did they produce? Did they produce life?

3. How does Goldilocks's paradox predict which planets might sustain life? What two limiting factors determine if life will occur?

4. Into what three categories can selection be classified, based on the agent of selection?

5. Describe how directional selection enhances an extreme trait to produce a shift in the distribution of that trait within a species. (Refer to Figures 2–8 and 2–9.)

6. How does stabilizing selection maintain the distribution of traits in Figure 2–8?

7. Compare and contrast branching and nonbranching evolution. (Refer to Figure 2–10.)

8. Define species. Do different races belong to different species?

9. What kinds of information are encoded in genes?

10. What is the difference between transcription and translation?

11. How is a Punnett square used to estimate how genes for a particular trait will be expressed? (Refer to Figure 2–17.)

12. What is the Hardy-Weinberg equation? What is this equation used to calculate?

13. What are two lines of evidence which indicate that evolution occurs by "tinkering" with the anatomy of adult species?

14. Define homologous and vestigial organs. Give examples of each.

15. Do microevolutionary changes produce new species? Why or why not?

16. Compare and contrast memes and genes.

17. Define and give examples of preadaptation.

18. Which is more rapid and efficient—biological or cultural evolution? Give several reasons to support your answer.

19. Give examples of localization within the brain. (Refer to Figure 2–30.)

20. Define four general levels of behavior. (Refer to Figure 2–31.)

Discussion Questions

1. What does the principle of mediocrity suggest about the possibility of life on other planets? Do you think there is life in other parts of the universe? Explain.

2. Briefly explain what is meant by "survival of the fittest." Do you think a more fit organism can be created by genetic engineering? What traits produce the most fit organism?

3. If genes, which encode particular traits, are passed on to offspring, why are siblings unique in appearance? What about identical twins?

4. Why was the "invention" of sexual reproduction one of the most important advances in evolution? Does selection explain why the "sex drive" exists in all organisms? Explain.

5. Do designed adaptations mean God exists? How does the blind watchmaker address this question? Does the theory of evolution conflict with or support your own views about the existence of God?

6. Is it likely that human culture and intelligence would evolve again if the human race was destroyed? Does this suggest that there is life on other planets similar to humans?

SUGGESTED READINGS

On physical evolutionary processes

FLOWER, P. 1990. *Understanding the Universe.* West, St. Paul.
SAGAN, C. 1980. *Cosmos.* Random House, NY.

On biological evolutionary processes

DAWKINS, R. 1976. *The Selfish Gene.* Oxford University Press, Oxford.
DAWKINS, R. 1986. *The Blind Watchmaker.* W. W. Norton, NY.
FUTUYMA, D. 1986. *Evolutionary Biology.* Sinauer, Sunderland, Mass.
GRANT, V. 1991. *The Evolutionary Process.* Columbia University Press, NY.

On human evolutionary processes

GRIBBIN, J. and M. 1988. *The One Percent Advantage.* Blackwell, NY.
NELSON, H., and JURMAIN, R. 1991. *Introduction to Physical Anthropology.* West, St. Paul.

A popular book on brain evolution

SAGAN, C. 1977. *The Dragons of Eden.* Random House, NY.

3

Evolution Past: History of Life

OVERVIEW: FROM BIG BANG TO BIG BRAINS

In Chapter 2, we discussed the processes that cause physical, biological, and cultural evolution. Now we discuss the products: what has been created by these processes through time. Two major themes, *time* and *continuity*, underlie our discussion of this history. The theme of time is impressed upon us by its sheer quantity: a human lifespan, and even the span of the human species, is dwarfed when compared to the billions of years spanned by the history of the universe, earth, and life itself. The theme of continuity refers to the continuity between the products of physical, biological, and cultural evolution. Thus, the history of the universe and earth is essential for understanding the history of life, which, in turn, is needed to understand the history of the human species. This is because events in the history of the earth have often had dramatic effects on the history of life. Changes in climate, sea level, and many other physical conditions have strongly influenced both the extinction and origination of species. Less obvious perhaps is that life has also influenced the history of the earth. For example, earth has had three atmospheres, with the modern oxygen-rich atmosphere being produced by plants over many millions of years.

We will begin with a history of the physical world, from the Big Bang that started it all to the formation of earth. This sets the stage for the history of life, which begins surprisingly soon after the earth formed. Finally, we discuss the history of our own lineage, which is a very recent occurrence in the history of life.

PRODUCTS OF PHYSICAL EVOLUTION

In our discussion of the processes of physical evolution, we noted that just a few basic forces caused matter to "clump" together to form larger, more complex forms of matter. The nuclear forces caused subatomic particles to combine into atoms. The electromagnetic force caused atoms to combine to form molecules. The gravitational force caused attraction among molecules, leading to the formation of stars and planets. We now see these processes in action, discussing how the universe originated as subatomic particles that eventually united, evolving into the myriad of galaxies, stars, and planets that now populate the universe.

This discussion has two parts:

Part 1: History of the Universe, Stars, and Planets
Part 2: History of the Earth

PART 1: HISTORY OF THE UNIVERSE, STARS, AND PLANETS

Origin and History of the Universe

The universe consists of billions of galaxies, each containing billions of stars. Most astronomers believe that many of these stars have planets. Trillions of miles separate these stars and galaxies. Our sun is only one of about 100 billion stars in a galaxy among billions of other galaxies.

But where did all this come from? In spite of the universe's awesome scale, we can simplify it considerably by pointing out that the universe is composed entirely of *matter* and *energy*. Matter makes up the physical universe and energy is what makes matter move. Therefore, to understand the origin of the universe, we must understand the origin of matter and energy.

THE BIG BANG. While there has always been much speculation about the origin of the universe, the first major evidence was discovered by Edwin Hubble in the 1920s. Hubble found that the light from other galaxies showed a **red shift.** This is, the light radiation is shifted toward the red end of the spectrum, indicating that virtually all the other galaxies are moving away from our own. He found that we live in what is now called an **expanding universe.** This does not mean that earth is at the center of this expansion. Instead, like raisins expanding in bread dough as it is baked, the galaxies are all expanding away from each other.

Hubble also found that the speed of galaxies moving away from us (determined by the amount of red shift) is strongly correlated to their distance: faster-moving galaxies are farther away (Figure 3–1). This relationship is called **Hubble's law.** It is what you would expect if all the galaxies started moving from the same point in space at the same time. The ones moving the fastest would now be the

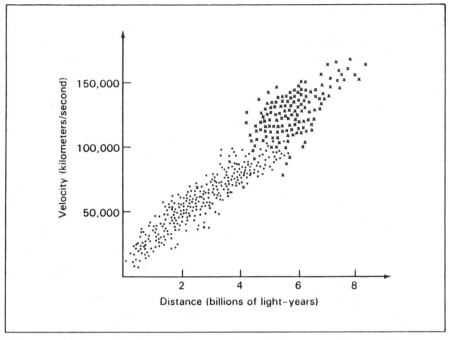

Figure 3–1. The Hubble diagram illustrates how galaxies are moving away from each other. Quasars represented by the x's are shown as the farthest. From E. Chaisson, *Universe* (Englewood Cliffs, NJ: Prentice-Hall, 1988), p. 294.

farthest away and the slowest would be closest to the starting point. In fact, given the current position, direction, and speed, it is easy to extrapolate "backwards" through time. When this is done, all galaxies and other forms of matter in the universe converge at one point in space. This point is now thought to have been the origin of all matter and energy. Considering how fast these massive galaxies are moving, it is clear that there must have been a cataclysmic explosion that hurled them outward. This explosion, which created all the matter and energy in the known universe is called (appropriately enough) the Big Bang. Based upon the current velocity and position of the galaxies, it is estimated that the Big Bang occurred about *15 to 18 billion years ago.*

Aside from the expansion of the universe, there is a second line of evidence for the Big Bang: cosmic background radiation. This was discovered in the 1960s when an annoying radio hiss was picked up by satellites of the telephone system. After it could not be accounted for on earth, it was found to be coming from outer space. This is now thought to be radiation produced by the Big Bang itself, a kind of "echo" of radiation that reverberates through space. We think this because: (1) it is a diffuse signal, coming from no single point, and (2) the characteristics of the radiation (the wavelength, frequency, and so on) could only have been produced by a fantastically high energy event, such as the Big Bang.

In science, the answer to one question usually raises more questions. The discovery of the Big Bang is certainly no exception. What existed before the Big Bang? Apparently it was literally "nothing." This is best seen if you consider the current universe as an expanding sphere. All the matter and energy in the universe lies within that expanding sphere. But what lies outside that sphere? The answer is: nothing. This "nothing" is not to be confused with the vast vacuum of space between stars and galaxies. There is actually a great deal of matter and energy in those places. Instead, if you could somehow travel to the edge of the expanding sphere of our universe, you would find a lack of any matter or energy at all beyond it.

Where, then, did the matter and energy come from? How does "something" originate from "nothing"? Many books have been written on this, by scientists, philosophers, and theologians, but the fact is that *no one really knows*. One key observation is that statistical fluctuations in energy can theoretically occur even when the average energy present is zero. For example, since a particle and its antiparticle can annihilate each other and become "nothing," it is also possible for the reverse to occur: A particle and antiparticle can be created from nothing. Like positive and negative numbers, they will always equally cancel out so that the net balance is zero. This is not only theory. Elementary particles have actually been observed to originate from literally "nowhere" during nuclear reactions. This has led many physicists to speculate that the universe arose as a "random" statistical fluctuation of some kind. Of course, even this observation leaves many questions unanswered, most of which boggle the mind. Yet even though many questions remain unanswered, knowledge in this area is growing rapidly.

AFTER THE BANG: THE TWO ERAS. In contrast to our near-total ignorance of events before the Big Bang, we know a great deal about what happened just after it. Using the laws of physics, scientists have been able to reconstruct in great detail the events that occurred just after the Big Bang and on through to the present. This is because we know a great deal about how matter behaves at various temperatures and pressures so that scientists need only calculate what would happen if all the matter and energy in the universe were brought together at different densities. For instance, we know that at early phases of the expansion, temperatures and pressures must have been incredibly high because of the high density of matter: All the matter in the universe was compressed into a much smaller part of space. By "running the expansion backward" (like reversing a movie), calculations show that the closer in time we get to the Big Bang itself, the greater was the compression. The compression (and density) would have been greatest immediately after the Big Bang, when all of the vast, current universe was squeezed into a tiny, pinpoint-sized area.

Table 3–1 shows the main events that occurred after the Big Bang. For convenience, astronomers group these into two eras: the **radiation era** and the **matter era.** Each of these eras is, in turn, subdivided into three epochs. The radiation era describes the first 100 seconds of the universe when temperatures and pressures were extremely high. During the first trillion-trillionth of a second after the Big

TABLE 3–1. The Major Epochs of the Universe in Order of Occurrence After the Big Bang. Also Shown Are Average Densities, Temperatures, and the Main Events of Each Epoch

EPOCH		TIME AFTER THE BANG	AVERAGE DENSITY (GRAMS/CUBIC CENTIMETER)	AVERAGE TEMPERATURE (KELVIN)	MAIN EVENTS
Radiation era	Chaos	$<10^{-24}$ second	$>10^{50}$	$>10^{20}$	Unimaginable Big Bang
	Hadron	10^{-24}–10^{-3} second	10^{30}	10^{15}	Annihilation of heavy elementary particles produces fireball of radiation
	Lepton	10^{-3}–100 seconds	10^{10}	10^{10}	Annihilation of light elementary particles continues to produce fireball of radiation
Matter era	Atom	100 seconds–10^{6} years	10^{-10}	10^{5}	Fireball diminishes; matter dominates and clusters into hydrogen and helium atoms
	Galaxy	10^{6}–10^{9} years	10^{-20}	300	Galaxies, quasars, and galaxy clusters form
	Stellar	$\gtrsim10^{9}$ years	$\sim10^{-30}$	~3	All galaxies have formed; many stars have formed; other stars are still forming, some accompanied by planets and life

Source: E. Chaisson, *Universe* (Englewood Cliffs, N.J.: Prentice-Hall, 1988), p. 322.

Bang (the chaos epoch), there must have been an incredibly hot and dense fireball of vast energy. At this time, *only energy (radiation) existed.* As it expanded, both temperature and density began to drop, causing matter to form from the energy. (Einstein's famous equations $E = MC^2$ shows that matter [M] and energy [E] can be converted into one another under unusual conditions; C is the speed of light). At first, in the hadron epoch, heavy subatomic particles ("hadrons") formed; examples are neutrons and protons. As cooling continued, lighter particles ("leptons") began to form; electrons are an example of this type of matter. However, energy (radiation) was still extremely abundant, which is why this first hundred seconds is called the radiation era.

The matter era describes a much longer time, from after the first 100 seconds until today. It is called the matter era because matter became much more common as temperature and pressure continued to decrease. Again there are three epochs. During the atom epoch, very simple atoms originated from protons, neutrons, and electrons. For example, hydrogen atoms, consisting of just one proton and one electron, formed as positively charged protons were attracted to negatively charged electrons. Helium atoms, consisting of two protons and two electrons, formed from pairs of hydrogen atoms. Since helium does not easily combine with other atoms, these two elements were all that formed at this time. It was not until stars formed and began to fuse hydrogen and helium atoms into more complex atoms that all the other elements in the universe were created. Even today hydrogen and helium (the two simplest elements) make up the major part of the universe, comprising nearly 90% and 10%, respectively.

As expansion continued, temperature and density dropped further, allowing the gravitational force to cause attraction between hydrogen and helium atoms. This marked the beginning of the fifth epoch in Table 3–1, the galaxy epoch. As local clusters grew, each began to have an even greater gravitational pull on surrounding atoms, causing a "snowball" effect (Figure 3–2). Eventually vast amounts of matter became grouped in separate islets, that were to become individual galaxies. As each grew, it began to spin and flatten due to the law of conservation of momentum (that is, as things contract, they spin faster and faster, such as ice skaters pulling in their arms). Today, there are about 100 billion galaxies in the universe. None are forming now and it is estimated that none have formed in the last 5 billion years.

Our own galaxy, the **Milky Way,** was named for the milky band we see in the sky when looking through the central plane of the galaxy. (Interesting, too, is that "galaxy" is derived from the Greek word for "milk.") The Milky Way is about 12 billion years old and is a fairly typical spiral galaxy, with about 100 billion stars arranged in a flattened, spiral shape that spins about the galactic core.

This brings us to the final phase of the matter era, the stellar epoch (Table 3–1). This began about 14 billion years ago with the formation of the first stars, and goes through the present as stars continue to form today.

Figure 3–2. Schematic diagram showing how changes in the average density of matter could have led to the formation of galaxies. Modified from E. Chaisson, *Universe,* p. 341.

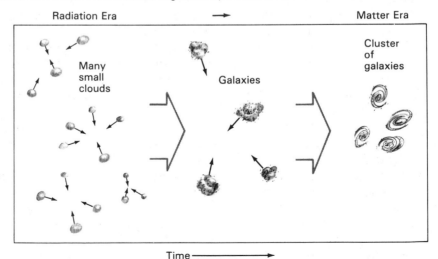

History of Stars

Stars, including our sun, are composed mainly of hydrogen and helium. These are the two simplest and abundant elements in the universe. When stars first form, they consist almost entirely of hydrogen, but the helium begins to accumulate as hydrogen atoms are fused to form helium. We know a great deal about the origin and evolution of stars not only because their simple composition allows us to predict events from physical laws but also because we can see stars in the process of forming.

Stars go through a very predictable series of *three stages*. These can be characterized as: contraction (birth), main sequence (adulthood), and fuel exhaustion (death). The first and last stages are relatively short compared to the main sequence, where a star spends most of its life.

CONTRACTION (BIRTH). Stars are born from huge clouds of hydrogen that were created in the Big Bang and still exist in many parts of many galaxies. Gravity causes the clouds to contract around a center to form a single gaseous object, as shown in Figure 3–3. Frequently, two or three local centers may dominate in different parts of the cloud. This results in two or three stars forming in close proximity. In fact

Figure 3–3. Artist's conception of interstellar cloud changes during early evolutionary stages. (a) Stage 1 represents an interstellar cloud; (b) stage 2, fragmentation; (c) stage 3, smaller, hotter fragments; (d) stage 4, protostar. (Not drawn to scale.)

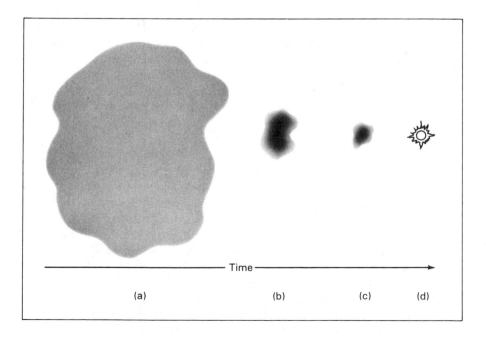

Time

(a)   (b)   (c)   (d)

such multi-star systems are more common in the galaxy than those with only one star.

MAIN SEQUENCE (ADULTHOOD). As hydrogen accumulates, the ever-growing mass will assume a spherical shape, since the matter will accrete symmetrically around the common center of gravity. The accumulating hydrogen exerts ever greater pressure on the core of the sphere, and temperature increases. Eventually, the temperature (at least 10 million degrees centigrade is needed) and pressure will become so great that **nuclear fusion** occurs, causing hydrogen atoms to fuse, forming helium atoms:

$$H + H \rightarrow He$$

The nuclear fusion within the star essentially begins the "adulthood" phase of its life. The energy generated stops the gravitational collapse of the sphere because the inward pull is counterbalanced by the outward pressure of very hot gas. As the energy diffuses out from the core it eventually radiates out into space as the light and other forms of radiation that we receive on earth.

All stars are not the same size. Larger stars have more hydrogen fuel but burn it up faster because the temperature and pressure at their cores is greater. They therefore give off more light and have a higher surface temperature, but have a shorter lifespan. Smaller stars burn fuel more slowly, giving off less light and having cooler temperatures. This relationship is shown in Figure 3–4. The band on the graph is called the **main sequence,** indicating that stars fall on that band during their "adulthood." Stars in the process of forming or dying will fall outside that band. Our sun falls in the middle of the band. This indicates it has a fairly average luminosity and surface temperature. This is because it is a roughly medium-sized star, burning fuel at neither very fast nor very slow rates.

FUEL EXHAUSTION (DEATH). *The moderate size of the sun is an important prerequisite for life.* Large stars burn hydrogen so fast that they use it up in less than a billion years. This is probably not long enough for intelligent life to evolve. On earth, it took nearly 5 billion years for humans to evolve. Due to its moderate size, our sun has a lifespan of about 10 billion years. Thus it is now "middle-aged" and will last for about another 5 billion years. In contrast, very small stars would last much longer than the sun, but they also give off much less light. As the sun is the source of energy for nearly all life forms, small amounts of light would greatly impede the evolution of life.

Figure 3–5 summarizes how stars die, depending on their size. The smallest stars simply burn out. However, all stars larger than these very small ones first go through a **red giant stage** before they die. This is because they are large enough so that, when hydrogen is depleted, the temperature and pressure at the core is great enough to cause helium atoms to fuse to create carbon. Thus, they burn the by-product accumulated from the hydrogen fusion of the main sequence. This new nuclear reaction releases much energy and the star begins to expand to a huge size,

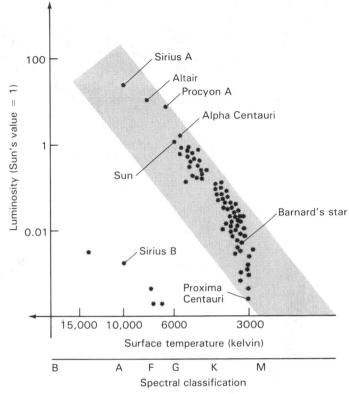

Figure 3–4. The majority of stars have what is called "normal" properties within the shaded region known as the main sequence. The values plotted here are for stars within a 20 light year region of the sun. From E. Chaisson, *Universe*, p. 210.

often giving off a reddish light. When the sun does this, it will become large enough to engulf the orbits of Mercury, Venus, and possibly even Earth and Mars.

Following the red giant stage, there are three pathways that a star will follow, depending on its size. These are illustrated in Figure 3–5. In medium-sized stars, such as our sun, the red giant stage is simply followed by a cooling off stage as the star gradually burns out to form a white dwarf and then a black dwarf. In larger stars, the core temperature and pressure will be greater (because of greater mass pushing on the core) so that the carbon atoms created by the helium fusion of the red giant stage will begin to fuse as well. Carbon fusion creates a still heavier atom, oxygen. Oxygen, in turn, will fuse to form still heavier atoms. Eventually, the dying star will become layered like an onion, with progressively heavier atoms at the core, as shown in Figure 3–6.

This sequence of progressive fusion of heavier elements can only occur up to the point where iron is reached because iron atoms will not fuse to form still heavier atoms. Iron begins to accumulate at the core and fusion begins to slow. As less

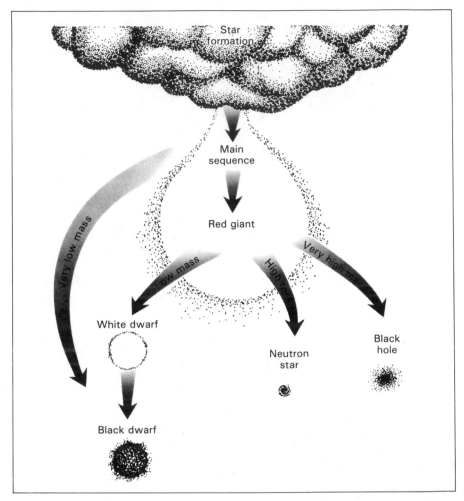

Figure 3–5. A summary of the evolutionary path of stars. From E. Chaisson, *Universe*, p. 392.

energy is produced at the core, the star begins to collapse on itself (implode), because there is no more pressure of hot gas to resist gravity. This collapse compresses the core material to phenomenal densities, causing some of it to rebound. This rebound expels enormous amounts of matter and energy into space with great speed. So much light is produced that stars not visible from earth before can become very bright. Hence, this violent event has come to be called a **nova** (Latin for "new"), since a "new star" is seen. It is called a **supernova** if the event is extremely violent.

Not all of the star material is ejected into space. Some of the highly compressed core remains behind. In some cases, this matter is so compressed that the electrons and protons in the atoms combine to form neutrons. Hence these remnants are

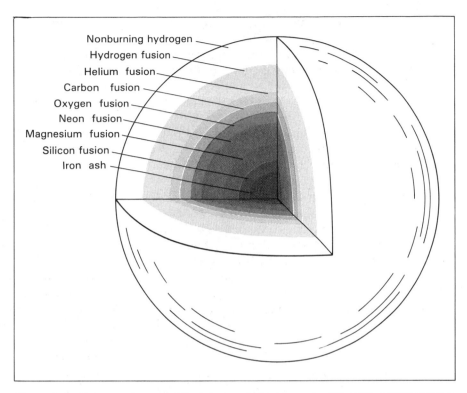

Nonburning hydrogen
Hydrogen fusion
Helium fusion
Carbon fusion
Oxygen fusion
Neon fusion
Magnesium fusion
Silicon fusion
Iron ash

Figure 3–6. Cutaway diagram of the interior of a highly evolved star with a mass greater than three solar masses. The interior resembles the layers of an onion. From E. Chaisson, *Universe*, p. 380.

called **neutron stars.** The density is so great that a single tablespoonful of neutron star material, if transported to earth, would weigh more than Mount Everest.

In very large stars, the implosion is even greater as the core is compressed to still greater densities. Even the neutrons are collapsed. It is as if the whole star were compressed to a pinpoint in space. At such density, the gravitational pull is so great that not even light can escape. Hence, these "ultimate" density remains of supermassive stars are called **black holes.** No one has ever seen a black hole (as expected from an object that emits no light), but there is much theoretical and indirect evidence of their existence. Not surprisingly, the physical properties of such a phenomenon are quite bizarre to everyday experience. Among the most unusual is that a black hole would "bend" space and time to the point that time would literally stop for anyone trapped inside of its gravity boundary.

STARS: MAKERS OF THE ELEMENTS. Star death is an essential ingredient for the evolution of life. *Most of the elements in the universe are created only during the death of large stars.* There are 92 naturally occurring elements on earth and only the two lightest (simplest) ones were made in the Big Bang. If only hydrogen and helium are in the universe, there would be no life or even any planets.

But there is something missing here. We said that dying stars fuse atoms only up to iron, yet iron is only element number 26 on the periodic table of elements. Where do atoms of heavier elements, that compose the rest of the periodic table, come from? They originate during the *nova and supernova* events. Only during these last, extremely violent episodes of the star's life is enough energy released at once to fuse atoms into still heavier ones than iron. One important result of this process of sequential fusion is that heavier elements are progressively less abundant in the universe than lighter ones.

History of Our Solar System

We said earlier that our galaxy, the Milky Way, is a typical spiral galaxy that is about 12 billion years old. It has about 100 billion stars and is about 100,000 light years in diameter (meaning that light takes over 100,000 years to cross it). Our sun and its planets are located far away from the galactic core, about two-thirds of the way out on one of the spiral arms (Figure 3–7). This is probably no accident. One reason is that star systems close to the galactic core would be subjected to massive gravitational disturbances. In contrast, in our area of space, the nearest star is over four light years away (which translates into trillions of miles).

Our solar system consists of the sun, 9 known planets with some 44 moons among them, asteroids, and comets (Figure 3–8). The sun has 1,000 times more mass than all the planets combined. The planets have nearly circular orbits that nearly all lie in the same plane. The four **inner planets**—Mercury, Venus, Earth, and Mars—have a much greater density (are "rockier") than the **outer planets,** which are highly gaseous and much larger in volume (except for Pluto, which is an escaped moon of one of the large gas giants).

ORIGIN OF THE SOLAR SYSTEM. Most astronomers agree that the solar system began, like the stars discussed above, from the contraction of gas and dust in space. But one major difference is that the parental cloud consisted of more than hydrogen and helium. The earth and all the other bodies orbiting the sun contain many other elements. The only source for such elements is the explosive remnants of other stars. Therefore, our sun is thought to be a second generation star and its planets to have condensed from supernova remains. These supernovas not only provide the raw material for systems such as ours, they also generate shock waves that initiate the contraction process.

As in the case of galaxy contraction, the contraction of the solar system cloud also results in a flattening and spinning. As shown in Figure 3–9, the main mass of the cloud forms in the center of the disk. The surrounding matter is thought to break into concentric rings as spinning continues. The matter in these rings will form the planets and moons and each ring's circular shape will determine a planetary orbit. This comes about because as local centers of gravity form in each ring,

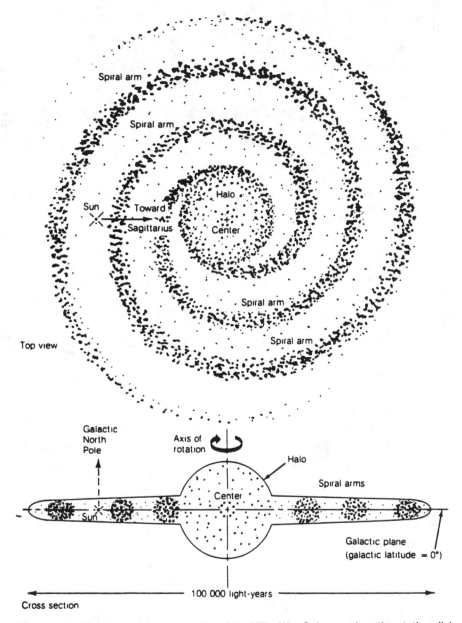

Figure 3-7. Top view and cross section of the Milky Way Galaxy, a gigantic rotating disk of stars, planets, gas and dust. The galaxy is about 100,000 light years in diameter and 5,000 light years thick, with a bulge (halo) in the center of mass. From W. Stokes, *Essentials of Earth History*, 4th ed. (Englewood Cliffs, NJ: Prentice-Hall, 1982), p. 142.

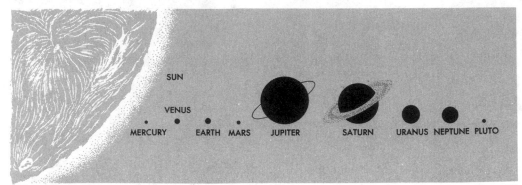

Figure 3–8. The relative sizes of the planets in relation to the sun. Satellites are omitted. From W. Stokes, *Essentials of Earth History*, p. 146.

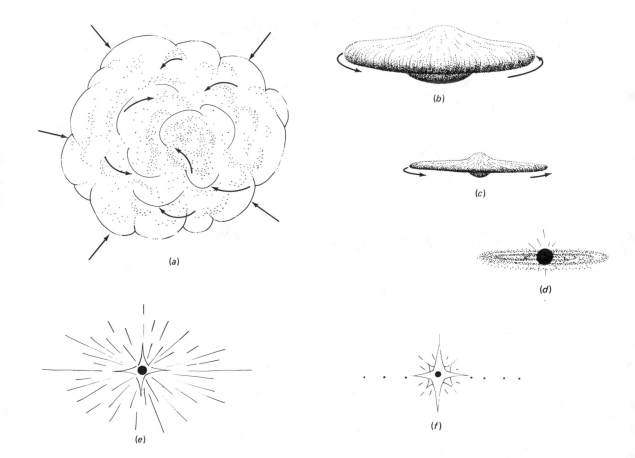

they will "snowball" and sweep through the ring to collect more and more material in it. Eventually, they will all collect into one body to form the planet.

Much evidence supports this model. For example, it explains the characteristics of our present solar system better than any other proposed explanation. The circular orbits of the planets in one plane result from their origin in the rings of the disk. Also explained is the revolution of the planets around the sun in one direction. This evidence has received even greater support in recent years by the use of computer modeling of how matter would behave in a second generation dust and gas cloud. More direct evidence has emerged very recently with the observation of solar systems apparently in the process of formation around nearby stars.

Our system is estimated to have begun forming about *5 billion years ago*. (Recall that the Milky Way is about 12 billion years old.) Computer models indicate that it would take about 100 million years for a spinning disk with rings to form and about 1 billion years for the planets to sweep up the debris. This agrees with observational evidence that all the planets underwent severe meteorite bombardment for about the first 1 billion years of their existence. This bombardment would be the last phase of accretion. Planets and moons with no atmosphere to erode them, such as our moon, still preserve the many craters created by the bombardment.

PART 2: HISTORY OF THE EARTH

The Geologic Time Scale

To discuss the next phase, the history of the earth, we first need a temporal framework so that we can order the sequence of events. That framework is a time scale that has grown over the last 200 years.

DEVELOPMENT OF THE GEOLOGIC TIME SCALE. In Chapter 1 we noted that humans have only recently become aware of the great age of the earth. In 1650, Archbishop Ussher estimated that the earth was created in 4004 B.C., based on the genealogy of the Bible. This age came under increasing doubt as geologists gained a better understanding of the rates of such processes as erosion and mountain building. By extrapolating present processes into the past ("uniformitarianism"), it was possible to estimate, for example, the length of time that river channels took to form, or for the sea to reach its present degree of saltiness. These estimates, made independently

Figure 3-9. Formation of the solar system: (a) A large dispersed nebula of interstellar gas and dust condensed under its own gravitational pressure. (b) As the nebula contracted, it began to rotate and flatten. (c) Material started to accumulate into the central protosun, and the nebula rotated faster and flattened further. Particles began to condense and accrete in disklike eddies. (d) The protosun continued to heat and became a star. Disk-eddies continued to consist mainly of gases, but planets began to form. (e) The contracting sun began to shine visibly. At this stage, solar winds drove off the remaining gases from the surrounding disk. (f) The sun began to fuse hydrogen, and planets and asteroids were the only remnants within the rotating disk. From D. Eicher and A. McAlester, *History of the Earth* (Englewood Cliffs, NJ: Prentice-Hall, 1980), p. 11.

by many observers, often yielded ages much older than the mere 6,000 years of Ussher. Charles Lyell (often called the "father of modern geology") published an extremely influential book called *Principles of Geology* in 1830. In the book, he presented many arguments to show that the earth was many thousands, perhaps even millions, of years old. Figure 3–10 shows how our view of the earth's age has grown exponentially since Ussher's initial estimates, until today when nearly all geologists agree that the *earth is about 4.6 billion years old.*

As important as estimating the overall age of the earth was the *ordering* of past events. There are three basic "laws," illustrated in Figure 3–11, that were crucial in establishing this ordering. The **law of superposition** says that whenever sediments are found layered one upon the other, the layers on top are younger than those on the bottom. This is because sediment is usually deposited in basins so that new sediment is deposited on top of older sediment. ("superposed" basically means "placed above.") The **law of original horizontality** says that these superposed sediments are deposited in horizontal layers, even though later events, such as mountain building, may have tilted the layers. Finally, the **law of original lateral continuity** states that the horizontal strata were laterally continuous when deposited, even though some may now be eroded.

By using such principles, geologists in the 1800s (mainly in Europe) were able to piece together rock layers across wide areas. Rock layers are called "strata," and **stratigraphy** refers to the study and mapping of such rock layers. The actual matching of the strata across long distances, even hundreds of miles, is called **correlation,** as shown in Figure 3–12. Long-distance correlation allowed geologists to reconstruct the order of many past large-scale geologic events, such as mountain building episodes and global sea-level changes.

Remains of past life, called **fossils,** within the sediments provided further evidence of the order of past events. For instance, strata containing dinosaurs always

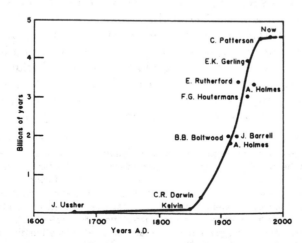

Figure 3–10. Changing views of earth's age. From P. Cloud, *Oasis in Space* (New York: W.W. Norton, 1988), p. 89.

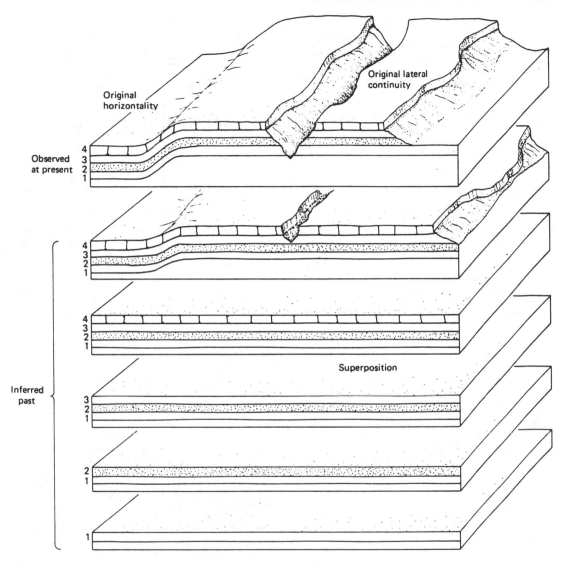

Figure 3–11. Application of the principles of original horizontality, original lateral continuity, and superposition in reconstructing geologic history. From D. Eicher and A. McAlester, *History of the Earth*, p. 37.

occurred earlier—beneath—strata containing modern mammals. Such observations allowed geologists to use fossils for correlation of strata across long distances, as shown in Figure 3–13. This use of fossils for correlation is called **biostratigraphy.** Because fossils of many species are often common in sedimentary rocks and each species varies in abundance, geologists could create highly refined fossil "range zones," such as shown in Figure 3–14.

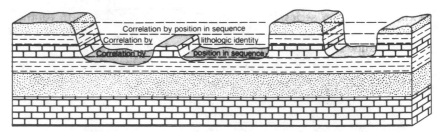

Figure 3–12. Correlation of rock layers across long distances. Modified from R. Wicander and S. Monroe, *Historical Geology* (St. Paul, MN: West Publishing, 1989), p. 70.

RELATIVE VERSUS ABSOLUTE TIME. While the *order* of past events across large areas could be reconstructed by the above methods, there was no way to know exactly when they occurred. In other words, the **relative time** of the events was known (dinosaurs came before humans, for example), but the **absolute time** (how many years ago dinosaurs lived) was not known. Even though it was possible to estimate that the earth was much older than 6,000 years, such methods as erosional rates were too crude to give any accurate dates as to just how much older, much less date individual events such as the age of the dinosaurs.

The major breakthrough in dating occurred in the early 1900s when the newly discovered phenomenon of radioactivity was found to have properties that allowed the absolute dating of rocks. Radioactive decay proceeds at a constant, predictable

Figure 3–13. The principle of fossil correlation. Rock strata from two geographic locations (A and B) contain similar fossils and are therefore considered to have been formed during roughly the same geologic time. From F. Racle, *Introduction to Evolution* (Englewood Cliffs, NJ: Prentice-Hall, 1979), p. 14.

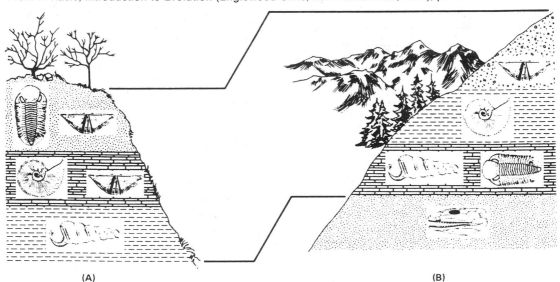

(A)   (B)

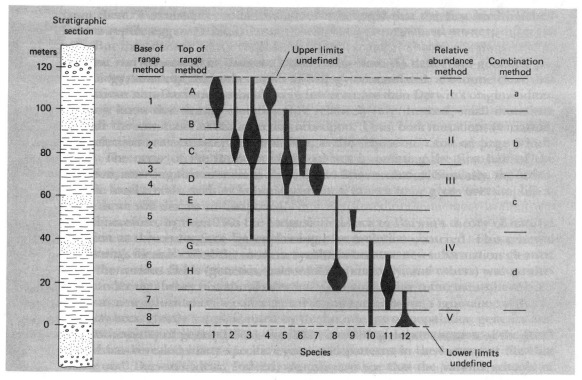

Figure 3–14. Four methods of recognizing fossil range zones in a rock sequence. Width of vertical areas indicates the relative abundance of specimens. From D. Eicher, and A. McAlester, *History of the Earth.* p. 157.

rate as the radioactive parent element decays to the more stable daughter element. Therefore, by measuring the *daughter/parent* ratio in rocks that naturally contain radioactive elements, it is possible to estimate how long the decay process has been occurring. This tells the geologist when the rock formed (having only the radioactive parent at the start). As shown in Figure 3–15, the **half-life** is the amount of time it takes one half of the parent to decay. For instance, uranium is a common element used in dating, since it is often found in rocks that form from magma. One type of uranium has a half-life of about 4.5 billion years. Therefore, if we were to start with 10 grams of this type of uranium, we would find only 5 grams 4.5 billion years later. The remaining 5 grams would now be daughter product. Such **radiometric dating** is possible not only because rocks naturally contain radioactive elements, but because the crystalline nature of many rocks allows the parent and its daughter product to be trapped in the crystals, allowing the ratio to be computed. Escape of parent or daughter would obviously invalidate the measurements.

There are many types of radiometric dating. As shown in Table 3–2, most of these use isotopes with half-lives of billions of years. These occur mainly in minerals found in granites and other rocks that form from the molten state. The key excep-

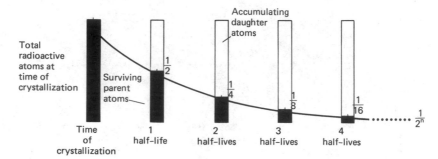

Figure 3–15. Exponential delay of radioactive minerals. Modified from D. Eicher and A. McAlester, *History of the Earth*, p. 54.

tion is carbon–14, which can be used to date fossils directly. Unfortunately, the half-life of carbon-14 is much shorter than the other isotopes so that it can only date materials less than about 100,000 years old. In fossils older than that, so many half-lives have passed that there is virtually no parent isotope left. It is therefore impossible to accurately calibrate the daughter/parent ratio.

THE MODERN GEOLOGIC TIME SCALE. By using radiometric techniques (and still other methods), geologists have been able to provide absolute ages for many events in the geologic timetable, now formally called the **geologic time scale.** Minor adjustments to the dates are continuously made as the scale is refined, but there is much agreement about the approximate age of most events. Most ancient events are known to within a few million years of error, with even less uncertainty about more recent events.

TABLE 3–2. Some of the Most Commonly Used Radiometric Dating Methods

PARENT ISOTOPE	HALF-LIFE (YEARS)	DAUGHTER ISOTOPE	PARENT ABUNDANT IN
Potassium-40	1.3 billion	Argon-40	Potassium-rich minerals (including feldspar, micas)
Rubidium-87	48.8 billion	Strontium-87	Potassium-rich minerals
Thorium-232	14 billion	Lead-208	Zircon and other minor minerals
Uranium-235	704 million	Lead-207	Uranium ores; zircon and other minor minerals
Uranium-238	4.5 billion	Lead-206	(same as uranium-235)
Carbon-14*	5730	Nitrogen-14	Organic matter; atmospheric CO_2; dissolved carbonate

* This method works somewhat differently from the other methods listed. With its very short half-life, carbon-14 cannot be used to date samples more than 50,000 to 100,000 years old. It is used principally for dating wood, cloth, paper, bones, and so on from archaeological sites.

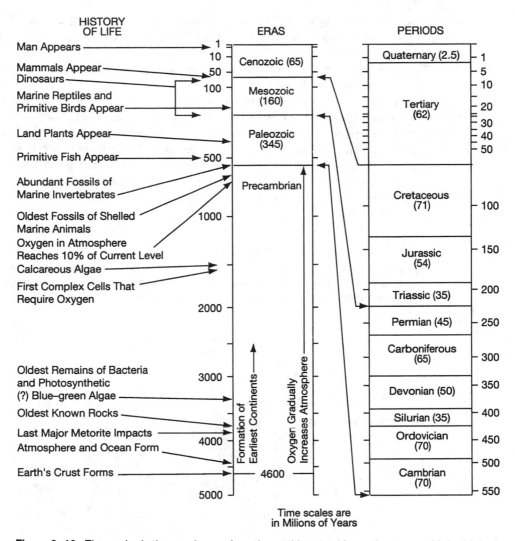

Figure 3–16. The geologic time scale runs from the earth's crustal formation to now. Major biological events are shown in conjunction with geological events. Duration shown in parentheses. Modified from M. Gross, *Oceanography* (Englewood Cliffs, NJ: Prentice-Hall, 1987), p. 381.

Figure 3–16 shows a modern version of the time scale. The four major subdivisions are **eras;** they serve as the main frame of reference for the rest of this chapter.

Cenozoic: 0–65 million years ago (mya)
Mesozoic: 65–225 mya
Paleozoic: 225–570 mya
Precambrian: 570–4600 mya

Notice that the eras become shorter as you move to the present. The vast Precambrian era composes nearly 90% of the whole scale. This is because we know more about more recent events.

The eras are subdivided into *periods*. These are generally named after the first major areas described with rocks of that age. For example, the Devonian period was named after rocks found around Devonshire, England. The derivation of the rest of the names are given in one of the review questions at the end of this chapter. Most of the era and period subdivisions are based on natural geological or biological events. For example, the ends of the Paleozoic and Mesozoic eras are both based on massive extinctions at those times. The end of the Devonian period is marked by major mountain building episodes. Such major events are used to define the beginning or end of an era or period because they can be recognized in rocks worldwide.

Figure 3–17. The earth's history placed within the boundaries of a conventional 24-hour time period. Within this time scale, life first appears around 5 A.M., but humankind doesn't appear until the final 20 seconds of the 24th hour.

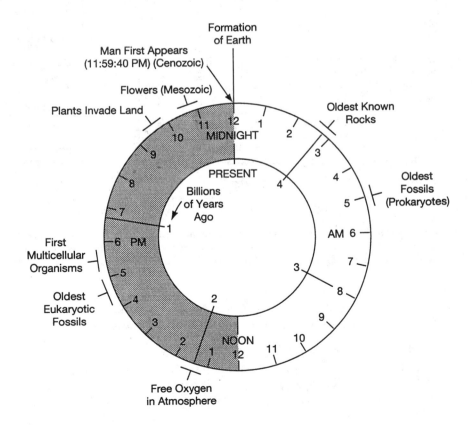

A better sense of this heritage is gained by converting the geologic time scale to a 24-hour clock (Figure 3–17) model. This gives us a more familiar reference. The main point is to illustrate the extreme *lateness* of human evolution. It is not until the *last minute* of the *last hour* that humans first appear. Recorded history is noted in the last few seconds. Even multicellular organisms do not become common until about 6 P.M. This is surprising given that the first microbes appear much earlier, after only about 5 hours. This early appearance of life, followed by a very long stasis where little happens, is of great interest, as we will discuss later.

History of Earth's Three Spheres: Land, Water, and Air

The earth's surface consists of three spheres, the *lithosphere, hydrosphere,* and *atmosphere.* These correspond to land, water, and air, respectively. A good way to remember this is that these three spheres represent the *three states of matter*: solid, liquid, and gas. In a solid the atoms are most tightly bound, whereas in a gas, they are least tightly bound. It is interesting that the ancient Greek philosophers believed that everything in the universe was made of one of four things: earth (land), air, water, or fire. If we consider life to be fire (which it basically is, since we live by "burning" calories metabolically), and add a fourth sphere, the **biosphere,** or life's sphere, then the list corresponds to the Greek's categorization of the world.

We first turn to a discussion of the evolution of the three physical spheres; we will consider biosphere evolution later.

EVOLUTION OF THE LITHOSPHERE. The **lithosphere** is the solid earth ("lithos" is Greek for "rock"). It includes rocks and their erosional products, sediments and soils. Figure 3–18 summarizes the evolution of the lithosphere.

The evolution of the lithosphere began about *4.6 billion years ago*, as the earth formed in its place as the third planet out from the sun. In condensing from the gas and dust cloud, earth began as an extremely hot molten sphere, forming in the manner discussed above with the origin of the solar system (see Figure 3–9). This heat came from three sources: the heat of collision from meteorite bombardment, the heat of collapse from gravitational shrinkage of the molten sphere, and heat from radioactive elements within the earth. A major result of this liquid state is that heavier elements, such as iron, sank toward the center of the sphere while lighter ones rose above them to the surface (Figure 3–19). This process of **differentiation** created three major layers: the **crust, mantle,** and the **core,** shown in Figure 3–19. The crust, on which we live, is only a few miles thick. It is not unlike the skin of an apple relative to the thickness of the other two layers. The composition of the crust is much higher in lighter elements such as aluminum, calcium, and sodium than the mantle and the core, which are richer in heavy elements like iron.

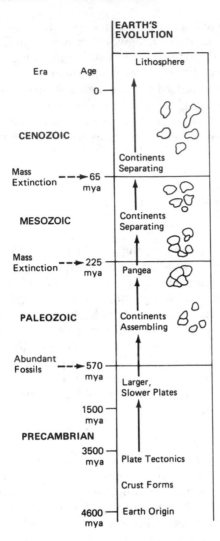

Figure 3–18. Geological time scale, showing history of the lithosphere. (mya = millions of years ago)

Even though differentiation of the deeper layers began soon after 4.6 billion years ago, it was not until about 4.0 billion years ago that the outer earth was cool enough for the crust to begin to harden. We know this not only from theoretical calculations and cooling experiments but also because a few mineral grains have been dated to about 4.0 billion years. At first, only small islands ("nuclei") of the crust began to harden. As these hardened, nuclei grew in size from continued cooling, and interior heat was trapped. To vent this heat, many volcanoes, which are essentially perforations in the crust, developed. However, much heat remained in the liquid, or more precisely "plastic," mantle under the surface, causing much

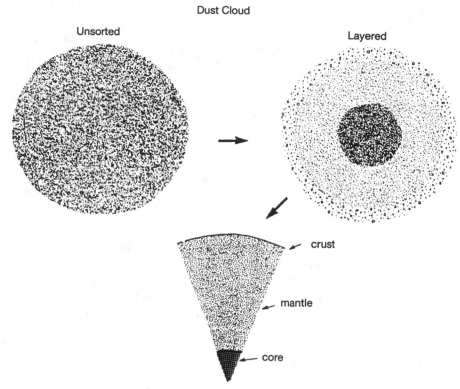

Dust Cloud

Unsorted

Layered

crust

mantle

core

Figure 3–19. Earth formed from a cosmic dust cloud. During its formation, it became stratified as the heaviest material sank to the center and the lightest rose to the top. Modified from T. van Andel, *New Views on an Old Planet* (London: Cambridge, 1985), p. 199.

turbulence. Under such conditions, **convection** occurs in a liquid (such as a cup of coffee) because liquid rises when hot and sinks as it cools (Figure 3–20). This convective movement of the underlying mantle caused movement in the hardening rigid material on top. This rigid layer soon broke into separate plates which moved about, floating on top of the more fluid mantle underneath. This movement of plates on the earth's surface is called **plate tectonics.** It is often called "continental drift" because continents, which ride on top of the plates, are also moving around. However, you should realize that it is the plates that move, not just the continents. This means that the land underlying the water in ocean basins moves as well.

Exactly when plates first formed and plate movement began is unknown. However, most calculations indicate that small plates ("microplates") must have formed not long after formation of the crust. These plates grew larger in size and fewer in number as the earth cooled and more molten magma hardened. Today, there are only about eight major plates. Most of these have continents riding on them, such as the North American plate. In addition, the rate of plate movement

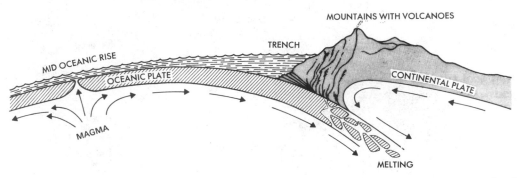

Figure 3–20. Geologic activity associated with subduction zones. Cross section (generalized) shows the relationship between plates as an oceanic plate is subducted beneath a continental plate. Modified from W. Stokes, *Essentials of Earth History*, p. 203.

has slowed as the earth has continued cooling. This is because cooling slows the convection currents that drive the plates along. Today, the plate movement is generally measured in inches per year, or about as fast as your fingernails grow.

There are just a few basic kinds of interactions that occur between the plates. At **divergent** boundaries, hot magma wells up and helps push the plates apart. This is shown as the mid-ocean rise in Figure 3–20. At **subduction** boundaries, a plate is subducted, or pushed down, underneath another plate. Often an oceanic plate is subducted under a plate with less dense continental material on it. This process has been very important on earth because subduction creates continental crust (in other words, "land" making up continents). As shown in Figure 3–20, when the subducted plate is pushed deep enough, it begins to melt. The less dense minerals (such as quartz and feldspar, which are rich in silicon and aluminum) in the melt separate and rise to the surface, often as lava extruded by volcanoes or as granite when intruded underground. These lighter rocks are "buoyed" up by their lower density to form continents. Note how this bouyant material is "welded" onto continental margins by this process (Figure 3–20). By about 2 billion years ago, about three-fourths of the modern continents had been formed in this way. The rest has been gradually added since, by the same process. Denser (richer in iron) crust rides lower on the mantle and forms the basins (oceanic plates) in which water (oceans) accumulates.

Aside from creating continents, another major effect of this plate movement has been to change the shape of the continents and seas many times. We are unsure of the exact changes in the Precambrian, but know quite a bit about changes in later times. Throughout most of the Paleozoic era, the continents were moving toward each other (Figure 3–18). This culminates at the end of the Paleozoic, 225 million years ago, to form one giant supercontinent called **Pangea,** shown in Figure 3–21. A series of massive mountain-building episodes occurred as Pangea was forming. A number of long mountain chains were thrust upward along continental margins as continents collided with one another. For example, the Appalachian Mountains

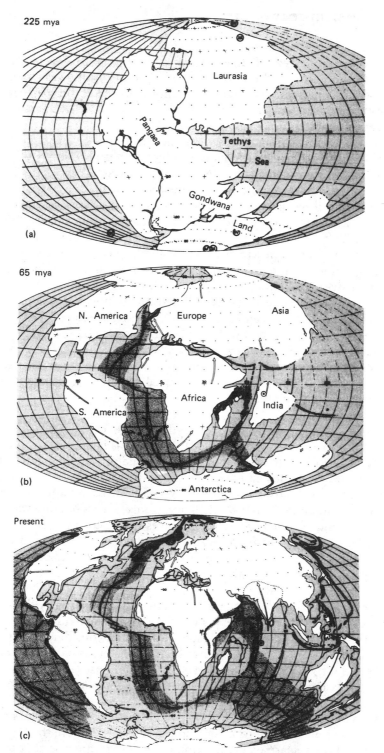

Figure 3–21. Continental drift from the late Paleozoic Era to the present. From M. Gross, *Oceanography*, p. 72.

of the eastern United States were formed when Africa collided with North America, as shown in Figure 3–21. Pangea began to break up in the early Mesozoic and this has continued through the Cenozoic up to present time. This breakup has led to the modern continental configuration shown in the lower part of Figure 3–21. For instance, North America and Europe continue to move apart, pushed by the divergent boundary in the middle of the Atlantic Ocean (mid-Atlantic ridge).

The overall influence of plate tectonics on the physical and biological evolution of earth is difficult to overestimate. Geologically, it has not only led to mountain chains and the great geographic diversity that characterizes earth, but plate motion has also affected climate, ocean currents, and virtually every other feature. For instance, we will see that massive global cooling has occurred a number of times because plates were moved over the polar regions, allowing ice to accumulate as glaciers. The importance of plate tectonics is especially clear when you compare the earth to most other planets in our solar system, which lack plate tectonics. The cold, lifeless, and unchanging surface of Mercury, for example, provides a stark contrast with the active evolution of the earth's surface.

The impact of plate tectonics on the history of life has been similarly profound. For example, migration routes were often created and lost. Even more influential were the broad changes that destroyed large areas of habitat. Indeed, the creation of Pangea caused the greatest mass extinction of all time, exterminating up to 96% of all species living then. However, before we discuss such events, we first complete our survey of the evolution of earth's physical evolution, looking at the hydrosphere and then the atmosphere.

EVOLUTION OF THE HYDROSPHERE. The **hydrosphere** consists of the earth's waters ("hydor" is Greek for "water"). About 97% of this is salt water in the oceans. The remaining 3% is fresh water, but the large majority of it is locked up as ice in glaciers. Therefore, only a tiny part of the hydrosphere is readily usable by humans as lakes, rivers, groundwater, and rain. Water is so essential for the maintenance of life that it composes 50% to 95% of all organisms. This is because water is a powerful solvent, meaning that matter easily dissolves in it, making it indispensable for biochemical reactions. For the same reason, water on earth has actively eroded the planet for billions of years, shaping the landscape many times over. If it were not for the continuing counteractive forces of plate tectonic uplift, the earth would have been worn flat eons ago.

The water on earth originated as water vapor expelled by the many volcanoes on the early earth. This process is called **outgassing.** Evidence for this is seen in gases released by modern volcanoes, which contain considerable amounts of water vapor. These gases are generated by chemical reactions in the magma beneath the volcano. As it rises, the water vapor cools and condenses in the atmosphere causing rain or snow. Upon falling, it will eventually collect in low-lying basins. The largest of these are basins between the continents, where oceans form.

Geologists estimate that earth's *oceans formed early in the Precambrian era*. Recall that there was much volcanic activity in the early earth from the high internal heat. This caused the release of great amounts of water vapor so that there must have been torrential rains for many thousands (perhaps millions) of years, filling up the ocean basins. Not only did oceanic *volume* reach modern proportions early in earth's history, but, perhaps even more surprisingly, the *composition* of those early oceans were probably modern as well. Our reason for inferring this is as follows. Today's oceans contain about 3.5% dissolved materials. The large majority of this is just "table salt"—sodium chloride. However, there is also a wide variety of other substances found in lesser amounts. Virtually all of these substances originally formed in rocks on land. Because water is a powerful dissolving agent, rain often reacts with rocks and sediment and carries off many materials in solution. The water finds its way to rivers, which then empty into the oceans. Since there is no outlet for water in ocean basins, these dissolved materials accumulate in the ocean water. However, when water becomes saturated with dissolved matter, any additional material will precipitate out as a solid. An analogy is the addition of sugar to iced tea until it no longer dissolves. Most geologists believe that the oceans became saturated very early in their history so that additional amounts of dissolved substances ("salts") brought in by rivers since then have tended to be deposited on the ocean floor. This has kept the composition of sea water roughly the same from its early history until today.

Much more changeable than volume and composition of the oceans has been the configuration of the oceans in how they covered the earth. Figure 3–21 shows how ocean basins have greatly changed in shape and location from plate tectonics. In addition, there have been major changes in sea level. As shown in Figure 3–22,

Figure 3–22. Generalized sea level and climatic conditions over the past 600 million years. Periods of low sea level correspond to periods of colder climate. Modified from M. Gross, *Oceanography*, p. 147.

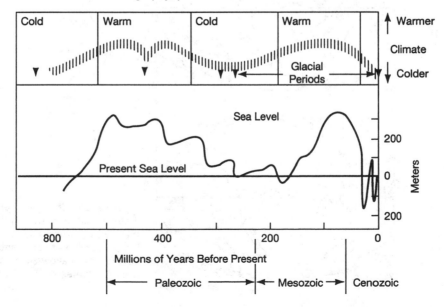

sea level was very high in the early Paleozoic. It gradually decreased throughout the Paleozoic until, by the end of that era, it was nearly what it is today. This drop in sea level is related to Pangea: As the continents assembled, shallow continental shelf area was squeezed out and rising mountain chains caused more land to be exposed. This is seen in Figure 3–23, which illustrates the changing distribution of oceans in the Paleozoic, as well as the formation of Pangea. Notice that in the early Paleozoic (500 million years ago), the oceans covered much of the continents. We

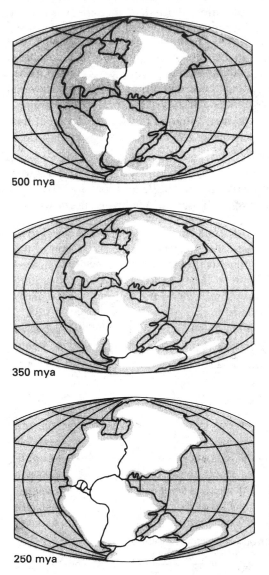

500 mya

350 mya

250 mya

Figure 3–23. Changing Paleozoic land-sea patterns. During this era, there was a progressive decrease in the proportion of continental surface covered by shallow seas. Modified from A. McAlester, *The Earth*.

will see later that these warm, shallow **epicontinental seas** were the home of much marine life. Indeed, the early Paleozoic was truly an "age of marine life."

After the Paleozoic, sea level rose once again until it was quite high at the end of the Mesozoic (Figure 3–22). This was the time of dinosaurs and many of the best dinosaur fossils are found along the shorelines of an epicontinental sea that covered western North America then. The Cenozoic has been a time of declining sea level. Indeed, at times sea level was lower than it has been for the entire preceding 600 million years. Even today, the present sea level is quite low compared to what it was for most of the past (Figure 3–22). The reason for this dropping sea level in the Cenozoic can be seen in Figure 3–21: the movement and isolation of the continent of Antarctica over the South Pole in the early Cenozoic (65 million years ago). The arrival of an isolated land mass allowed ice to accumulate at the Pole, causing a drop in overall sea level as water was taken from the oceans. Not surprisingly, this accumulation of ice is also associated with a temperature drop in global climate. There is a general correlation between cooling climate and lower sea level throughout earth's history. As shown in Figure 3–22, times of cooler climate tend to be associated with times of lower sea level.

HISTORY OF THE ATMOSPHERE. The **atmosphere** is the layer of air surrounding the earth ("atmos" is Greek for "vapor"). The gas molecules are held by gravity so that air becomes progressively less dense as it moves away from the earth. About 75% of the atmosphere is within 7 miles of the earth's surface.

Like the hydrosphere, the atmosphere is crucial for life. Not only does it provide oxygen, a very reactive element used in biochemical reactions, but it serves as a protective filter against harmful solar radiation (in the form of ozone). The current composition of the atmosphere is:

78% nitrogen
21% oxygen
1% carbon dioxide, argon, other trace gases

The other gases besides oxygen are also important to life. Nitrogen is a key element in all food chains. Carbon dioxide is a major determinant of global temperatures since it traps heat from solar radiation (the "greenhouse effect," which is discussed in Chapter 5).

Like the hydrosphere, the atmosphere originated from outgassing by the many volcanoes of early earth. However, unlike the composition of the ocean, the composition of the atmosphere has changed considerably through time. Most current theories indicate that earth has had three atmospheres. As shown in Figure 3–24, the first, "condensing atmosphere," was a short-lived envelope of gases that formed as the earth condensed from about 5 to 4.6 billion years ago. Since the gas cloud was hydrogen-rich, this first atmosphere was also *hydrogen-rich*, with not only pure hydrogen but also methane (CH_4), ammonia (NH_3), and other hydrogen compounds. However, as hydrogen is the lightest element, it soon trickled out into space

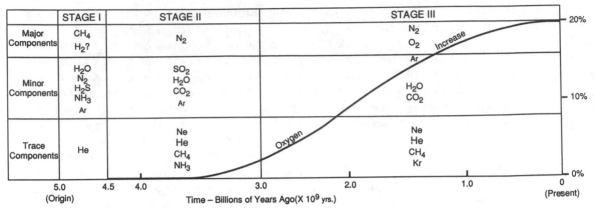

Figure 3–24. Probable evolution of composition of the earth's atmosphere from its origin until today. From R. Dott and R. Batten, *Evolution of the Earth*, 4th ed. (New York: McGraw-Hill, 1988), p. 117.

because the earth's gravity was not strong enough to retain it. It was replaced by gases produced by volcanic outgassing. This second, "outgassing atmosphere," was *nitrogen-rich*. While many gases are released by volcanoes, nitrogen was the major gas accumulating because it is relatively inert compared to other major gases. That is, nitrogen does not readily combine with other elements.

The third, "photosynthetic atmosphere," formed from the addition of oxygen to the reservoir of nitrogen gas. This *oxygen-rich* atmosphere was produced by the release of oxygen by plants. In **photosynthesis,** plants absorb carbon dioxide and combine it with water taken up by the roots. This produces food (CH_2O) and oxygen, both required by animals. This is summarized by the following simplified chemical equation:

$$CO_2 + H_2O \rightarrow CH_2O + O_2$$

Figure 3–24 shows that this oxygen accumulation was gradual at first, beginning at about 3.5 billion years ago when the first plants appeared. These were single-celled blue-green algae, as we discuss shortly. Even today, the major oxygen producers are single-celled marine plankton. Over the next 2.5 billion years, oxygen accumulation seems to have accelerated, until it approached the 21% oxygen level of today, at about 1 billion years ago (late Precambrian). Evidence for the oxygen curve of Figure 3–24 comes from minerals that formed during the Precambrian era. Minerals that form only in the absence of oxygen (or in low oxygen) are much more common in the early Precambrian. Conversely, oxidized minerals that form in oxygen-rich environments become progressively more common in the later Precambrian. Even so, the exact shape of the curve in Figure 3–24 is very hypothetical. Yet whatever the true shape, we can be sure that the increase was gradual enough to allow life to evolve adaptations to the additional oxygen. Recall in our origin of life discussion that the highly reactive nature of oxygen means that life could only have initially evolved in the absence of it.

Besides the composition of the atmosphere, the heat distribution and other climatic characteristics of the atmosphere have also changed greatly through time. Temperatures were very hot on the early earth, but there is evidence of a large-scale glaciation ("ice age") by about 2.5 billion years ago, so we know temperatures had cooled considerably. As seen in Figure 3–22, the early Paleozoic era was generally characterized by a warm global climate. As Pangea formed, temperatures dropped and two glaciations occurred. During the Mesozoic, climate gradually warmed so that the middle to late Mesozoic, which was the heyday of the dinosaurs, was characterized by very warm temperatures. During the Cenozoic era, the temperature began to drop again when Antarctica moved over the South Pole.

As the Cenozoic climate continued to cool, it led to a period of worldwide glaciation, often called the "ice age." Beginning about 2 million years ago, large masses of ice called glaciers advanced southward a number of times. However, this advance was not steady and was often interrupted by interglacial warming episodes wherein the ice temporarily retreated. A main cause of this alternating cycle of advancing and retreating is believed by many to have been the regular "wobble" that occurs as the earth spins on its axis (analogous to a spinning top). Even though it is very slight, this cyclical wobble, called the **Milankovich cycle,** would affect the angle at which sunlight strikes the earth. This is likely to have a major impact on global warming and climate.

At the peak of glacial advances, ice sheets over a mile thick covered more than 30% of the earth's surface. In North America, a nearly solid sheet reached down from Canada all the way south to Ohio (Figure 3–25). Many landscape features of the United States are testimony to this ice. Numerous lakes of the upper Midwest and Northeast, including the Great Lakes, occur in basins gouged out by the glaciers. Unsorted gravel deposits and huge boulders represent material carried by the advancing ice and left behind when it melted. Even the formation of the Mississippi River was strongly influenced by the vast meltwaters when the ice withdrew. The last withdrawal ended only about 10,000 years ago and now only about 10% of the globe is covered with ice (Figure 3–25).

Is our current climate just another temporary interglacial episode? Many scientists believe that it is. However, human warming of the climate, especially by adding gases into the atmosphere, will probably delay or maybe even prevent another glacial advance (see Chapter 5).

SUMMARY: PHYSICAL HISTORY OF THE EARTH. An overview of the history of the lithosphere, hydrosphere, and atmosphere is presented in Figure 3–26. Some main points to remember about this earth's evolution are as follows.

1. The earth condensed into a molten sphere from a huge cloud of gas and dust about 4.6 billion years ago. As the sphere cooled, it differentiated into layers. The outermost layer, the lithosphere, hardened and became mobile, driven by the underlying hot magma. These *plate tectonics* led to the assembly of continents during the Paleozoic, culminating with the supercontinent Pangea 225 million years ago. Throughout the last 225 million years of the Mesozoic and Cenozoic, Pangea has been breaking up.

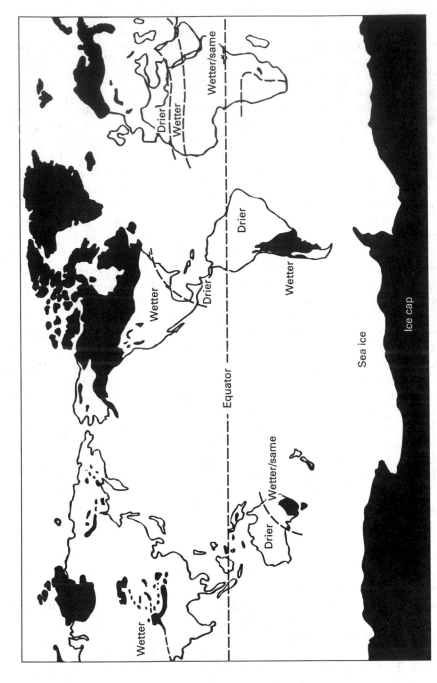

Figure 3–25. Ice Age glaciers (dark shaded areas) covered 30% of the land. Also note that rainfall patterns were much different. From R. Wicander and S. Monroe, *Historical Geology* (St. Paul, MN: West Pub., 1989), p. 510.

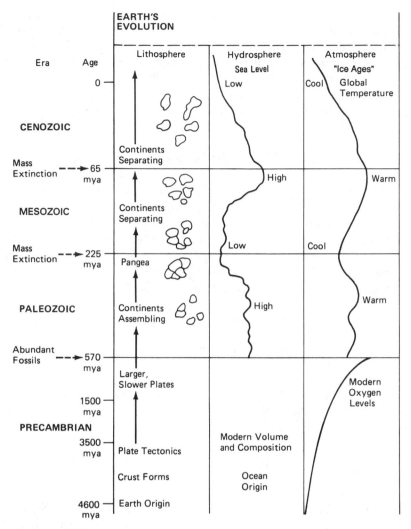

Figure 3–26. Evolution of the earth's lithosphere, hydrosphere, and atmosphere.

2. The hydrosphere originated from water vapor outgassed by volcanoes that condensed and filled up the ocean basins. It probably reached modern volume and composition early in its history. Sea level has fluctuated. It was high throughout most of the Paleozoic (a time of epicontinental seas, rich in tropical marine life) and decreased greatly when Pangea formed. The second period of high sea level was during the late Mesozoic (the time of the dinosaurs). Throughout the Cenozoic sea level has dropped as Antarctica moved over the South Pole, allowing ice to build up and withdrawing water from the ocean.

3. Earth has had three atmospheres: hydrogen-rich (from planetary condensation), nitrogen-rich (from outgassing), and oxygen-rich (from photosynthesis). Near-modern oxygen levels (21%) were apparently reached about 1 billion years ago. Low global temperatures have generally correlated with low sea level as cold periods result in ice accumulation. The most recent temperature decline culminated in the "ice age," a series of glacial advances and retreats that began about 2 million years ago. The last retreat was about 10,000 years ago.

PRODUCTS OF BIOLOGICAL EVOLUTION

We turn now to the history of the biosphere. We will see that life first appeared in very simple form, as primitive single-celled microbes, surprisingly soon after the earth became habitable. There was a very gradual evolution toward more complex cells, and then multicellular life. Most of this evolution occurred in the ocean (hydrosphere). However, land (lithosphere) and air (atmosphere) were eventually colonized, as natural selection of variation caused life to branch out into new ways of living. We will study life's evolution in this order:

Part 1: Evolution Toward Multicellular Life
Part 2: Evolution of Life in the Water
Part 3: Evolution of Life on Land
Part 4: Evolution of Life in the Air

Before recounting the actual history of life, we first review some basic facts about fossils. As the only direct evidence of the history of life, an understanding of fossils is essential.

THE FOSSIL RECORD

Fossils are the remains of prehistoric life. The word "fossil" is derived from the Latin word to "dig up"—these remains are usually preserved in rocks. The vast majority are preserved in **sedimentary rocks.** Sedimentary rocks form when erosional products (sediments) are deposited by water in basins. Common examples include sandstone and limestone.

How Fossils Form

We tend to think of fossils as "bones," but there are actually many kinds of fossils. **Paleontologists,** people who study fossils, recognize two basic categories of fossils: body fossils and trace fossils. **Body fossils** are the remains of the whole body or, more often, parts of the body. Frozen carcasses of mammoths—whole bodies—have been found in Siberia and other areas that have stayed frozen since the last glaciers retreated. Body fossils have also been found where the chemical conditions under water (especially in swamps) are just right to preserve organic matter and the complete remains, including soft parts, of organisms. Much more common are body fossils that represent only part of the body. Not surprisingly, these are usually the *hard parts* of the body, such as the shells of clams or the teeth of mammals (teeth are much harder than bones). Sometimes the hard part is preserved in its original composition (such as with calcite seashells). However, very often the original fossil is altered in a number of possible ways; some of these are shown in Figure 3–27. Three basic categories of fossilization can be identified in the figure: *replacement, molds and casts*, and *recrystallization*. In replacement, the original mineral is replaced by other minerals that seep in from surrounding sediment. In some cases this seepage will even fill into tiny pores, preserving fine details of the original anatomy. A familiar example of this is "petrified wood," which is wood that has had its pores filled with the mineral silica. Sometimes the whole body will dissolve and the remaining cavity will be mineralized, leaving a *cast*, or a *mold* where only the wall of the impression is preserved. In recrystallization, the original shell is destroyed and is often preserved only as an outline.

Trace fossils preserve *indirect evidence* of past life, such as burrows, tracks, or trails. In this case, the soft sediment that bears these impressions is often mineralized. While they cannot tell us much about the anatomy of the organism, trace fossils are very useful in providing information about behavior. Most often, trace fossils are associated with marine animals (clams, worms, snails, and many other animals) that burrow or crawl in the soft sediment. An example of the diversity of marine trace fossils is shown in Figure 3–28. These traces are very useful in reconstructing water depth and other paleoenvironmental conditions. For instance, winding, surface trails on the sediment tend to indicate deeper waters because the sediment is nutrient-poor and the animal must search wide areas for food particles. In contrast, deep, tubelike burrows tend to form in shallow-water sediments, in part because the animal can "mine" deeper into the nutrient-rich sediment. Trace fossils of land life are also well known. Dinosaur footprints are known from a number of localities, usually fossilized in the soft sediment of river beds. These have been very useful in estimating how fast the dinosaurs ran, providing evidence of herding and other clues to behavior.

How Complete is the Fossil Record?

There are an estimated 5 to 50 million species of life on earth today. Yet, since life first originated over 3.5 billion years ago, an estimated 1 to 3 *billion* species have come and gone. (This estimate is based on the fact that average species duration is about 5 to 10 million years, as measured in the fossil record; see Chapter 4.) This means that *over 99% of all species that ever existed are extinct.* They will be found only in the fossil record, if they are to be found at all. The question is, How complete is this record? Unfortunately, it has proved disappointing. Of the estimated 1 to 3 billion species that have existed, only one to a few million, or *less than 1%,* have

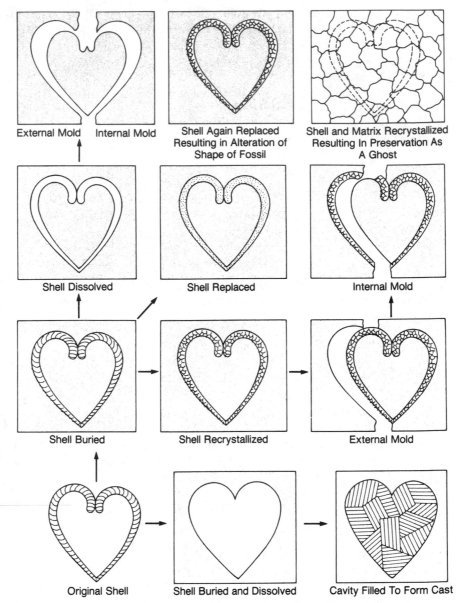

External Mold   Internal Mold

Shell Again Replaced Resulting in Alteration of Shape of Fossil

Shell and Matrix Recrystallized Resulting In Preservation As A Ghost

Shell Dissolved

Shell Replaced

Internal Mold

Shell Buried

Shell Recrystallized

External Mold

Original Shell

Shell Buried and Dissolved

Cavity Filled To Form Cast

Figure 3–27. Diagram showing how a shell may be preserved by replacement, recrystallization, or by the formation of molds and casts. The arrows show possible sequences of preservation.

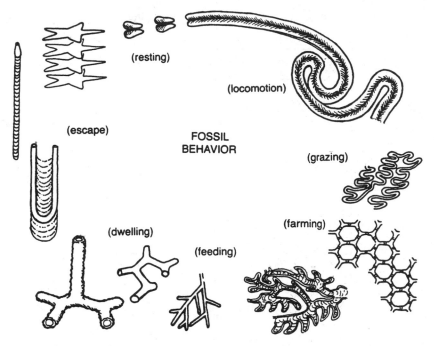

Figure 3–28. Schematic diagram illustrating some of the diverse behavior traces made on or in sediment by various invertebrates. Modified from J. Cooper, R. Miller, and J. Patterson, *A Trip Through Time*, 2nd ed. (Columbus, OH: Merrill, 1990), p. 77.

probably been fossilized. Of those fossilized, only about 10% have so far been discovered and described by paleontologists over the last 100 years.

Aside from the lack of overall preservation, another problem is the *biased preservation* of the record. Some groups have a much greater tendency to be preserved than others. Two major factors promote fossilization in a group: (1) hard parts in the anatomy, and (2) rapid burial. Thus, shelled marine invertebrates, such as clams, snails, and corals, have a relatively good fossil record because they not only secrete hard mineralized shells or other body parts, but they live in an environment of relatively rapid deposition (sediment is usually deposited underwater). Mammals and other land vertebrates have a fair record because bones and teeth are very durable and are often deposited in rivers, swamps, sinkholes, or other watery environments.

Virtually unrepresented as fossils are organisms that have only soft tissues and live in environments that have little deposition. Insects fall into this category; they make up an astounding 75% of living animal species yet are extremely rare in the fossil record, making up less than 1% of the described species. Not only do insects lack mineralized hard parts, but they live on land, with little chance of rapid burial. One interesting exception involves insect fossils found in the gemstone amber,

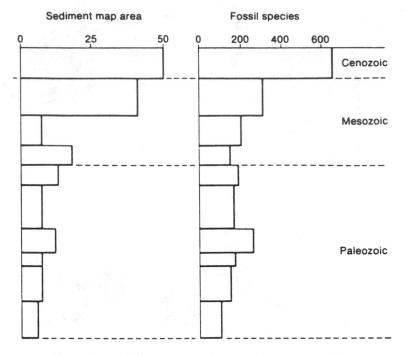

Figure 3–29. Sedimentary rock and fossil diversity through geological time. All data compiled through worldwide sampling. Modified from D. Raup and S. Stanley, *Principles of Paleontology* (New York: Freeman, 1978, p. 9.

which is fossilized tree sap. In such cases the insect is trapped and literally sealed off, preserving very fine details such as antennae and wings. Amber-preserved insects as old as 50 million years often bear a striking resemblance to living ones, providing invaluable evidence on the pace of evolution. Other soft-bodied groups with a poor record include worms, jellyfish, and many kinds of plants. Fortunately, many plants have durable pollen grains or spores that preserve well. Also, there are very rare conditions where soft-bodied impressions are preserved in sediment. Usually these involve deposition in oxygen-poor environments, such as swamp waters, which inhibit bacterial decay of the soft tissue.

Aside from anatomy (hard parts) and habitat (rapid burial), a final key influence on an organism's chance of fossilization is *when it died.* As shown in Figure 3–29, the preservation of sediment (and hence fossils within it) improves as it becomes younger. This **pull of the recent** occurs because the longer a deposit of sediment and fossils is around, the greater the chance that it will be eroded, deeply buried, or otherwise rendered inaccessible to humans. Thus, younger species have a progressively better chance of being preserved and found by paleontologists than older species.

PART 1: EVOLUTION TOWARD MULTICELLULAR LIFE

Virtually everyone knows that the evolutionary record of life shows an increase in complexity through time. Much less well known are the three major steps that were taken in the evolution of increasing complexity. The three steps involve the development of these life forms, from the earliest to the most recent:

1. simple single-celled microbes (prokaryotes)
2. complex single-celled microbes (eukaryotes)
3. multicellular organisms (metaphytes, metazoans)

Each of these advances, like the origin of life itself, occurred *in the ocean.* As shown in Figure 3–30, the first step, the origin of the simplest microbes, occurred in the early Precambrian, while steps 2 and 3 occurred in the late Precambrian. This is one of the most important observations in evolution: the early appearance of life on earth followed by an extremely long (billions of years) period of little change.

Figure 3–30. Distribution of life remains found in Precambrian rocks. Eukaryotic fossils are not found until late in the Precambrian, while prokaryotic and stromatolite remains are found throughout the Precambrian fossil record. From D. Eicher and A. McAlester, *History of the Earth*, p. 125.

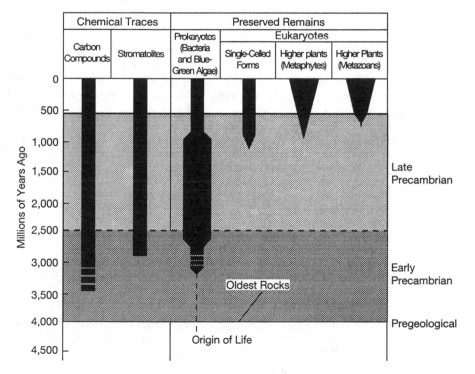

Once change does occur, to more complex cells, the pace of evolution increases progressively faster thereafter.

Simple Single-Celled Microbes (Prokaryotes)

The first known life form is evidenced by microscopic fossils. Even microbes are vastly more complex than nonliving matter; therefore, it has been a major surprise to find these first fossils in rocks 3.5 billion years old. This is much sooner after the earth cooled than most would expect. Further, given the vagaries of fossil preservation, we would hardly expect these earliest fossils to represent the exact moment of life's origin. We would expect instead that life began earlier than 3.5 billion years ago. Evidence for this is provided by chemical traces of certain carbon compounds in rocks older than 3.5 billion years (Figure 3–30). The isotopic composition and other characteristics of this carbon indicate that it is likely associated with biochemical processes.

The earliest microfossils belong to a group of organisms still alive today, the **prokaryotes.** These consist of bacteria and blue-green algae, the simplest independent life forms known. (Viruses are simpler, but they are parasitic on more complex life forms.) Prokaryotes differ from more complex life in having cells that lack a true nucleus. They also lack specialized internal organs (called organelles), such as mitochondria, which act as centers of energy conversion. All other life on earth today belongs to the **eukaryotes.** These have a true nucleus, specialized organs, and other differences shown in Figure 3–31. Eukaryotes do not appear until much later (Figure 3–30).

Early prokaryotes are found mainly in two parts of the fossil record. These are deposits dating from about 3.1 to 3.5 billion years ago in South Africa and Australia. These early deposits contain only a modest assemblage of different types. A much richer diversity of prokaryotes is found in deposits in the Great Lakes area of North America. These are much younger, dating to nearly 2 billion years ago.

Figure 3–31. Differences between prokaryotic and eukaryotic cell structures. Notice the lack of organelles and nucleus organization in the prokaryotic cell. From D. Eicher and A. McAlester, *History of the Earth*, p. 118.

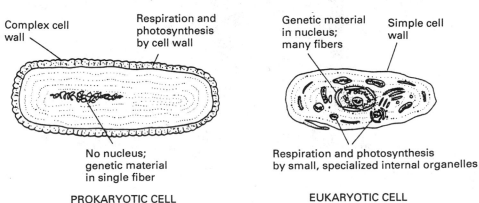

Complex cell wall

Respiration and photosynthesis by cell wall

Genetic material in nucleus; many fibers

Simple cell wall

No nucleus; genetic material in single fiber

Respiration and photosynthesis by small, specialized internal organelles

PROKARYOTIC CELL

EUKARYOTIC CELL

Thus, although there was some increase in diversity, no major evolutionary changes occurred in over 1.5 billion years. This was a very long time for so little change when compared to later evolution, when many changes occurred in just a few million years.

In all cases, the deposits with prokaryotes are sedimentary rocks such as chert (similar to flint), in which the cells have been mineralized by replacement. Such microfossils are best studied by cutting the rocks into very thin slices (called "thin sections") and transmitting light through them under a microscope. These show that the early prokaryotes had a variety of shapes, from spheroidal algal bodies to filamentous colonies (Figure 3–32). Often these colonies of algae formed large

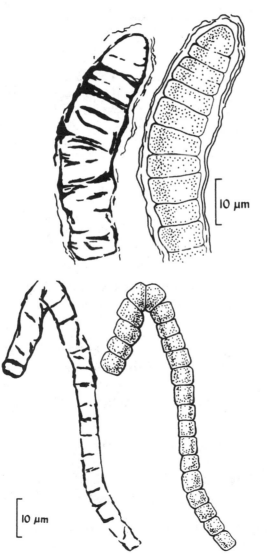

Figure 3–32. Schematic drawings of fossil algal bodies from the early Precambrian. (μm = 1 millionth meter.) Modified from R. Wicander and S. Monroe, *Historical Geology*, p. 221.

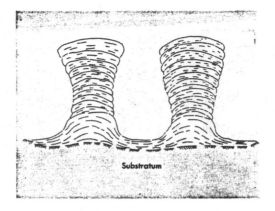

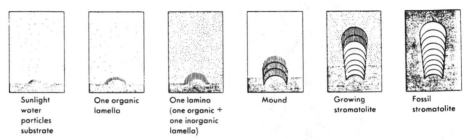

| Sunlight water particles substrate | One organic lamella | One lamina (one organic + one inorganic lamella) | Mound | Growing stromatolite | Fossil stromatolite |

Figure 3–33. Vertical sketch of algal laminated sediment, algal mats localized as small mounds which grow upward as successive layers of sediment are trapped by successive generations of mat formation.

layered mounds called **stromatolites.** These mounds, shown in Figure 3–33, have the important distinction of being perhaps the first common, large structures produced by life to appear on the earth.

THE FIRST ECOSYSTEMS. The simplicity of early life was not limited to cellular structure. Organisms also interact with one another and their physical environment to form an **ecosystem.** The basic foundation of any ecosystem are the **primary producers,** which are usually plants, taking sunlight and using the energy to produce food via photosynthesis:

$$CO_2 + H_2O \rightarrow CH_2O + O_2$$

Animals are **consumers,** meaning that they cannot synthesize their own food from sunlight and must eat it directly. They also take in oxygen to metabolize the food molecules.

We can see therefore that the plant-animal relationship is a *cycle.* Animals excrete H_2O and exhale CO_2 to provide raw materials for the plants, which use the material for photosynthesis. As we will see, increasing diversity has modified this simple relationship into complex food webs: Some ecosystems today have many different kinds of plants as primary producers and many levels of consumption among animals. However, when only prokaryotes were alive, this system was much simpler. Blue-green algae were the sole primary producers, while bacteria were the consumers. Yet even here we can see signs of increasing ecosystem complexity: While many bacteria fed directly on the algae (and are thus called "primary consumers"), no doubt some of them also fed on other bacteria, the "secondary consumers."

Complex Single-celled Microbes (Eukaryotes)

The next major advance was the appearance of eukaryotes. As noted, this type of cell, unlike the prokaryotic cell, has a true nucleus and specialized organelles (Figure 3–31). Eukaryotic cells are also larger in size. All organisms, except the bacteria and blue-green algae, have eukaryotic cells.

ORIGIN OF EUKARYOTIC CELLS.   The most widely accepted view on the origin of eukaryotic cells is the **symbiosis theory.** This theory proposes that eukaryotic cells arose from a cooperative relationship (symbiosis) between different prokaryotic cells. As shown in Figure 3–34, some single-celled prokaryotes engulf other single-celled

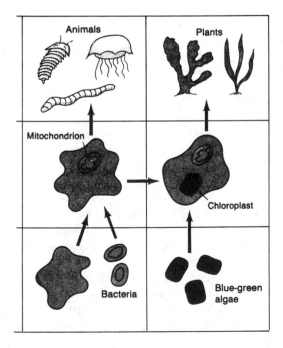

Figure 3–34. The probable sequence of major events leading to multicellular animals and plants: engulfed prokaryotes become organelles in eukaryotes. Modified from S. Stanley, *Earth and Life Through Time* (New York: Freeman, 1986), p. 287.

prokaryotes. It may be that the engulfed cell became a permanent "organelle," performing specialized metabolic and other tasks. This theory explains a number of things, most importantly why organelles often have separate genetic information from that in the nucleus. This would be expected if the organelle was originally another organism.

DIVERSIFICATION OF SINGLE-CELLED EUKARYOTES. Eukaryotic cells apparently first appeared in the fossil record shortly after 1.5 billion years ago (Figure 3–30). However this is only an estimate because it is difficult to distinguish prokaryotic from eukaryotic cells in the fossil record.

Once they appeared, eukaryotic cells evolved much more rapidly than the prokaryotes. A main reason is that eukaryotes generally reproduce by *sexual* means. In sex, there is an exchange of genetic material (DNA) between two organisms so that offspring have a combination of the DNA of both parents. Sex greatly accelerates evolution because it drastically increases the variation in the offspring. In contrast, prokaryotes reproduce asexually meaning that one parent almost always produces exact copies ("clones") of itself. Obviously this greatly limits the amount of variation of individuals and would tend to slow down changes into new species. The primary way for any variation to occur is by a gene mutation, which usually kills the offspring and, in any case, takes much longer than sex to produce variation.

This key role of variation in evolution was discussed in detail in Chapter 2. Suffice it to say here that the "invention" of sex by eukaryotes is probably the main reason that evolution was so slow in the prokaryotes compared to that of the eukaryotes. As we shall see, a veritable "explosion" of evolutionary change occurred not long after eukaryotes appeared. It began with a diversification of the single-celled eukaryotes to produce protozoa and other single-celled groups. However, evolution accelerated even more as the single eukaryotic cells began to group together and specialize to form multicellular organisms.

Multicellular Organisms (Metaphytes, Metazoans)

ORIGIN OF MULTICELLULAR ORGANISMS. The much more rapid evolutionary change by eukaryotic cells soon led to a major innovation: integrated groups of cells joining together to form a single "organism." Not surprisingly, there is no direct evidence in the fossil record of how this came about. The first such organisms must have had only a few cells and had not yet evolved shells, teeth, or other hard parts that form most fossils. Nevertheless, clues have been observed in living single-celled eukaryotes. These have led to a number of suggestions on how single-celled organisms could have evolved into multicellular forms. Two of the most prominent are the symbiotic and colonial hypotheses (Figure 3–35).

The **symbiotic hypothesis** says that *different kinds* of single-celled organisms grouped together in a cooperative unit to form one organism. This is much like the theory on the origin of eukaryotic cells. The problem with this idea is that multicellular animals generally have the exact same set of genes in all their cells, although

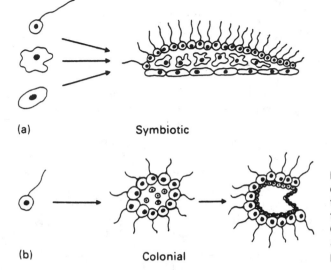

(a)   Symbiotic

(b)   Colonial

Figure 3–35. Possible routes for the evolution of animal multicellularity from single-celled animals. Modified from R. Barnes, P. Calow, and P. Olive, *The Invertebrates: A New Synthesis* (London: Blackwell, 1988), p. 13.

they are expressed differently in different cells (see Chapter 2). Furthermore, when they reproduce, half of this same gene set is transmitted to the offspring. It is very difficult to see how this situation could have evolved from a group of different kinds of single-celled organisms, since each kind would have had a different set of genes.

The **colonial hypothesis** states that multicellular organisms evolved when reproduction in eukaryotic cells produced a number of *identical offspring cells*, which stayed together in a colony. This production of identical offspring is not uncommon in single-celled eukaryotes and is similar to the production of much more complex human identical twins, which have identical cells. This hypothesis solves the problem of identical gene sets, since all cells are identical. Therefore, the colonial hypothesis is the most accepted one, although it is still debated.

Most of us think of ourselves as individuals. Yet in reality, we are each composed of trillions of individual cells. Our "colonial" composition is much more easily seen when we consider that we (and most other multicellular organisms) begin life as just one cell, a fertilized egg, that divides into ever greater numbers of cells. We owe our existence to the advantages that accrued to individual cells that tended to group together and cooperate. Recall from Chapter 2 that our embryological development repeats much of our evolution. This is seen not only in the multiplication of many cells from one, but also in that these cells in the developing embryo migrate, communicate, and even form a hollow ball very similar to that shown in Figure 3–35. Such hollow structures are common in both colonies of cells and constitute an early stage in the developing embryo.

Obviously an animal as complex as a human is more than just a group of cells. There are two main differences between us and a simple colony: Our cells are much more *specialized* and *integrated*. Specialization refers to the variation among cells that allows them to perform different tasks. In the human body there are over 100 kinds of specialized cells, such as nerve cells, liver cells, skin cells, and so on. This specialization allows the cells to become much more effective and efficient at what they do than if they had to remain capable of doing many things. In contrast, the cells in a colony are all one kind. By integration we mean that there is communication among the cells, usually by electrical and chemical means. This allows the specialized cells to perform their separate tasks, but also to interact with the group as a single unit and share the results of their specialized labors.

While the differences between a complex multicellular organism like ourselves and a simple colony of cells may seem too great to be related, we must realize that we are comparing endpoints in a long evolutionary sequence. In reality, the specialization and integration must have been gradual and occurred in increments over millions of years. The best understanding of how multicellularity arose is to look at living intermediates that have not evolved much past the colonial stage. The classic example is the sponge. Here we have an animal with specialized cells for circulating and filtering water, structural support, and other tasks. Yet when the sponge "animal" is squeezed through a fine mesh and broken up into individual cells, the cells will reunite in a few hours and the sponge animal will again exist. Obviously the cells in a human being will not reunite after such an experiment. Our cells have become so specialized and dependent on one another that they cannot survive alone.

Finally, note that multicellularity evolved more than once. **Metaphytes** are multicellular plants with ancestral ties to single-celled plants, beginning perhaps with colonies of algae. **Metazoans** are multicellular animals; they originated with protozoans (single-celled animals). Indeed, even within the metaphytes and metazoans, it is likely that some groups arose from different single-celled ancestors. However, many researchers think that most of the major groups of animals arose from just one single protozoan ancestor. As shown in Figure 3–36, primitive multicellular animals (such as jellyfish and sponges) gave rise to still more complex forms, as new cell types were added and more closely integrated with the evolving cell assemblage. For example, while insects and mammals (at the top of Figure 3–36) have many kinds of well-integrated cells, sponges and jellyfish have relatively few kinds, and these are less integrated. Note the two-fold branching of the "trunk" of this family tree: arthropods, mollusks, worms, and other invertebrates branched off from the vertebrate line very early. What is the evidence for this branching and, indeed, for the tree itself? Primarily the evidence is embryological. As just noted, embryological development gives important clues to how a group evolved: More closely related groups tend to have more similar development. For instance, mammals have cell cleavage and other embryological patterns that are much more similar to birds than to insects.

FOSSIL RECORD OF MULTICELLULAR EVOLUTION. The first direct evidence of multicellular plants and animals dates from about ½ billion years ago (see Figure 3–30). Fossils

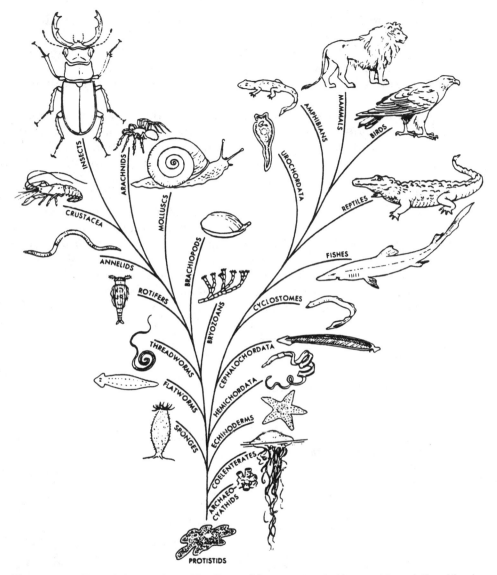

Figure 3–36. This phylogenetic or "family tree" is reconstructed to consider relationships between organisms. Note two major branches of complex life: the "insect branch" and the "vertebrate branch." From W. Stokes, *Essentials of Earth History*, 4th ed., p. 109.

of these early multicellular forms are rare because they had soft bodies; mineralized hard parts were a later specialization. Thus, wormlike or jellyfishlike organisms were probably among the early metazoans. For this reason, the first fossil evidence of metazoans are their trace fossils. The appearance of small tracks, trails, and burrows about a billion years ago is our first clue for primitive and perhaps wormlike organisms. Such traces are much too large to have been made by microscopic life.

By about 700 million years ago (very late in the Precambrian), multicellular life had evolved into a wide variety of forms. This is best known from a very famous fossil locality in Australia called the **Ediacara Formation.** This is a highly unusual deposit because it has the preserved remains of many soft-bodied animals. Rather than relying on their traces, we can directly see the remains of the animals themselves. These provide a fascinating glimpse of the earliest multicellular life to evolve on the planet earth. Living in a shallow marine environment, some of the animals in the formation are recognizable as relatives of living groups, such as jellyfish and segmented worms (Figure 3–37).

However, many late Precambrian fossils are unfamiliar because this was a time of great evolutionary experimentation. The unfamiliar forms represent unsuccessful experiments whose basic designs have now become extinct. For example, one organism has a body with threefold symmetry. This is completely unknown in any living organism. Most animals have twofold symmetry. A few, like the starfish, are based on fivefold symmetry. As organisms with certain basic body designs became more common, other basic designs were eliminated. Therefore, later evolution has certainly involved much change, but it is more modest. It consists mainly of minor variations of the more basic anatomical themes established during these early times. More basic changes, such as the creation of new major groups, were not possible because existing ecological space became filled up: Evolutionary "experiments" had a progressively lower chance of success as competitors and predators became more effective and increased in number.

Figure 3–37. Artist's rendition of Ediacara animals as they may have looked in Precambrian seas. From A. McAlester, *The History of Life*, 2nd ed., p. 24.

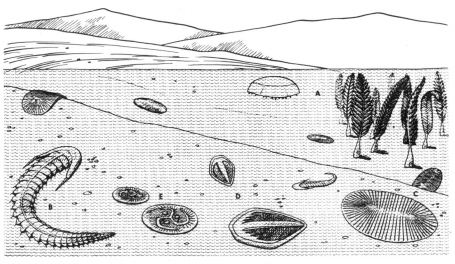

In addition to their novelty, one is also impressed by the complexity of these metazoans, with their highly differentiated organs and structures, compared to single-celled animals. This, (plus the diversity of animals found) indicates that multicellular evolution had been proceeding for quite some time. Whether it began sooner than the 1 billion years indicated by trace fossils is uncertain.

THE EARLY PALEOZOIC "EXPLOSION" OF LIFE. In sharp contrast to earlier rocks, rocks formed after about 570 million years ago are rich in fossils. This "explosion" of fossils is used to mark the end of the Precambrian and initiate the Paleozoic era on the geologic time scale. Just what caused the sudden appearance of fossils at this time is uncertain. It may be that organisms were there before, but had not evolved hard parts, such as shells and teeth, which would easily fossilize. Or it may be that some physical threshold was reached, such as enough oxygen in the atmosphere to support complex life. (See Figure 3–24.) This may have allowed large numbers of groups to evolve very rapidly.

Whatever the cause, this early Paleozoic explosion marks the last major "experimental" phase of life's evolution. By about 500 million years ago, representatives of all the basic groups of life had appeared. These basic groups are called **phyla,** such as phylum Echinodermata and phylum Mollusca (phylum is Greek for "tribe"). As we have said, all evolution after this consists of variations on these basic phylum body plans that had originated by 500 million years ago. Thus, many new snail and clam species evolved (and still do), but they all represent variations of the basic mollusk body plan. No new phyla have appeared in the last 500 million years. In his book *Wonderful Life* (1989), Stephen Jay Gould gives a very colorful description of this last great phase of experimentation and its profound implications for evolution.

PART 2: EVOLUTION OF LIFE IN THE OCEANS (HYDROSPHERE)

All that we have discussed so far—the origin of life, its evolution into multicellular complexity, and life's first major diversification (the early Paleozoic "explosion")—occurred in the ocean. Not long after this, life began to radiate into the other two environments—land and, later, the air. However, before we discuss this radiation into new realms, we must complete the story of life in the hydrosphere over the next 500 million years.

As shown in Figure 3–38, the evolution of life in the oceans can be divided into three main stages:

1. Rise of sessile invertebrates and fishes (Paleozoic)
2. Rise of mobile invertebrates and marine reptiles (Mesozoic)
3. Rise of marine mammals (Cenozoic)

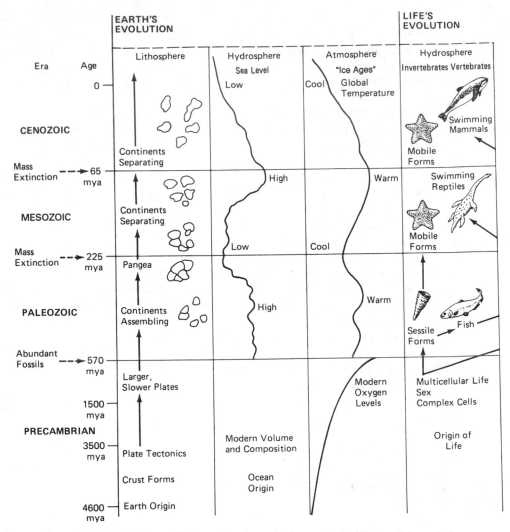

Figure 3–38. Evolution of life's major stages in the context of the earth's evolution.

These stages represent times when the groups were first prominent. This does not necessarily mean that they faded thereafter. For instance, fish and mobile invertebrates continue to be extremely diverse and important today even though they first appeared much earlier.

Note that these three stages can be roughly correlated with the three latest eras of earth history (Figure 3–38). We will see that stages in life's evolution onto land and air also correlate well with these eras. This is no accident since the eras were originally defined by major fossil changes. For example, we have already seen that the Paleozoic begins with the explosion of life. In fact, Paleozoic means "old animals," Mesozoic means "middle animals," and Cenozoic means "new animals," each referring to the groups of animals that were abundant when they first developed. Thus, "new animals" abundant in the Cenozoic include mammals, while the "middle animals" of the Mesozoic include dinosaurs. The eras are separated by mass extinctions that often ended the dominance of animals, such as dinosaurs, in the preceding era (Figure 3–38).

Rise of Sessile Invertebrates and Fishes (Paleozoic)

The Paleozoic world had extensive seas and a warm climate (Figure 3–38). These provided many habitats for extremely abundant and diverse marine life. Even today warm shallow areas, although much reduced in size, are the most diverse and productive parts of the marine environment. This is because shallow waters: (1) receive nutrient runoff from land, and (2) are areas of sunlight penetration for photosynthesis, the base of the food chain. Below a few hundred feet of water depth, sunlight becomes too diffuse.

SESSILE INVERTEBRATES. **Invertebrates** are animals without backbones. Included in this group are clams, snails, starfish, worms, and many other familiar animals. Invertebrates have always been much more common than vertebrates. During the Paleozoic, an observer would have viewed a sea floor with invertebrates that showed mainly **sessile** ways of living. By sessile we mean stationary: They stayed in one place and trapped microscopic food (such as plankton and organic debris) from the water. This trapping of food is usually done with filters or sticky mucous secretions. As shown in Figure 3–39, these include the following specific groups that were the most abundant then: **stalked echinoderms** (such as sea lillies), **brachiopods** (lamp shells), **corals,** and **bryozoans** (tiny encrusting colonies of "moss animals"). In contrast, mobile invertebrates (burrowers, crawlers, and swimmers), while present, were not as common until later times. The notable exception to this are the **trilobites** (segmented ancestors of the insects), but even these diminished in abundance and became progressively less important after the early Paleozoic. A typical Paleozoic community is shown in Figure 3–40, illustrating the predominance of sessile life.

While we focus here on the marine hydrosphere because it constitutes the largest part of the hydrosphere, we should point out that some invertebrates made the transition to fresh water in the Paleozoic. Among the most successful have been the mollusks, especially clams and snails. Other groups have never been able to make the transition. One such group is the echinoderm, which includes starfish, sea urchins, and their kin. Apparently the physiology of some groups is difficult to adapt to the chemical and physical properties of fresh water. A large part of the difficulty is that because life originated in the ocean, the chemistry of living organisms (including our own) is much closer to that of ocean water than to fresh. It is therefore more difficult to maintain this internal chemistry in fresh water (a problem even more difficult on land).

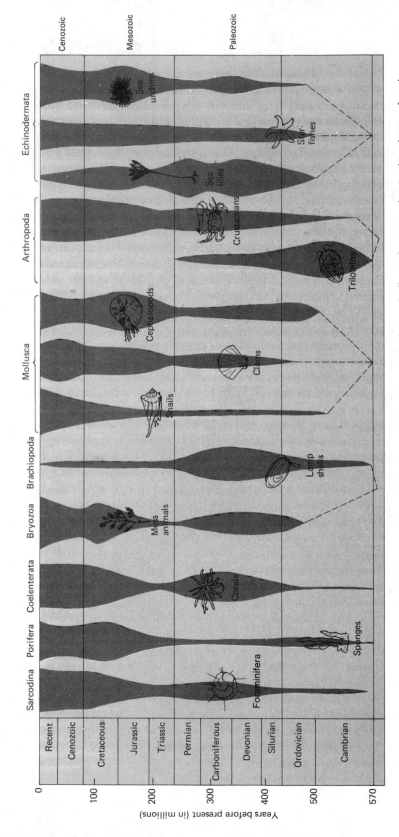

Figure 3–39. Geologic record of major groups of marine invertebrates. Width of the vertical areas indicate the approximate abundance of each group and dashes show likely evolutionary relationships between groups. Modified from D. Eicher and A. McAlester, *History of the Earth*, p. 237.

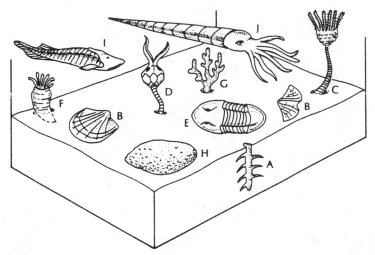

Figure 3–40. Prominent fossils from the Paleozoic: (A) Spiral burrow; (B) Brachiopods; (C) Crinoid; (D) Cystoid; (E) Trilobite; (F) Solitary horn coral; (G,H) Bryozoan colonies of branching and massive species; (I) Jawless fish; and (J) Nautiloid cephalopod. Modified from N. G. Lane, *Life of the Past* (Columbus, OH: Merrill, 1989), p. 144.

FISH. Fish are of great evolutionary importance because they were the first major **vertebrates** (animals with backbones). The backbone was a key evolutionary innovation because it led to an *internal* skeleton, which has a number of advantages over shells and other external skeletons. Among the advantages of an internal skeleton are: (1) greater mobility (due to a more flexible body), and (2) larger body size (due to greater structural strength of internal support). Both of these advantages were crucial in the evolution of large mobile land animals: Amphibians, reptiles, and mammals (including humans) evolved from fish ancestors.

Four main groups of fishes evolved by the middle Paleozoic: *jawless fishes, placoderms, sharks,* and *bony fishes.* Only the last two are abundant today, as shown by the geologic range chart of these groups (Figure 3–41).

The **jawless fishes** were the first fishes to appear. They were small, only a few inches long at most. They lived by "vacuuming" food from the sea bottom through an opening under the skull. Jawless fishes were moderately diverse, although they were not very common. They appear strange to the modern observer not only because of their jawless appearance but because of the heavy armor, especially around the head (Figure 3–42). They were soon replaced when fish with jaws evolved, probably because of the advantages of jaws in acquiring and manipulating food. However, a few jawless fish have survived until today by adopting a parasitic life habit. The modern lamprey uses its jawless mouth (equipped with rasping teeth) to attach itself and eat the insides of other fish. Such specialization and parasitism is a common mode of survival among old groups that have been evolutionarily superseded.

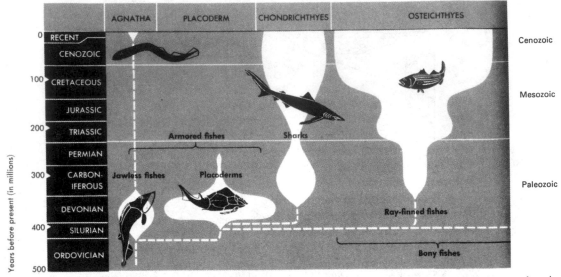

Figure 3–41. Evolutionary history of fish. Width of white areas indicates the approximate abundance of each group during its existence. Dashes show probable evolutionary relations of the groups. Modified from D. Eicher and A. McAlester, *History of the Earth*, p. 275.

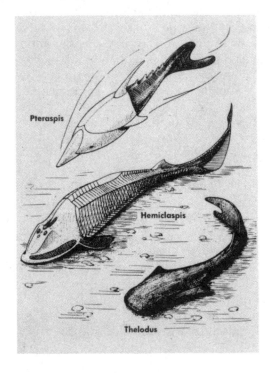

Figure 3–42. Armored jawless fishes lived during the middle Paleozoic. These fish were bottom feeders with characteristic shovel-like mouths. From A. McAlester, *The History of Life*, 2nd ed., p. 80.

Figure 3–43. This Paleozoic, carnivorous fish reached a length of 30 feet and had a mouth several feet wide. From A. McAlester, *The History of Life*, 2nd ed., p. 82.

The most primitive jawed fish were the **placoderms.** These were ferocious-looking fish that also had bony armor (placoderm means "plated skin"). The jaws were equipped with sharp beveled teeth (Figure 3–43). Even more intimidating, placoderms grew up to 30 feet long. In spite of these qualities, the placoderms were replaced by the cartilage-bearing and bony fishes. The reason may be that the heavy armor was an impairment to swimming agility and speed. The function of the armor is not known, though it may have been protection against predators such as giant "sea scorpions." These scorpions were similar to scorpions today, but they were much larger and lived in the water.

Sharks (including the rays) are still abundant today. They have skeletons of *cartilage* instead of bone. Cartilage has some advantages: It is more flexible and, unlike most bone, continues to grow throughout life so that sharks and rays can become very large. Nevertheless, it is the fourth major group, the **bony fishes,** that have come to dominate the seas, making up over 90% of all fish today. A main reason for this dominance is the evolutionary flexibility of having a skeleton of bone. Even though bone stops growth, it is much harder than cartilage and capable of being modified into many shapes. Most of the bony fishes belong to a group called the **ray-finned fishes** (Figure 3–41). A second group, the **lobe-finned fishes,** is much less numerous but has great evolutionary importance: They are the direct ancestors of land vertebrates. Unlike ray-finned fish, which have only small bones in the fins, lobe-finned fish have a large, fleshy lobe that reinforces the fins. This lobe was an important precursor to legs, for crawling on land.

ORIGIN OF FISH AND JAWS. As the first vertebrates, the origin of fish is of special interest. However, in looking for their origin, the casual observer is faced with an appar-

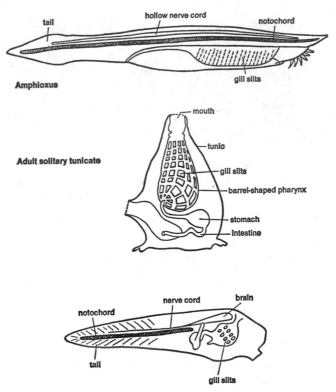

Figure 3–44. Vertebrate features, such as the muscular tail used for swimming, are visible in the *Amphioxus* fish. These same features are also evident in the young tadpole (larva) of the tunicate but not in the bottom-dwelling adult. From A. Romer, *The Procession of Life* (Highland Springs, VA: Anchor Books, 1972), p. 157.

ent discontinuity because no invertebrates seem to bear any resemblance to fish. But this is because we are focusing on the adults. In fact many invertebrates produce offspring that swim as juveniles ("larvae"). Upon finding a suitable habitat, the larvae settle and undergo metamorphosis. Some invertebrates have larvae with a strong resemblance to primitive fish. Some of these larvae even have a nerve cord, gill slits, and a tail (Figure 3–44).

But how can a juvenile give rise to a new group? **Heterochrony** is a common occurrence whereby the developmental process of an organism undergoes a change in timing ("hetero" = different, "chrony" = time). This often occurs when slight changes are made in the hormonal signals that stimulate the onset of a stage, such as sexual maturity. For instance, a mutation in a regulatory gene (Chapter 2) can cause the hormonal signals for maturation to occur early. Thus, even though it is still *anatomically* a juvenile, the individual is *sexually* an adult, able to reproduce itself. This has been observed in a number of living animals such as some salamanders—juveniles may stay and reproduce in the water instead of metamorphosing into adults and leaving the water. Thus, it is hypothesized that the first fish arose through heterochrony: The swimming larval stage of an invertebrate, complete with nerve cord, gill slits, and so on, became an "adult," never reaching the settling stage. Since it was able to reproduce, the settling stage was permanently lost and the swimming way of life was passed on.

Which major invertebrate group led to fish? Here we turn to evidence from living animals. When we examine their biochemistry and cell cleavage patterns during development of the embryo, we find that vertebrates show striking similarities with *echinoderms*; animals such as starfish and sea urchins are echinoderms. In contrast, mollusks, insects, and other invertebrates have clearly distinct embryologi-

Figure 3–45. Origin of the vertebrate jaw. Fish jaws (b) evolved through enlargement of the anterior gill arch supports of jawless fish (a). From R. Wicander and S. Monroe, *Historical Geology*, p. 334.

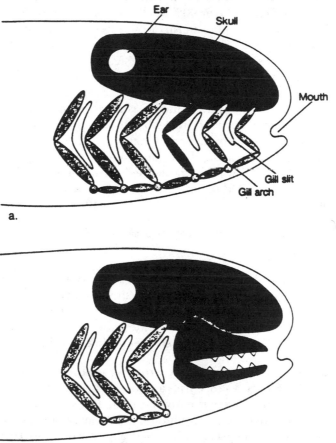

cal development, and thus occur on the second major branch of the "family tree" of animal life (see Figure 3–36).

Clues to the origin of jaws are also found in observing the development of living fish. During embryonic development, modern fish go through a stage where gill slits become supported by small bony structures called *gill arches*. However, as shown in Figure 3–45, the first three arches back from the mouth subsequently become modified by continued growth and develop into the large bones making up the jaw. Jaws apparently arose in fishes by mutations (and by subsequent selection favoring them), which altered bone growth in the forward gill arches to form the jaw bones.

Whether fish originated in the ocean or fresh water has been debated for decades. The question remains unresolved because the earliest remains, from early Paleozoic rocks, are very fragmented and hard to precisely date. Either way, fish were soon adapted to both marine and fresh waters.

THE END—PALEOZOIC MASS EXTINCTION. The Paleozoic ended with the largest mass extinction of all time. Over 90% of all species were apparently wiped out. Trilobites, primitive corals, and placoderms would never be seen again. Many of the groups that survived would never recover. Brachiopods and stalked echinoderms would never become as abundant or diverse again.

A large part of the cause of the extinction must have been the uniting of all the continents into the supercontinent Pangea. This would have literally squeezed out many marine habitats around the continental margins, as shown in Figure 3–23. Adding to this was the withdrawal of the sea from continents, removing even more habitats of the marine organisms. Finally, global weather and ocean current patterns were also altered, causing disruption to ecosystems in general. Unfortunately, the same withdrawal of waters that helped cause the extinction also led to fewer sediments being deposited so the geological and fossil record of this interesting time is not well preserved.

Rise of Mobile Invertebrates and Marine Reptiles (Mesozoic)

It was not until the middle and late Mesozoic that the seas again covered the continents and marine life began to flourish as before. Diversification was further encouraged by the breakup of Pangea, which created more habitats along the continental margins (see Figure 3–38). Mesozoic sea life may be roughly characterized by the rise of (1) mobile invertebrates and (2) marine reptiles.

MOBILE INVERTEBRATES. In the Mesozoic, we see the rise of *mobile* invertebrates. Referring back to Figure 3–28, corals and bryozoans are important, but brachiopods and stalked echinoderms are greatly reduced. The corals are not the primitive kind of the Paleozoic but have a considerably different biology, such as a sixfold symmetry. Indeed, many biologists think that the modern corals reevolved in the Mesozoic from a different, noncoral ancestor.

Increasing in abundance and diversity during the Mesozoic (Figure 3– are *burrowers*, such as **sea urchins** and **clams,** and the *crawlers*, such as **snails** an **crustaceans** (crabs, lobsters, shrimp, and other relatives). In addition, swimming invertebrates increase, in the **cephalopods.** These are relatives of the squids and octopi that swim by jet propulsion, squirting water out of a funnel. They grasp food with tentacles and eat it with a powerful crushing beak (like a parrot's beak). A typical Mesozoic sea floor community, illustrating these animals, is shown in Figure 3–46. Compare it with the sessile community that a visitor would have seen in the Paleozoic (refer to Figure 3–40).

The marine invertebrates of today have changed little since the Mesozoic. Therefore, we may characterize the Mesozoic as the rise of modern marine invertebrates. Most of these mobile animals (clams, snails, cephalopods) belong to one phylum, the **mollusks.** Thus, a shoreline observer of a modern ocean will generally see a preponderance of members of this group, making it the most familiar group to the average shell collector today. This change from sessile to modern mobile invertebrates is often called the **Mesozoic marine revolution.**

What is the cause of this "revolutionary" rise of mobile animals? It is not the end-Paleozoic mass extinction because these mobile forms all existed before it. Also, there is some evidence that sessile forms were beginning to decline before the

Figure 3–46. A generalized Mesozoic marine community picturing organisms common during the Era: (A) Dwelling burrow; (B) Burrowing bivalves; (C) Burrowing echinoids; (D) Partially burrowing oyster; (E) Nestling bivalve; (F) Reef-forming rudistid bivalves; (G) Gastropods; (H) Coiled and straight-shelled ammonoids with complex sutures; and (I) Bony fish. From N. G. Lane, *Life of the Past*, p. 145.

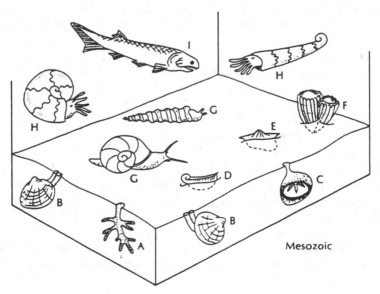

ended. Instead, it seems to have resulted from evolving interactions
organisms. For instance, more efficient predators, such as snails with
lls, appear. Another major predator are the crustaceans (for example,
and lobsters) with huge crushing claws. Mobility would be an important means
escape from such predators. Thus, even filter feeders, such as clams, could no
longer remain in one place, but developed the ability to dig deep into the sediment,
using a long tube to breathe and eat with. (A rarer, but more striking means of
escape evolved with the familiar scallop. It claps its shell together to swim rapidly
away.) Sea urchins either evolved long spines or the ability to burrow, like the sand
dollars. This mutual "escalating" co-evolution between predator and prey has been
appropriately called an **arms race.** Further, the initial need to become more mobile
to escape predators also caused a positive feedback ("snowball effect") that led to
even more mobility: Mobile organisms tend to stir up the sediment and this suspended sediment clogs the filtering and breathing mechanisms of many sessile
organisms, making the environment even more inhospitable for them.

MARINE REPTILES. Like the mammals after them, some reptiles returned to the sea
after evolving on land. Also like the marine mammals (whales and porpoises), swimming reptiles attained large sizes. This is because the bouyancy of water offsets the
pull of gravity, allowing large sizes without the animal collapsing under its own
weight, as it would on land. This same principle works with invertebrates and they
also have evolved giant sizes as swimmers. Today, giant squids over 20 feet long
have been known to attack whales. However, most invertebrates live on the bottom
and a very large size is not advantageous for burrowing, crawling, or filter-feeding
lifestyles. In spite of their impressive sizes, large organisms are not nearly as abundant or diverse as smaller organisms. This is normally the case for animals at the
top of food chains because much energy is lost at each link in the chain: When
eating a smaller organism, much of the energy is lost as heat during digestion and
metabolism.

There were four major kinds of swimming reptiles: ichthyosaurs, plesiosaurs,
mosasaurs, and turtles. While they lived at the same time as dinosaurs, they are *not*
dinosaurs. Nor are they closely related to each other.

The **ichthyosaurs** adapted best to water, having a porpoiselike appearance
with fins and streamlined body (Figure 3–47). Their development provides a classic
example of **convergent evolution,** which means that two unrelated groups develop
similarities—in this case, in appearance and way of life. Ichthyosaurs bear a remark-
able resemblance to porpoises (which are mammals) and some fish. ("Ichthyosaur"
means "fish-lizard.") This is due to the physical properties of water, which demand
a certain hydrodynamic design for fast swimmers. Also like porpoises, ichthyosaurs
had live births, unlike most other known reptiles. Ichthyosaurs grew quite large,
up to 30 feet long.

Plesiosaurs were much slower swimmers, equipped with four paddlelike flippers on a larger, less streamlined body. Instead of pursuit they relied on a long

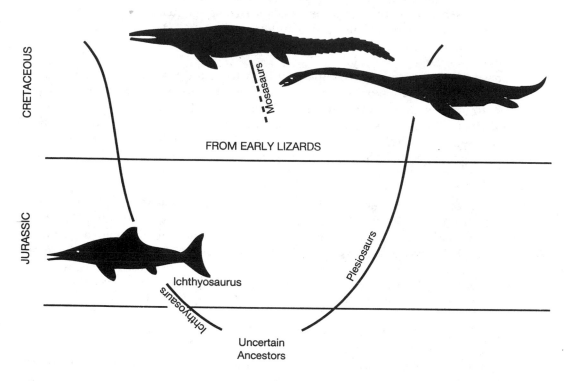

Figure 3–47. A generalized family tree showing evolutionary relationships in the three major groups of marine reptiles of the Mesozoic era. Modified from H. Levin, *The Earth Through Time*, p. 476.

snakelike neck to pluck fish from the water, much like some shore birds. They grew up to 50 feet long. If plesiosaurs look familiar it (see Figure 3–47) is because they have been widely suggested as the celebrated "Loch Ness Monster." However, there is no documented evidence for this and it seems especially doubtful considering that no fossils of plesiosaurs are known since at least 65 million years ago. In addition, the cold climate of Scotland is a most unlikely place for a large reptile to live.

Mosasaurs were giant lizards. They were similar to alligators in appearance, with elongated bodies and heads, but short necks (see Figure 3–47). They swam like alligators, in a sinuous side-to-side motion using their flattened tails. Also like alligators, they probably spent only part of the time in water. Mosasaurs grew to over 30 feet long.

Finally, giant turtles were very abundant in Mesozoic seas, using flipperlike limbs for propulsion. They were very similar to sea turtles today, only larger, growing up to 12 feet long. Of the four groups, turtles are the only survivors today. As we often see in evolution, ferociousness and size is no guarantee to evolutionary success. Indeed being inconspicuous and/or well protected, as with turtles, often seems to help most in increasing a group's longevity.

MODERN PHYTOPLANKTON. Microscopic plankton form the base of nearly all marine food chains, so we should not overlook these tiny floaters. Primitive phytoplankton ("phyto" means "plant") were the photosynthetic base of Paleozoic and Precambrian food chains. In the Mesozoic, modern phytoplankton became common: especially diatoms and coccoliths. Coccoliths are perhaps best known because their abundant microscopic fossil shells form many chalk deposits, such as the famous "white cliffs of Dover" in England.

END-MESOZOIC MASS EXTINCTION. The mass extinction ending the Mesozoic is best known for killing off the dinosaurs. It was a major event, with over 50% of all species on earth dying out. However, it also affected marine life. The large swimming reptiles (except turtles) died out, although only cephalopods, reef-forming clams, and plankton suffered great losses among the rest of the marine ecosystem. What caused this extinction? Many scientists believe that a massive meteorite impact was involved, as we will discuss later with the demise of the dinosaurs.

Rise of Marine Mammals (Cenozoic)

As temperatures gradually cooled throughout the Cenozoic, sharper and more numerous temperature zones formed between the poles and the equator. In response, gradients of marine communities evolved that were adapted to them. This, plus the continuing fragmentation of Pangea to form more diverse habitats, led to much diversity of marine life during this time. However, the sea has generally remained withdrawn, so there have been no large epicontinental seas.

As noted, there have been no major changes in invertebrate or phytoplankton evolution since the Mesozoic, although a number of minor changes have occurred. More conspicuous has been the replacement of large swimming reptiles by large swimming mammals.

MARINE MAMMALS. The **whales** and their close relatives the **porpoises** appeared fairly early in the Cenozoic, a few million years after the demise of the large swimming reptiles, which happened about 65 million years ago. Early whales are relatively common fossils in rocks about 40 million years old. As seen in Figure 3–48, these early whales were already highly adapted for sea life, with flippers, tail, and no hind limbs. Many of them were up to 70 feet long. Given this adaptedness to water, we may assume that the whale and porpoise ancestor had left the land much earlier. This was probably a gradual process, where the animal spent progressively more

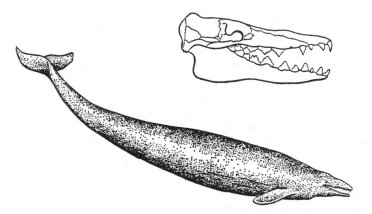

Figure 3–48. An early Cenozoic whale. Note simple teeth not found in later whales. Modified from A. Romer, *The Procession of Life*, p. 320.

time in the water. Interestingly, some "evolutionary throwbacks" have been observed wherein whales today are born with the remnants of hind-limbs. The exact land ancestor of the whales and porpoises is a matter of debate. This is not surprising because, as we see later, early Cenozoic land mammals were in a very "archaic" state, meaning very generalized anatomically. The familiar, rather specialized, groups that we see today (cats, cows, bats and so on) had not yet evolved. However, the bulk of the evidence indicates that the most direct ancestor was also an ancestor to deer, sheep, and other cloven-hoofed animals.

Two basic groups of whales have evolved from this ancestor. The oldest group are the flesh-eating *toothed whales*. This includes the sperm whale (the whale in *Moby Dick*), killer whales, and porpoises. As you would expect, these whales have rows of sharp teeth. Early whales were also carnivorous because even the early Cenozoic whales have large, sharp teeth (Figure 3–48). The second group of whales evolved away from the flesh-eating lifestyle about 15 million years ago. These are the *baleen whales*. They are called baleen whales because they strain tiny plankton through the baleen bone (a finely divided network of bone). Baleen whales not only include the largest whales but are the largest known animals to ever exist. The blue whale, the largest, reaches over 100 feet long and weighs over 130 tons. (For comparison, the largest land animal today, the elephant, weighs less than 10 tons and even the largest dinosaurs, such as *Brontosaurus*, rarely exceeded 30 tons, and certainly never reached more than 70 or 80 tons).

Whales are of special interest not only for their sheer size but also for their intelligence. Porpoises (and killer whales), in particular, have an extremely large relative brain size. Indeed, the porpoise is the only animal with a relative brain size on par with our own. Whales and porpoises have a highly advanced social structure and communicate in a complex language using sound waves. So far, attempts to understand this language and assess what kind of "minds" these animals possess have been largely unsuccessful. Mainly this is because their environment is so vastly different from our own.

Although mammals (porpoises) have "re-invented" the ichthyosaur way of life, and alligators have retained the mosasaur way of life (marine crocodiles can reach huge sizes), evolution has never replaced the third extinct group of giant marine reptiles, the long-necked plesiosaurs (Figure 3–47). It is true that seals and walruses (distant relatives of dogs and bears) also use flippers and eat fish. But they actively swim in pursuit of fish, are much smaller, and lack the long neck of plesiosaurs. Wading birds use the "long-neck" principle, but they are obviously much smaller in size and are nonswimmers. This illustrates how the "tinkering" nature of evolution, discussed in Chapter 2, is incapable of replacing some patterns if the biological raw materials are not available. Perhaps some long-necked fish-eating mammal, such as a long-necked seal, might eventually evolve. Indeed, plesiosaurs started off with short necks and less flipperlike limbs and took many millions of years to develop the very long necks. Since long necks were advantageous for fish eating in large animals at one time, is there any reason why they would not be now?

PART 3: EVOLUTION OF LIFE ON LAND

Earth before land life must have been much like the barren, rocky surface of Mars, with great hills of rolling sand dunes, unbroken by trees, grasslands, or any living thing. Unlike the sea, the land was not occupied by multicellular life in the very early Paleozoic. Fossils of the first known land life are not found for at least 100 million years after the "explosion" of life at the beginning of the Paleozoic. It was not until the middle Paleozoic that land life became abundant. Even then it was confined to moist, near-shore environments. This long delay in colonizing land occurred because, having evolved in the sea, life found land to be a hostile environment.

There were three major problems for life on land: (1) *drying out*, (2) a much *higher gravity*, and (3) a need to *breathe oxygen* as a gas instead of a liquid. Plants, invertebrates, and vertebrates each solved these problems differently. Not surprisingly, the shift toward freedom from water was gradual in all cases. The earliest land plants (seedless plants), and the earliest land animals (invertebrates and amphibians) needed to live close to water to reproduce.

Given this adversity, why did life colonize land at all? Any time there are unused resources, life will eventually find a way to use them because individuals who can do so will flourish in the absence of competition and predators. (In other words, natural selection favors such individuals.) In the case of plants, which must have been the first land life because they are the base of all land food chains, those that could tolerate the open air would have relatively unimpeded sunlight. Not only would plants be free of competition from other plants, but air blocks less light than water. Similarly, once plants had adapted, any plant-eating animals that could adapt to land would have a rich source of food, without competition or predation. However, predators would soon follow as the early meat-eating animals pursued this new source of food. Eventually, as more and more competitors and predators evolved, a balance would be struck and the land ecosystem, like the ocean ecosystem, would have a complete food web, with plants, herbivores, and carnivores. Later evolution would create new actors in this ecological drama, but the basic ecological roles would change much more slowly.

As with marine life, the evolution of land life can be divided into three stages, which include both plants and animals:

1. Rise of seedless plants and amphibians
2. Rise of seed plants and reptiles
3. Rise of flowering seed plants and mammals

As with sea life, these phases can be roughly correlated with the Paleozoic, Mesozoic, and Cenozoic eras, as summarized in Figure 3–49.

Rise of Seedless Plants and Amphibians (Paleozoic)

The extensive seas of the Paleozoic did not completely cover the continents. As the continents collided during the gradual unification of Pangea, land was thrust upward, creating mountains and higher ground. These islands served as the homes of the first plants. Initially, these plants were single-celled, such as algae. Microbial animals, such as bacteria, were probably the first life to feed on the plants. Thus, the first land-based food webs probably consisted of such structures as algal mats that were grazed upon by bacteria. After millions of years, multicellular plants arose from colonies of single-celled plants. Exploiting these new plant resources were more animals, including such invertebrate animals as scorpions and millipedes. Vertebrates in the form of amphibians probably did not arrive until millions of years after these first invertebrates.

ORIGIN OF LAND PLANTS. The first multicellular life on land was probably some kind of seedless plant. Most investigators believe that the first land plants arose from **green algae.** Not to be confused with blue-green algae (a prokaryote), green algae is a eukaryotic single-celled plant which lives not only in the sea, but has become adapted to fresh water. This would have made the transition to land much easier because available water would usually be fresh. Aside from its fresh water adaptation, two other lines of evidence suggest green algae as the ancestor. One is the multicellular colonies that these algae form. These would serve as the same raw material for both integration and specialization of colony members, discussed above for the origin of multicellular animals. The second line of evidence is the unusual biochemical similarity between living land plants and this kind of algae. In particular, they share a characteristic way of carrying out photosynthesis.

Land plants solved the problem of gravity with the development of a stem that was strengthened by "woody" cells of tough cellulose. The problem of drying out

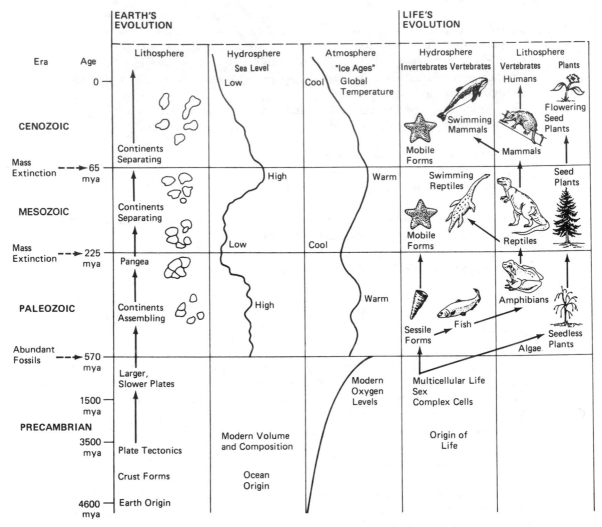

Figure 3–49. Evolution of land life in context of other events.

was solved by development of (1) roots, which penetrated the soil to absorb the water molecules trapped therein, and (2) a **vascular system.** This system is composed of conduits that transport water from the roots and up the stem to the rest of the plant. ("Vascular" is from Latin for "small vessel," such as a blood vessel). Finally, the third problem, that of breathing air, was solved by pores on the plant that open and close to exchange gases with the atmosphere. Recall that in photosynthesis, plants take in carbon dioxide and expel oxygen during this exchange.

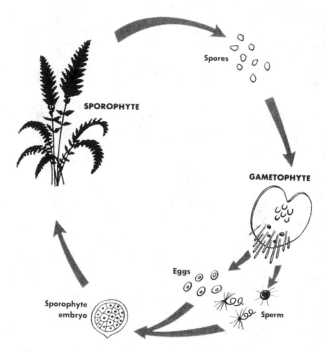

Figure 3–50. Life cycle of a seedless land plant. Fertilization of the egg cannot take place except in the presence of water. From A. McAlester, *The History of Life*, 2nd ed., p. 94.

SEEDLESS PLANTS. When we say these first land plants were seedless, we mean that they had **spores** in their life cycle instead of seeds. Living examples are ferns. Whereas seeds are fertilized plant embryos that simply grow into adults, spores are unfertilized cells which grow into **gametophytes.** As shown in Figure 3–50, this is a small specialized plant that produces both sperm and egg cells. These cells are dispersed into a watery environment where the sperm moves about until it finds an egg to fertilize. (Each gametophyte releases its sperm and egg at different times to avoid fertilizing itself). Once an egg is fertilized, it grows into a **sporophyte.** This is the larger spore-producing plant, and the cycle starts anew.

   This method of reproduction in seedless plants varies from that of seed plants, which evolved later, in two ways. One, it has alternating generations, the gametophyte and sporophyte. Two, seedless plants require a moist environment for the sperm to move about. Usually this amounts to no more than a thin film of water, but this requirement has limited the spread of seedless plants on land. Even today the surviving seedless plants (such as ferns) are restricted to moist environments such as forest floors.

   There were four main kinds of seedless plants in the Paleozoic: psilopsids,

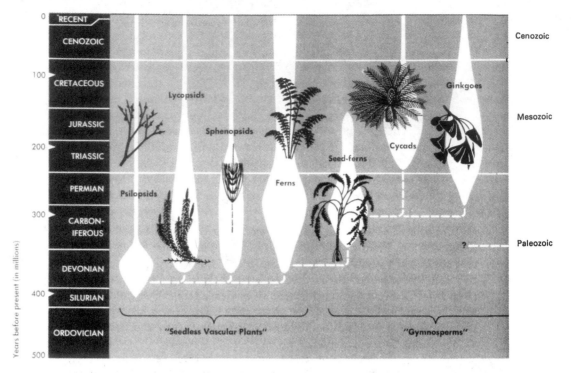

Figure 3–51. Evolutionary history of land plants up through gymnosperms. Dashes show the probable relationships between groups, while the width of the white areas indicates the approximate abundance of each group. Modified from A. McAlester, *The History of Life*, 2nd ed., p. 92.

ferns, sphenopsids, and lycopsids (Figure 3–51). The first to arise were the **psilopsids.** Spores of these plants have been dated as far back as about 400 million years ago, making them the earliest known multicellular life form on land. Psilopsids grew in an underground horizontal network that put up vertical "shoots." These shoots were small, only a few inches high. As you might expect, these plants, which still exist today, lack many of the specializations of later plants. For instance, they lack leaves, relying on pigment in the stem to capture light for photosynthesis.

By the middle Paleozoic, the other three groups, constituting the more advanced seedless plants, appeared (see Figure 3–51). These all had true roots and leaves. Leaves were important in increasing the surface area to capture light. **Ferns** are familiar to most of us because they are still common today, especially in warm moist areas, such as forest floors. As shown in Figure 3–52, **sphenopsids** had radiating bushy whorls at regular intervals along the trunk, while **lycopsids** are recognizable by the scaly bark and concentration of leaves and branches only at the top of the tree. The "scales" formed when old leaves dropped off as the tree grew upward. Sphenopsids and lycopsids are rare today, but in the middle and late Paleozoic they formed vast forests with trees reaching over 100 feet tall. Underneath this canopy

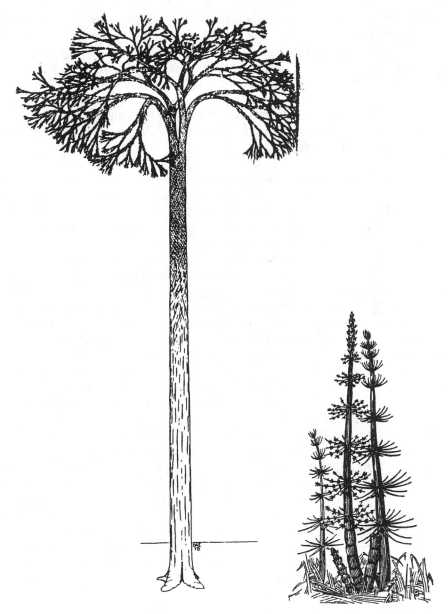

Figure 3–52. Seedless vascular plants of the Paleozoic. Left: A Lycopsida. Right: A sphenopsid. Modified from H. Levin, *The Earth through Time*, p. 390.

the ferns grew to over 10 feet tall, trying to absorb what light escaped the leaves of the taller trees.

These vast swamps of the middle and late Paleozoic, composed of huge ferns, sphenopsids, and lycopsids, are of practical interest. When these plants died, some of them did not completely decay, becoming buried instead. As the organic matter continued to accumulate, the bottom matter was compressed and **coal** was formed. Many of the major coal deposits of the world, especially of the eastern United States, were formed at this time from these trees. Impressions of these plants are very common in the coal.

THE FIRST LAND ANIMALS: INVERTEBRATES. The arthropods are a phylum of animals characterized by jointed body segments and legs. ("Arthro" means "jointed.") This includes spiders, crustaceans, and insects. The first known land animals to follow the land plants were arthropods (still abundant today), the **millipedes** (multi-legged relatives to centipedes), and **scorpions.**

This early appearance on land is not surprising considering that the marine adaptations of arthropods prepared them well for the rigors of land life. A hard **exoskeleton** served not only as support against the stronger gravity but helped retain body moisture. Breathing oxygen as a gas was also readily accomplished in the scorpions in that their ancestors absorbed oxygen with **book gills.** These are leaflike organs that, if kept moist, can be easily modified to absorb oxygen in air. Millipedes developed another method of breathing based on diffusion through pores in the exoskeleton. Millipedes are of special interest in that they, or a close relative, gave rise to the insects, the most abundant life form on earth today. (As many insects are primarily adapted for flight, we discuss insects later, in the colonization of the atmosphere.) In short, their preadaptations for land were so effective that they were not only the first animal pioneers on land, but arthropods have become the most diverse of all land organisms.

Other types of invertebrates have adapted to land. Snails, which are mollusks, are one of the most successful nonarthropods. Again the adaptations of the marine ancestors provided preparation. The molluscan shell initially served as a place to conserve moisture and keep the gills wet even on land, much like the scorpion breathing apparatus. It also served as protection for an otherwise slow-moving and defenseless animal. Eventually some snails adapted to land without a shell. Slugs are snails that have lost the shell, secreting instead a thick mucous to retain moisture and as partial protection. Support and locomotion were provided by the muscular "foot" on which they crawl.

THE FIRST LAND VERTEBRATES: AMPHIBIANS. The first fossil amphibians date from the middle Paleozoic, indicating the earliest occupation of land by our vertebrate ancestors. We often hear that the fossil record provides few or no "missing links." Yet these first amphibian fossils are good examples of just such transitional forms, for they bear a strong resemblance to lobe-finned fish, from which amphibians surely

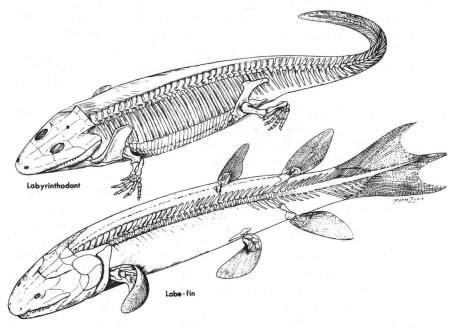

Labyrinthodont

Lobe-fin

Figure 3–53. Bottom: Paleozoic lobe-fin fish. Top: An early amphibian (middle Paleozoic).

evolved. Even to the nonspecialist, the anatomical similarity of early amphibians to these fish is obvious (Figure 3–53).

Lobe-fins are a type of bony fish equipped with fleshy, muscular fins. In swimming, the fins are used for navigation and perhaps moving on the sea bottom. These fish have never been especially common, but they have two key preadaptations for *locomotion* and *breathing* on land. Regarding locomotion, the muscles in the lobed fins are: (1) strong enough to provide locomotive power, and (2) the fins swivel freely at the point of attachment for control. In time, these fins evolved into legs. Indeed, reptiles and mammals still retain many of the same bones in their limbs. The breathing problem was also resolved by their ancestral biology since these fish have rudimentary lungs. Most fish have air bladders used in controlling bouyancy in water. In the lobe-fins and their relatives, the bladder has been converted into a chamber that not only contains but absorbs air into the bloodstream (in other words, a lung). In some living relatives of the lobe-fins, this lung is useful when a lake dries up—they are able to breathe air until water is again available. The problem of water loss was partially resolved by development of a thicker skin.

However, *amphibians never fully adapted to land.* Their legs did not adapt well to land—they were clumsy, with a slow sprawling gait. While their lungs were adequate enough, their heart and circulatory system, which transports the oxygen absorbed by the lungs, was relatively weak. Water loss was still considerable. Thus, as adults, many still returned to the water during much of their life. Finally, and perhaps most notably, amphibians needed water to reproduce since eggs must be laid there and the offspring need water to develop as larvae. This cycle of life on both water and land gives amphibians their name ("amphi" is Greek for "both," and "bios" for "life"; hence "both-life"). However, it meant that they would never be extremely successful in either environment because they could not compete with vertebrates that specialize in living completely in water (fish) or completely on land (reptiles, mammals). Nevertheless, amphibians were crucial in the vertebrates' transition to life on land. From amphibians evolved the reptiles, which became the dominant vertebrates in the Mesozoic.

Remaining amphibians living today are the frogs and salamanders. However, these familiar forms evolved later, in the Mesozoic, and are *not typical* of Paleozoic amphibians. Rather, we once again see that the survivors of groups replaced by later ones become highly specialized to life habits in inconspicuous or marginal environments. Thus, if we look at the Paleozoic amphibians, before the evolution of reptiles, we see large alligatorlike animals that grew up to 10 feet long (see Figure 3–53). Many of these were plant eaters, feeding on the abundant ferns and other seedless plants in the lush late Paleozoic swamps. Others were meat eaters, consuming fish or other amphibians. The general name for all of these Paleozoic amphibians is **labyrinthodonts** ("labyrinth" is Greek for "maze," "dont" refers to "teeth"; hence "maze-teeth" because they had teeth with many ridges).

Rise of Seed Plants and Reptiles (Mesozoic)

The formation of Pangea at the end of the Paleozoic caused a great drop in sea level and increased climatic aridity. This seems to have contributed to the demise of spore plants and amphibians because of their need of water to reproduce and incomplete adaptation to land in general. In their place came their descendants, the seed plants and reptiles. However, the huge end-Paleozoic mass extinction did not, by itself, cause the demise of seedless plants and amphibians. In both cases, their descendants, the seed plants and reptiles, had begun to replace them many millions of years before the end of the Paleozoic. The mass extinction may thus simply have served as a *coup de grace* that accelerated an already ongoing replacement. This culminated, by the middle and late Mesozoic, in the great dinosaurs, undoubtedly the most famous of all fossils. These often giant beasts inhabited widespread tropical swamps that surrounded great inland seas, which occurred when the sea level rose again in the middle Mesozoic (refer to Figure 3–49).

SEED PLANTS. The second great radiation of land plants consisted of primitive plants with seeds, usually called the **gymnosperms.** Their most familiar representatives are the cone-bearing plants, such as pine trees. In gymnosperm reproduction the gametophyte stage is eliminated (Figure 3–54). Instead, in each generation there is a female organ, such as the pine cone, which contains the egg. This egg is fertilized by pollen (plant sperm cells) produced by the smaller male cone. The pollen is

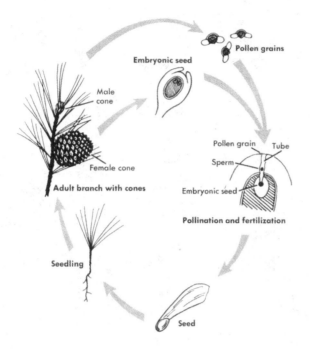

Male cone

Embryonic seed

Pollen grains

Female cone

Adult branch with cones

Pollen grain   Tube

Sperm

Embryonic seed

Pollination and fertilization

Seedling

Seed

Figure 3–54. The life history of a gymnosperm. Female cones contain embryonic seeds; male cones produce sperm-bearing pollen grains. When pollen is carried to the female cone by wind, fertilization is accomplished by movement through a tube which grows from the pollen grain. From A. McAlester, *The History of Life*, 2nd ed., p. 101.

released and carried by the wind to the female cone, which is designed to catch the pollen. Reliance on wind for transport is obviously a chancy method, so huge quantities of pollen must be released. Once a pollen grain enters the female cone, a moist tube grows to the egg allowing the sperm cell itself to travel and fertilize the egg. Upon fertilization, a **seed** is formed, which will grow into a new generation, given proper conditions. Thus, a seed is a fertilized cell, while a spore, used by seedless plants, is an unfertilized cell. The term gymnosperm means "naked seed." This is applied because, unlike seeds of flowering plants, which evolved later, gymnosperm seeds do not have a fleshy covering.

We may identify four basic kinds of gymnosperms in the Mesozoic: the seed ferns, cycads, ginkoes, and conifers (Figure 3–55). The **seed ferns** were the earliest to appear, in the late Paleozoic. They apparently evolved from seedless ferns and are very similar to them. **Cycads** and **ginkoes** were very prominent trees in the early Mesozoic. Cycads resemble palm trees but are completely unrelated. Ginkoes have a fan-shaped leaf. None of these three groups has fared well since the Mesozoic. Seed ferns are extinct and cycads and ginkoes have few living species. However, the fourth group, the **conifers,** has fared better. Conifers, such as the familiar pine trees, have needlelike leaves and differ in other ways from the other three groups of gymnosperms, indicating no clear relationship to them. Today there are a moderate number of species that flourish mainly in cold, dry climates. This is partly because the conifer needles minimize loss of moisture, permitting them to

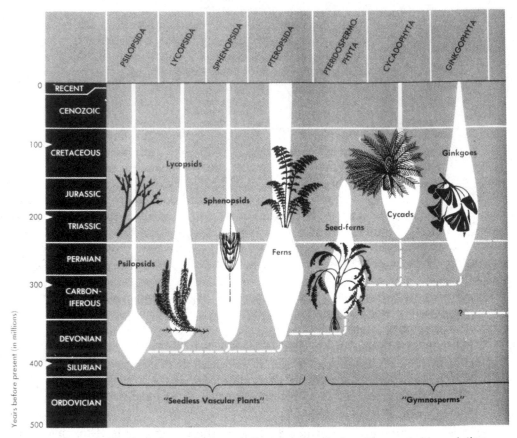

Figure 3–55. Evolutionary history of all land plants. Dashes show probable evolutionary relationships between the groups. The width of the white areas indicates the approximate abundance of each group. Modified from A. McAlester, *The History of Life*, 2nd ed., p. 92.

thrive where broad-leaved trees would lose too much moisture through evaporative processes.

REPTILES. Reptiles first evolved in the late Paleozoic, from an amphibian ancestor. In 1989, the earliest known reptile was reported (found in Scotland) and dated to about 340 million years ago. The first reptiles are known as **archaic reptiles** and had a number of general reptilian features. Of major importance is the **hard-shelled egg,** very similar to a bird's egg. This key feature allowed reptiles to more fully inhabit land, removing the obstacle that had barred the amphibians. By allowing air to diffuse in while retaining nutrients and body fluids, the egg allowed the embryo to develop on land. Reptiles were freed from the need to return to water to reproduce. Reptiles are also better adapted for land in that they have a stronger

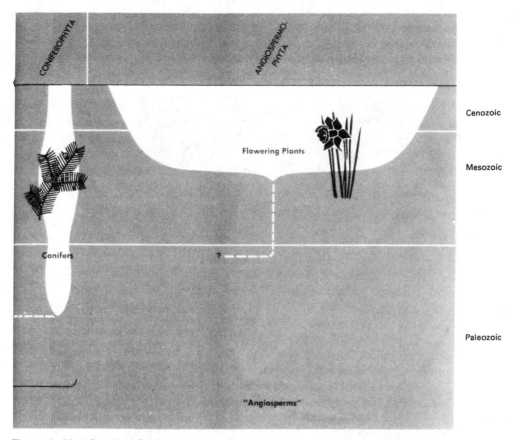

Figure 3–55. (*Continued*)

heart and body musculature better suited for locomotion on land. But eggs and soft tissues are seldom preserved in the fossil record. How, then, do we know when the first reptiles evolved? Mainly this is based on anatomical details of the skull and backbones that differ between amphibians and reptiles.

Archaic reptiles gave rise to a number of more specialized reptiles (Figure 3–56). We will group these into three broad categories: mammallike reptiles, nondinosaurs, and dinosaurs. The **mammal-like reptiles** were the ancestors of the mammals and will be discussed later. The **nondinosaurs** include three modestly successful groups: the turtles, lizards/snakes, and crocodiles (Figure 3–56). While these never became as prominent or as large as the dinosaurs, they have outlasted them. Turtles first appeared in the early Mesozoic. Much of their success is owed to the protective shell made of interlocking bone. They have changed little in 200 million

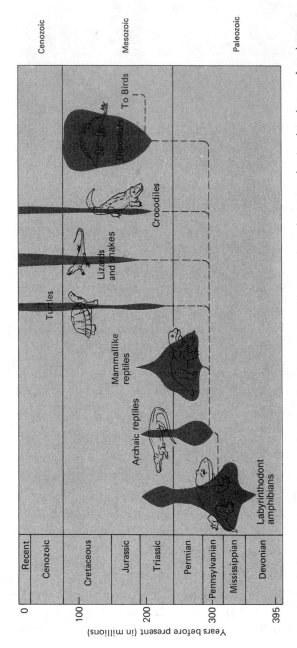

Figure 3–56. Evolutionary history of reptiles. The width of vertical areas show the approximate abundance of each group of organisms and the dashes show the most likely evolutionary relations between groups. From D. Eicher and A. McAlester, *History of the Earth*, p. 347.

years, aside from developing better ways of withdrawing the head and limbs. Instead of retracting their head and limbs, early turtles simply laid the head and limbs sideways, under the ledge of the shell.

Lizards also appeared in the early Mesozoic and have been generally successful at a secretive, often burrowing lifestyle. However, they can be large. The Komodo Dragon lizard of the Pacific can grow to over 10 feet and is an intimidating meat eater. Recall also the huge lizard that took to the sea, the mosasaur. Sometime in the late Mesozoic some lizards lost their legs and evolved into snakes. This leg loss is not uncommon among lizards even today. Burrowing lizards often have little need of legs and have reduced or no legs at all. The secretive lifestyle has been very advantageous—snakes are an extremely abundant group. Contributing to their longevity was the development of poison glands in some of the more recently evolved snakes. However, some of the older groups of snakes, such as the constrictors (e.g., boas and pythons), are also very successful. In their case, prey is killed by suffocation.

Crocodiles (including alligators) are more closely related to the dinosaurs than the other nondinosaurs. They have changed little since the middle Mesozoic, living a largely aquatic, flesh-eating existence. The longevity of this group may in part be attributed to a dependable food supply (fish) and lack of major predators, at least until humans. Also, unlike the extinct large marine reptiles, crocodiles can inhabit fresh water, which seems to have been a major factor in their survival. Perhaps the additional geographic range onto land has helped buffer them against oceanic catastrophes.

DINOSAURS: OUR FAVORITE REPTILES. **Dinosaurs** are the most widely known group of extinct animals by far. Much of this fascination is due to their sheer size, which often was many tons. (The name "dinosaur" means "terrible lizard"; in fact, they were not lizards, but "cousins" to them, as seen in Figure 3–56.) Aside from their size, dinosaurs also owe their dominance to a key feature of their anatomy. They had legs positioned directly under the body as opposed to the sprawling limbs of many other reptiles (e.g., the alligator). This allowed dinosaurs to move quickly, even with their great weights. This was true for both four-legged and two-legged dinosaurs.

Dinosaurs arose from reptiles called **thecodonts** (meaning "socket-tooth"), which flourished in the early Mesozoic. It was not until the middle and late Mesozoic eras that dinosaurs became common. As shown in Figure 3–57, there were two basic groups of dinosaurs: the lizard-hipped and the bird-hipped groups, based on whether the hip bones resembled those of living lizards or birds. The **lizard-hipped** dinosaurs included two subgroups, the sauropods and theropods. The **bird-hipped** dinosaurs included four subgroups: the ornithopods, stegosaurs, horned dinosaurs, and ankylosaurs (see Figure 3–57). One way to distinguish the two basic groups is to remember that the bird-hipped group consisted of moderate-sized herbivores with specialized defenses, including horns and plates, while the lizard-hipped group consisted of meat eaters (theropods) and very large herbivores (sauropods).

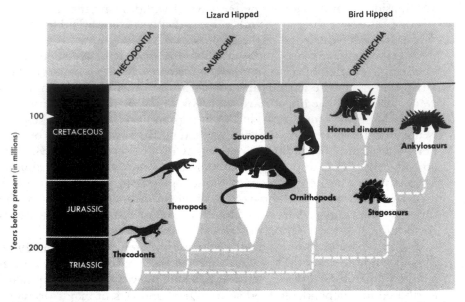

Figure 3–57. The evolutionary history of dinosaurs is divided into two groups, the lizard-hipped and the bird-hipped. The dashes show the most likely evolutionary relations of the groups. The width of the white areas indicates the approximate abundance of each group. Modified from A. McAlester, *The History of Life*, 2nd ed., p. 120.

In the lizard-hipped group, **theropods** were the carnivorous subgroup. The most familiar species of this group was *Tyrannosaurus rex* (meaning "terrible lizard king"); this dinosaur stood 20 feet tall and weighed about 8 tons (Figure 3–58). These awesome animals are the *largest meat eaters* to ever exist on land. Unlike most large meat-eating animals today, these dinosaurs walked on two legs and had small front limbs that were basically useless. They were no doubt fast runners who used the long claws on their powerful legs to great effect. Also used in killing prey were long, saberlike teeth—they were as long as 6 inches. Not all theropods were as large, but most were fast moving and employed similar methods, running on two legs.

What defenses could be used against such a dangerous predator? In the case of their lizard-hipped relatives, the **sauropods,** sheer size may have helped. For example, lions rarely attack adult elephants because there is much easier game to kill. Sauropods are the largest known dinosaurs and, in fact, the largest land life to ever exist on earth. This group includes the well-known *Brontosaurus* and its relatives. Sauropods had a long neck and tail, with thick legs for support, as shown in Figure 3–58. Some species grew to truly huge sizes. The recently discovered *Seismosaurus* was over 110 feet long (one-third the length of a football field), while the also newly discovered *Ultrasaurus* was shorter but heavier, weighing perhaps 140 tons. (For comparison, the average adult elephant weighs a mere 7 tons.) A major puzzle is how such huge organisms could support and feed themselves. The thick legs and wide feet helped distribute the weight, but the gravitational stress still must have been enormous. Probably some of the stress was relieved by a life of wading in the water of lush swamps. Even aside from structural stress, feeding such an enormous mass was also a challenge, considering that sauropods were plant eaters and plants are generally low in calories compared to meat. Thus, a cow spends many more hours of its life eating than does a lion. Apparently, this problem was relieved by consuming large amounts of vegetation without chewing. This allowed food to be eaten faster. Food was broken down by swallowed rocks, which stayed in the stomach and ground up the food like mill stones. Also, it may be that

Figure 3–58. Representatives of two main types of lizard-hipped dinosaurs. Adult human for scale. Modified from D. Lambert, *The Dinosaur Data Book* (New York: Avon Books, 1990), various figures.

LIZARD-HIPPED

Theropod

Sauropod

the relative metabolism (the rate at which food is burned) of these massive creatures was relatively reduced compared to the relative mammalian rate.

All four subgroups of the bird-hipped dinosaurs were plant eaters and therefore also needed defense against the theropods. In their case, size was not a factor because most were about the size of the larger theropods, being about 20-30 feet long as adults. Instead, specialized defense mechanisms were often used. The **ceratopsians** included the well-known *Triceratops* and its relatives. These had a number of sharp bony horns (three in *Triceratops*) extruding from the head, seen in Figure 3–59. These, plus a thick bony shield around the neck, served as protection. **Stegosaurs** had a series of alternating plates running down the back (Figure 3–59). These were not used for protection, as may first appear. They were suffused with vessels which carried blood to the surface for cooling. These plates were apparently *heat radiators* used to disperse heat generated by the massive body volume. These animals had long spikes on their tails that were their defensive mechanisms. **Ankylosaurs** were also short-legged, low-built forms like the stegosaurs, to whom they were related. Instead of spikes, they had a heavy club on the tail. Ankylosaurs also had armor, with much of the upper body being encased in heavy, bony plates (Figure 3–59). The **ornithopods** were two-legged forms who often lived near lakes or other bodies of water. They are often called "duck-billed" dinosaurs. As shown in Figure 3–59, these dinosaurs had complex, often ornate crests on their skulls, which contained hollow cavities connecting to their sinuses. Apparently these were used as resonating chambers for vocalizations. Many paleontologists believe that these were often highly social animals, perhaps with herding behavior. The presence of *offspring care* is best documented for ornithopods: Evidence of nests for large offspring indicates that mothers took care of their young.

MORE ABOUT DINOSAURS. Dinosaurs have been enormously popular since their discovery in the 1800s. Let us discuss some general points about them. To begin, there are some major misconceptions to clear up. First, it is common to suppose that all dinosaurs were gigantic. It is true that many were large: Estimates indicate that over half of the dinosaurs weighed more than 2 tons. (In contrast, only about 2% of modern mammals reach this size.) However, this means that about half were smaller than this. In fact, *some dinosaurs were quite small*, with species ranging down to dog and even chicken-sized animals. Many of these smaller dinosaurs were very active two-legged forms.

A second misconception is that dinosaurs were slow, clumsy creatures. In fact, there is now considerable evidence that even many of the larger dinosaurs were *fast* and *agile*. This image problem comes from our experience with living reptiles, which often are slow moving and have sprawling limbs compared to living mammals. The slowness usually comes from their low metabolism and "cold-bloodedness," meaning that they become less active when it gets cold because they cannot generate much internal body heat. However, many scientists believe that at least some dino-

Figure 3–59. Representatives of four main types of bird-hipped dinosaurs. Adult human for scale. Modified from D. Lambert, *The Dinosaur Data Book.*

BIRD-HIPPED

Ceratopsian

Stegasaur

Ankylosaur

Ornithopod

saurs had a relatively high metabolism and could at least partly regulate their body temperature. Further, as already noted, they had limbs more oriented under their bodies, unlike their living relatives.

A third misconception is that dinosaurs were relatively stupid. However, there is now considerable evidence that they were at least *moderately intelligent*. It is true that some dinosaurs had astonishingly small brains. For instance, *Stegosaurus* controlled a 20-ton body with a small brain, roughly the size of a walnut. For comparison our own 100 to 200 pound bodies are controlled by a much larger, 3 to 4 pound brain. However this is misleading because the dinosaur's brain was not as localized as ours. It had a number of secondary brains that swelled out along the nervous system, performing many of the functions our single brain must accomplish. The secondary brains helped organize and synthesize sensory information and transmit motor (movement) signals. *Stegosaurus* had a particularly large secondary brain in its pelvis. The reason for this more "diffuse" brain control in many dinosaurs is probably related to their large size. Nerve signals travel at about 225 miles per hour. While this may seem fast, big animals have a vast amount of nerve information to process. Even a few microseconds' delay in detecting a predator and moving away from it could be fatal, so speeding up information processing would be important. (Consider a nerve signal from a sauropod's brain telling its legs to move; it would have to travel over 40 feet.) Secondary brains processed information closer to local areas before it was sent on to the major brain center and organized outgoing signals to the body areas.

There was also much variation in brain power between dinosaur groups. As shown in Figure 3–60, stegosaurs and sauropods had the smallest brains, when body size is taken into account. The smartest dinosaurs were meat eaters, such as the theropods. This is true of most groups, including mammals, because meat eaters must generally outwit their prey in order to eat. The plant-eating ornithopods have the second-largest relative brain size, thus substantiating behavioral evidence of social behavior, communication, and parental care that seem to have been more advanced in the ornithopods than any other herbivore group.

Moving away from misconceptions, we turn to a more basic question: Why were many dinosaurs so big? Not only were there so many large ones (recall that over half were at least 2 tons), but the largest dinosaurs far exceeded the largest mammals. The largest sauropods were at least 80 and perhaps 140 tons, while the largest land mammal, a hornless rhinoceros that became extinct a few million years ago, weighed a mere 25 tons. Unfortunately, we do not know for certain why dinosaurs were so big. However, at least three contributing factors can be cited. First is the exceptionally warm, stable climate of the Mesozoic. Even today reptiles grow largest in warm tropical environments. This is because there is ample food and no seasonal cold periods during which growth is reduced or stopped. (Cold periods are less of a problem for mammals because their body temperature is less affected by the environment.) Second, while the dinosaurs may have been more active than reptiles today, they still probably had a lower metabolism compared to mammals. As a result, larger animals could be supported with the same amount of

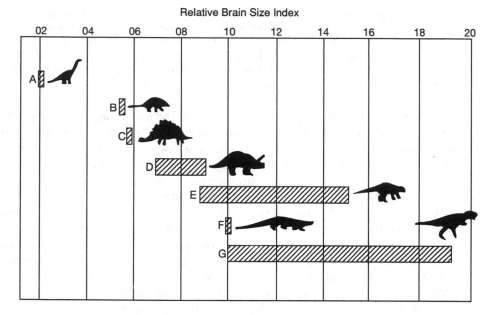

Relative Brain Size Index

Figure 3–60. This diagram plots the range of brain sizes of dinosaurs against that of their living relatives the crocodiles. Encephalization quotient (EQ) numbers take account of body size. The higher the EQ the greater the assumed brainpower. (A) Sauropods; (B) Ankylosaurs; (C) Stegosaurs; (D) Ceratopsians; (E) Ornithopods; (F) Crocodiles; (G) Carnosaurs (carnivores). Modified from D. Lambert, *The Dinosaur Data Book*, p. 183.

food because they burned it more slowly. Third, large size was advantageous because it allowed dinosaurs to maintain a more constant body temperature: Large bodies retain heat better. Small animals have a higher amount of surface area relative to their body mass, providing relatively more area to dissipate heat. Thus, daily or overnight temperature drops would have less effect on large dinosaurs. This so-called "inertial body heat" may explain why dinosaurs were so active in spite of having relatively low metabolisms.

EXTINCTION OF THE DINOSAURS. It is fitting that such a fascinating animal as the dinosaur should even intrigue us in death: The extinction of these creatures is one of the major and controversial enigmas in life's history.

Let us begin with some of the basic facts about dinosaur extinction.

1. No dinosaur fossils are definitely known in rocks after the Mesozoic. Therefore, the end of the Mesozoic about 65 million years ago is used to demarcate the end of the dinosaurs.
2. Some dinosaurs began to wane before others. For example, stegosaurs declined by the late Mesozoic (refer to Figure 3–57).

3. Other organisms died out, besides dinosaurs, at the end of the Mesozoic. Most notable were the large marine and flying reptiles, but some marine invertebrate groups, especially plankton and reef-building corallike clams called "rudists," died out as well.

4. About 70% of all species died at the end of the Mesozoic and the surviving animals were all small (a few pounds or less). This included our own ancestors, the mammals.

5. Some plants also seem to have been decimated at the end of Mesozoic.

These and other lines of evidence have led to numerous suggestions about the cause of this so-called *end-Mesozoic extinction*. Many of these suggestions are now generally dismissed. One suggestion that was once very popular is that the "superior" mammals outcompeted the clumsy, stupid, large reptiles. However, we have already noted that at least some dinosaurs were neither large, clumsy, nor stupid. More important, this superior mammal idea cannot be true because mammals originated early in the Mesozoic and *coexisted with dinosaurs* for over 100 million years. Yet mammals never became larger than a housecat and remained inconspicuous and relatively rare during this entire time. Thus, it was clearly mammals and not reptiles who were being suppressed. Mammals diversified rapidly only when the large reptiles no longer existed. This is called **ecological release,** when a dominant group is removed, releasing another group to diversify.

A second discarded suggestion is that the appearance of flowering plants (discussed on page 180) near the end of the Mesozoic altered the food sources of dinosaurian plant eaters and they were unable to adjust. However, this does not explain the disappearance of the many flying or marine reptiles (the fish eaters) or the widespread extinction of many other marine organisms such as plankton.

Yet a third suggestion is that climatic cooling occurred and the "cold-blooded" dinosaurs were unable to adjust to the change in temperature. However, while climate probably was a factor, cooling alone was not sufficient to account for the massive number of deaths that occurred. There were some downward temperature fluctuations at the end of the Mesozoic, but they were relatively minor and the overall climate did not cool dramatically. Indeed, very warm global temperatures continued into the early Cenozoic (see Figure 3–49). Furthermore, there is growing evidence that dinosaurs could withstand cold temperatures, even if they had dropped radically. A number of fossil dinosaur species have been found in areas (e.g., Alaska and Antarctica) that were quite cool during their lifetimes.

Yet a fourth commonly advanced theory on the dinosaur's demise is that disease wiped them out. The problem with this theory is that diseases tend to be very specific in their effects. It is thus very difficult to see how an entire group of animals as diverse and widespread as the dinosaurs could have been wiped out this way. More important, some victims almost always survive—it is to the virus's or bacteria's advantage not to eliminate all of the potential hosts. Disease has literally wiped out entire groups only when the group has been isolated, never exposed to the disease before, and therefore never had a chance to adapt to it slowly. (An example is the large number of American Indian tribes eliminated by European diseases.) Since dinosaurs were so diverse and widespread, such a situation could not have existed.

Until recently, the theory that climate change caused the dinosaur's extinction was probably the most popular among researchers. However, in the last 10 years, another theory has been gaining increasing acceptance: that a large meteorite struck the earth at the end of the Mesozoic. This possibility has been suggested repeatedly over the last 50 years. However, there was no sound evidence for it until 1980, when a very famous scientific paper was published by Luis Alvarez and his associates. In performing chemical analyses of sediment in Italy deposited at the end of the Mesozoic, they reported finding an abnormally high amount of the element **iridium** (Figure 3–61). This is indicative of a meteorite impact because iridium is extremely rare on the earth's surface but is abundant in meteors and some other bodies in the

Figure 3–61. One possible explanation for the dinosaur extinction. Stratigraphy of the Mesozoic-Cenozoic boundary shows a 2.5 centimeter thick boundary of clay and an abundance of the element iridium. Iridium is rare on earth, but is abundant in solar debris (such as meteors). Although iridium is measured only in parts per billion (ppb), there is still a marked increase in its presence at this boundary, in rocks all over the world. Modified from R. Wicander and S. Monroe, *Historical Geology*, p. 423.

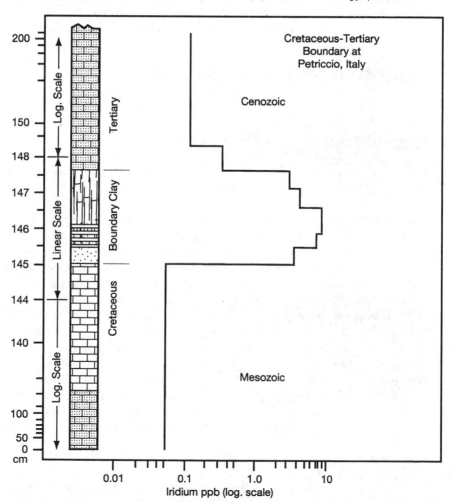

solar system. (Because of its chemical and physical properties, most of the iridium migrated into the mantle and core during earth's differentiation.) Thus, this group of researchers suggested that the iridium "spike" at the end of the Mesozoic represents dust, rich in iridium, that settled out when a large meteorite struck the earth. Further chemical testing of very late Mesozoic sediments all over the world has substantiated their initial findings: There is an enrichment of iridium in over 70 locations from New Zealand to New Mexico. Even now, as this is being written, there are many articles appearing in scientific journals (especially *Nature, Science,* and *Paleobiology*) that report further analyses and arguments, pro and con. Some report not only iridium "spikes" but also **shocked minerals** in end-Mesozoic rocks from many parts of the world. These are deformed mineral grains, especially quartz, interpreted to have been created by the impact of a massive object at high speed.

Some scientists have proposed that the high iridium and shocked minerals may indicate widespread volcanic activity instead of a meteorite. They point out that some lavas on earth are enriched in iridium and that cataclysmic eruptions could cause the shocked minerals. There is indeed evidence for much volcanism at the end of the Mesozoic. However, there are a number of problems with inferring volcanism as the main cause of the end-Mesozoic extinction. One of the most important is that the abundance and size of shocked minerals are greatest in North America, whereas the greatest amount of end-Mesozoic volcanism was in Asia, especially India. Another important problem is that detailed analyses indicate that even the most iridium-rich lavas have an iridium content well below that needed to generate the pronounced iridium "spike" over a worldwide area. For these and other reasons there is a growing consensus among most specialists that a large meteorite better explains the current data.

Assuming that the meteorite hypothesis is correct, let us examine how a meteorite impact, such as shown in Figure 3–62, could cause widespread extinction of dinosaurs and many other organisms. Upon impact, the meteorite would create a huge explosion, in the form of a mushroom cloud much like an atomic bomb. The meteorite would disintegrate, dispersing its atoms, including the iridium atoms, into the atmosphere in the explosion. Judging from the amount of iridium, most estimates indicate that the end-Mesozoic meteorite was about 7 miles (10 kilometers) in diameter. Such meteors are not uncommon in space and astronomers agree that the earth has been bombarded by such large meteors many times in the past (Chapter 4). Based on the number and location of meteors, the best estimate is that one or two such large impacts would occur every 100 million years. If such an event were to occur, a vast amount of energy would be released, comparable to that of a many megaton thermonuclear weapon. Everything within the immediate area would be obliterated, creating a crater about 100 miles (150 kilometers) in diameter. If you were a few hundred miles away you would see the mushroom cloud reaching far into the atmosphere. The ground would shake and you would hear a deafening blast as the sonic waves spread out as shock waves in the surrounding air. (You could hear the echo on the other side of the world, like rolling thunder.) The results

Figure 3–62. *Tyrannosaurus rex* cower before a force of spectacularly immense power—an asteroid striking the planet.

would be the same whether land or ocean was hit. Astronomers estimate that a large meteorite would hit the earth at a speed approaching 60,000 miles per hour. At such speed, even deep water would be punched through like paper, striking the earth on the sea bottom.

As dramatic as these effects are, they would not cause a mass extinction of global proportions. Only those animals and plants within a few hundred or thousand square miles would be immediately affected. Instead, according to computer simulations, the main cause of mass extinction would be the long-term atmospheric effects. A huge dust cloud would be raised high into the atmosphere where the fine particles would remain for an estimated 6 to 24 months. Earth would be cut off from much sunlight for this entire period and plants would die from lack of photosynthesis. As plants form the base of any food chain, the animals that eat the plants would die, as would the predators that eat the plant eaters. In addition, temperatures would drop and many organisms would die of the unaccustomed cold. Simulations, such as shown in Figure 3–63, indicate that a drastic temperature drop would occur even in tropical regions, lasting for days or weeks. Such modeling of this event led scientists to realize that a small nuclear exchange could cause similar dust clouds. The models are very tentative, but they indicate that a so-called **nuclear winter** could occur, wherein relatively small nuclear explosions could cause temperature decreases similar to those seen in Figure 3–63. This would cause massive agricultural losses, with calamitous consequences in today's overpopulated world (Chapter 5).

As if weeks of darkness (loss of photosynthesis) and cold were not enough, the meteorite impact simulations indicate that a host of other effects could also occur. Table 3–3 lists some of these: high winds (over 300 miles per hour); tsunamis ("tidal waves"); global wildfires ignited by the impact; pyrotoxin and acid rain from poisons and acids cast into the air by the burning forests; and other effects. The time scale of these effects varies, from hours to thousands of years (millennia). While these effects are, in part, inferred from computer simulations, there is direct evidence for some of them. For instance, a layer of soot has been identified in end-Mesozoic sediments. This layer is sufficiently thick and widespread that it apparently represents the remains of a global wildfire. Evidence of a tsunami has been found in end-Mesozoic rocks around and within the Caribbean. This evidence consists of very thick layers (6 meters, or 20 feet) of jumbled rocks that seem to have been deposited by the large energy of the crashing "tidal wave."

How did anything live through such a catastrophe? **Extinction selectivity** refers to the process whereby groups in an extinction event are affected differently because of their different traits. For instance, many plant species could do quite well in a short-term catastrophe because seeds and spores are fairly resilient. Hence

----

Figure 3–63. Surface temperatures from computer models of various meteorite impacts. Times of maximum cooling are shown. Shaded areas are below 0 degrees Celsius and hachured areas are below 10 degrees. Modified from V. Sharpton and P. Ward, eds., 1990, Geological Soc. of America Special Paper 247, Fig. 2, p. 266.

Control case 30-day average

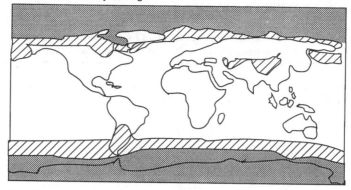

Small Impact case days 14-17

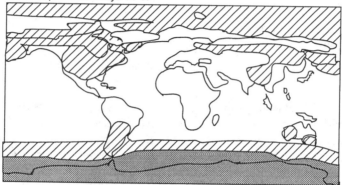

Medium Impact case days 2.5-5.5

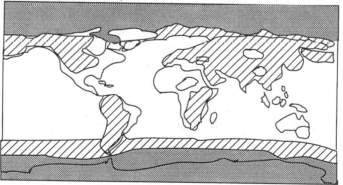

Alvarez Impact case days 10-20

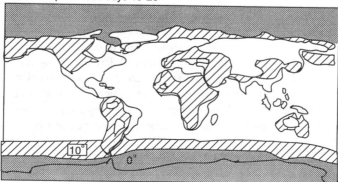

TABLE 3-3. Environmental Stresses Caused by a Meteorite Impact, Which May Have Killed Off the Dinosaurs. Note Variety of Time Scales During Which Stresses Persist

STRESS	TIME SCALE
Darkness	Months
Cold	Months
Winds (500 km/h)	Hours
Tsunamis	Hours
H_2O-Greenhouse	Months
CO_2-Greenhouse	Decades
Fires	Months
Pyrotoxins	Years
Acid rain	Years
Destruction of ozone layer	Decades
Impact-triggered volcanism	Millennia?
Mutagens	Millennia

Source: Geological Society of America, Special Paper 247, p. 398.

they would rebound in just a few years even if the adults died. Indeed, this is what the evidence seems to show: Fossil leaves indicate that adult plants were temporarily devastated by the wildfires and other possible effects, but entire plant species do not generally become extinct as do animal species. Selectivity is also seen within the animals. Small animals reproduce very rapidly and need relatively little food. In addition, many of them are very generalized omnivores and scavengers (such as rodents). Indeed this pretty well describes our ancestors, the mammals in the late Mesozoic. Perhaps they even benefited from all the dead and decaying matter.

In contrast, large species, such as most of the dinosaurs and other reptiles, are almost always the first to die from environmental disturbances. Large animals have at least three disadvantages during stressful times:

1. Each individual requires more food to live than a small animal, so the large animals die more easily when resources are disrupted
2. Large species have lower population densities (because they need more food), which means there are fewer individuals to start with
3. They reproduce and grow more slowly, so large animals take longer to recover from stress. (In the time it takes for one elephant to mature, many thousands of rats and millions of insects could be produced.)

These reasons explain why small organisms, such as the mammals, would preferentially survive the end-Mesozoic extinction. They also explain why so many large species are endangered today: rhinos, gorillas, large cats, and so on, because humans cause environmental disturbances. A final kind of selectivity seen at the end-Mesozoic extinction is that against tropical organisms. Reef-builders and other tropical organisms appear to suffer disproportionate losses. Again, this would be expected from the meteorite scenario because the drastic cooling would have its greatest effect on organisms most accustomed to warm temperatures.

Figure 3–64. Map showing possible impact site of meteorite that killed the dinosaurs. From D. Norman, *Dinosaur!*, p. 153.

Has a meteorite crater from this impact been found? The earth has many such craters, but for a decade, none could be found that was the right size and age. Then, in 1991 a crater was discovered in the Yucatan Peninsula of Mexico, on the western edge of the Gulf of Mexico (Figure 3–64). Detailed studies indicate that the crater is indeed about the right size (170 km. diameter) and age (65 million years old). Furthermore, the location of the crater closely matches that expected from evidence already found in end-Mesozoic rocks in the region: (1) shocked minerals are most abundant and largest in North America and nearby areas, (2) tsunami deposits are known from a number of sites around the edge of the Gulf of Mexico, and (3) bodies of melted rock called tektites have been found in Haiti and other Caribbean islands, indicative of melted rock ejected by a nearby impact.

Rise of Flowering Plants and Mammals (Cenozoic)

As the continents comprising Pangea continued to move apart, the plants and animals on them became progressively isolated from one another. Antarctica moved over the South Pole, accumulating ice that caused global temperatures to become cooler. Thus, widespread tropical life in the early Cenozoic was replaced by cold-tolerant life at higher latitudes in the late Cenozoic. The global cooling culminated in the advance of glaciers ("ice age") about 2 million years ago. The loss of water to glaciers led to dropping sea levels when glacial advances occurred. These created "land bridges," which periodically reconnected some of the continents, such as North America and Asia, allowing cross-migrations.

FLOWERING PLANTS. The third and final great land plant diversification consisted of flowering plants, which are also called **angiosperms.** This diversification occurred mainly in the Cenozoic, although it actually began in the preceding Mesozoic era (refer to Figure 3–55).

Flowering plants reproduce with seeds, but unlike gymnosperms, they do not rely on the vagaries of the wind to carry pollen. Instead, the flowers produced on the adult plant are used to attract animals that carry the pollen from the male organ (**stamen**) to the female organ (**pistil**) of another plant (Figure 3–65). The "flowers," as we call them, are actually modified leaves with pigments that reflect light at wavelengths attractive to certain insects. This method of pollination is obviously much less wasteful. Aside from attractive colors, flowering plants may produce attractive tastes (nectar), smells, and other stimuli to draw animals, especially insects. As we will see, the evolution of insects and land plants is intimately connected.

The reproductive effectiveness of flowering plants does not stop with less wasteful fertilization. Members of this group also produce a seed that usually has a fleshy covering. Often this is edible (as on apples, citrus, and many other fruits) so

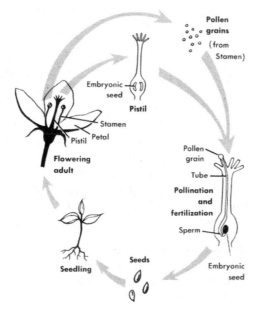

Figure 3–65. Life history of an angiosperm. Pollen is produced by the stamen and seeds are developed in the pistil of the flower. From A. McAlester, *The History of Life*, 2nd ed., p. 105.

that animals will eat them and disperse the seeds in their feces. (The seed itself is designed not to be digested.) This is a major advantage in dispersal, considering that this part of the life cycle is a plant's only opportunity to be dispersed. Yet another advantage is double fertilization. This means that the egg is fertilized twice so that while one fertilization forms the offspring embryo, the second forms nutrient tissue used for food by the developing embryo. This allows faster growth.

In sum, there are three major advantages to the flowering plant mode of reproduction over nonflowering plants: (1) more efficient pollination, (2) greater dispersal of seeds, and (3) seeds that grow into plants faster. These allowed flowering plants to readily displace most other plants after they appeared in the late Mesozoic. Today, about 96% of all land plants are flowering plants. Of particular economic importance are grasses, such as wheat and other grains. Grasses became widespread during the middle Cenozoic as the climate cooled.

RISE OF MAMMALS. Mammals evolved from mammal-like reptiles in the early Mesozoic. These reptiles were sprawling, sometimes large, animals, and there were a great many of them at this time (refer to Figure 3–56). Some of these reptiles showed certain trends toward mammalian traits. One of the most important of these mammalian traits is specialized teeth. Reptiles generally have simple peglike teeth for tearing and do not process their food before swallowing. In contrast, mammals have incisors for biting, canines for tearing, and molars for grinding. Other mammalian traits recognizable in fossils are the single-boned jaw (reptiles have a number of bones locked together to form the jaw) and three-boned ear.

Unfortunately, many mammalian traits are made of soft tissues and so do not fossilize. We can only assume that they were present in these early mammals. These include hair and a stronger heart than reptiles. Both of these, along with the specialized teeth, are part of a more fundamental adaptation of mammals: a roughly constant body temperature, more loosely called "warm-bloodedness." Unlike cold-blooded animals such as reptiles, amphibians, and insects, mammals are able to generate heat when needed to regulate their body temperature. This permits a generally more active existence since cool external temperature (such as at night or in winter) need not lead to a drop in body temperature and consequent lethargy. Yet there is a cost for this regulating ability, that of consuming many calories of food. On average, a mammal must eat about *10 times* more food than a reptile of the same size. We now can see the importance of specialized teeth, hair, and the stronger heart. The teeth are required to more efficiently preprocess the greater amount of food before swallowing. The hair is important as an insulator to retain the heat so expensively produced. The stronger heart is related to the higher metabolism, carrying more nutrients and oxygen to the more active cells. A final mammalian trait related to warm-bloodedness is the birth of live young. This reflects embryonic development within the mother because of greater control of her internal environment compared to that of an egg laid outside. Also, parental care is extended beyond birth, including the feeding of milk from the mammary gland (from which the name mammal derives).

These traits allowed mammals to diversify and become very successful, but they do not mean that mammals are "better adapted" than the dinosaurs. The fact that mammals were suppressed for so long in the presence of dinosaurs emphasizes this. What seems to have happened is that the elimination of dinosaurs by a "chance" catastrophe created an opportunity for mammals to diversify using their own kinds of adaptations. As noted above, such opportunities to diversify, created by the elimination of a dominating group, are called ecological release.

RADIATION OF MAMMALS. Before the end of the Mesozoic, mammals had evolved into three basic groups: monotremes, marsupials, and placentals (Figure 3–66). All of these still exist. The most primitive group are the **monotremes.** These are the egg-laying mammals such as the duck-billed platypus. They are primitive not only in

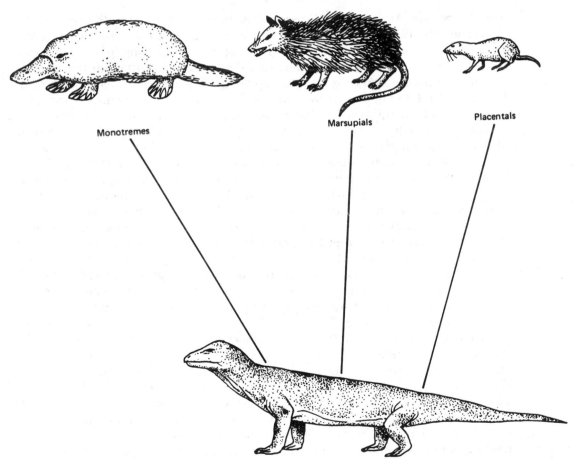

Figure 3–66. Evolution of three basic mammalian forms from mammal-like reptiles. Modified from F. Racle, *Introduction to Evolution*, p. 108.

egg-laying but in other respects as well, such as poor temperature regulation. They are a very small group and are found only in Australia. The **marsupials** are the pouch-bearing mammals (which is what marsupial means). Most of these live on Australia, including such familiar groups as the kangaroos and koalas. (The opossum is one of the few marsupials to thrive outside of Australia.) Marsupial offspring are forced to leave the mother's womb at a very early stage. The tiny animal, less than two inches long, must climb out of the mother's womb and into the pouch on the mother's abdomen, where it attaches to a nipple and resumes growth. The reason for this is that the mother's immune system attacks the offspring as it grows older. Since half of its genes belong to the father, it is rejected as "foreign" tissue.

Placental mammals have solved the rejection problem with the evolution of the placenta, which is an organ that partly separates the developing embryo from the mother's circulatory system. It allows nutrients to flow into the embryo, but acts as a barrier to the mother's immune system and other components in the bloodstream. Whether this makes placentals better competitors than marsupials is a matter of debate, although there is some evidence that it does. The only place where marsupials (and monotremes) have fared well is on Austrialia, which has been isolated from most of the rest of the world since the early Cenozoic due to continental drift. Furthermore, the introduction of placentals to Australia has eliminated marsupials in many areas, indicating they are not as competitive. The same thing happened about 4 million years ago when South America, which also had been isolated and contained many marsupials, was connected to North America by the creation of a land bridge in Central America.

Placental mammals constitute over 90% of all living mammals. There are many different kinds of placental mammals, including the swimming whales and porpoises already discussed, and flying forms, the bats, to be discussed later. Most of this evolutionary diversification of mammals occurred in the late Mesozoic and early Cenozoic when many lines arose from **insectivores.** These were small, generalized, insect-eating mammals with long snouts—rather ratlike in general appearance. Such distinct forms as bats, whales, carnivores, hoofed-mammals, rodents, and others all separated at this time (Figure 3–67). Evolution after this has consisted mainly of progressive modification of these more basic lines. For instance, the carnivore lineage was initiated by a group of archaic carnivores with teeth for tearing flesh and other basic meat-eating adaptations. As the Cenozoic progressed, their ancestors separated to form the more familiar cat and dog branches. Even later, part of the dog branch separated as bears.

Another important line is the hoofed mammals ("ungulates"), plant-eating animals adapted for running fast. This ability is seen in the development of long, stiff legs with hooves, derived from toenails. Early forms ran on paws much like a dog. Eventually, all but one or two of the toes was lost. The remaining toe or toes became elongated and the toenail greatly enlarged to form the hoof, a tough projection to protect the soft tissue. **Archaic ungulates** featured a variety of hoofed animals that would seem strange to us today. One of the most interesting forms, shown in Figure 3–68, was one of the only known plant-eating animals to have

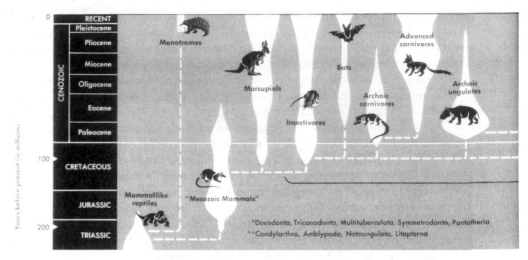

Figure 3–67. Evolutionary history of mammals. The dashes show the white areas indicates the approximate abundance of each group. From

saber teeth. It is unclear whether these teeth were for defense against predators or for use in fighting others of its own species, such as in male combat for females. The archaic ungulates eventually became extinct, but they gave rise to two separate lines that are very important in today's world. In one line, the **odd-toed ungulates,** only one toe came to support the weight. These include the horses, tapirs, and rhinoceroses. This group was much more diverse in the early and middle Cenozoic (see Figure 3–67). Many of these were very large, including the largest land mammal ever known, a relative of the rhinoceros that grew up to 18 feet high and weighed as much as 30 tons. Figure 3–68 shows another once-common odd-toed rhino relative; this one had unusual horns that may have been used by swinging the head sideways. In the middle Cenozoic the odd-toed group was replaced in dominance by the **even-toed ungulates** which have two-hoofed feet, also known as cloven feet (see Figure 3–67). This includes pigs, cattle, deer, camels, and other familiar animals. The reason for the replacement is that even-toed hoofed mammals have a more effective digestive system. Their stomachs have a number of chambers (sometimes called "rumin"; hence the name ruminants), which allow food to be digested more completely. This was important in the middle Cenozoic because grasslands became widespread and grass is extremely tough and low in nutrients. As with the odd-toed forms, many interesting ancestors existed earlier in the Cenozoic, including a deer with "antlers" growing from the top of its nose.

In addition to carnivores and even-toed ungulates, other groups of mammals that have expanded in the late Cenozoic include: bats, primates, and especially rodents (Figure 3–67). Primates, a group adapted for life in the trees, include monkeys, apes, and humans. But it is mainly monkeys who have expanded, with the number of apes and humanlike species decreasing in number in the last few

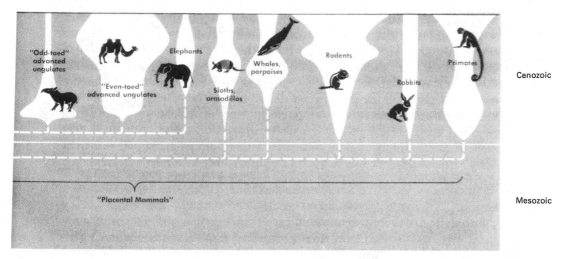

most likely evolutionary relations between the groups. The width of the
A. McAlester, *The History of Life*, 2nd ed., p. 132–133.

million years. Rodents, a group characterized by gnawing teeth, include not only the familiar rats and mice, but also squirrels and other less obvious relatives. These have been spectacularly successful because of their ability to adapt to many situations, their small size, and their potential to speciate readily.

ICE AGE MEGAFAUNA. As the cooling climate culminated in the "Ice Age" (more technically called the Pleistocene epoch), beginning about 2 million years ago, many very large mammals, called **megafauna,** became common over much of the world. The reason for this size increase is probably that large mammals retain body heat better than small ones. This is because they have less surface area exposed relative to their body volume. (Mathematically speaking, as any object increases in size, its weight increases much faster than its surface area.) Since heat loss occurs through exposed surface area, larger animals have less relative heat loss. Even today, rabbits, bears, and other mammals are larger toward the poles than in warm areas. This relationship, of increasing size with cooler climate, is called **Bergmann's rule.**

A number of megafauna are illustrated in Figure 3–69. Large hoofed mammals in the Ice Age included the "Irish Elk" (it was really a deer), with a rack of antlers 10 feet across that weighed many pounds. Elephants were very common, such as the woolly mammoths and mastodons. They lived in cold regions and had a thick coat of hair with three inches of fat for insulation. They were huge, impressive beasts, up to 14 feet tall, with long curved tusks formed from greatly enlarged teeth, as in elephants today. Many other kinds of mammals were much larger than today, such as huge armadillos, bison, and ground sloths. Recently, fossils of a giant kangaroo dating to this period have even been discovered. Preying on these large plant eaters were large predators, such as the massive "dire wolves" and large cats.

Figure 3–68. Early Cenozoic plant-eating mammals. Modified from G. Simpson and W. Beck, *Life* (Orlando, FL: Harcourt, Brace, & World, 1969), p. 505.

Best known of these was the saber-toothed cats, which had long fangs for penetrating the thick neck flesh of their prey. This allowed them to reach the jugular blood vessels and the windpipe.

Virtually all of the megafauna died out by about 10,000 years ago for reasons that are not entirely clear. At least part of the cause seems to be the *warming climate* that occurred as the ice sheets retreated. This would make large size disadvantageous if the large animals had trouble dissipating heat. However, more important would be the major effects of changing climate on plant life. This would have drastically altered the base of the food chain, with radical effects to the large herbivores (and thus their predators). While warming climate is accepted by many, others question it as the only cause. They note that there had been a number of similar warming periods before in the Ice Age (called "interglacials") and there was no similar megafauna extinction. Therefore, they argue for a second possible cause:

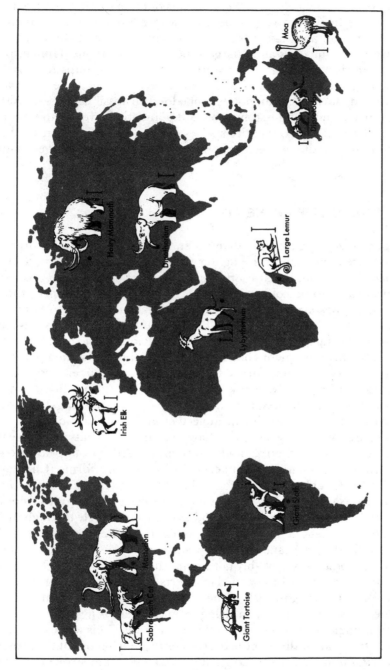

Figure 3-69. Some giant mammals, birds and reptiles of the Pleistocene "ice ages." The bar near each animal represents about 3 feet. From W. Stokes, *Essentials of Earth History*, 4th ed., p. 424.

the theory of *human overkill*. This theory points out that by 10,000 years ago, human technology had finally evolved to the point where humans were a dominant force in the environment and had spread over nearly all parts of the globe. They had highly effective stone tools and a large repertory of weapons. Human population densities were rising rapidly in many areas and human groups from Asia had recently colonized the Americas. There is much evidence that the megafauna were indeed hunted, such as spear points embedded in fossil carcasses. Critics of the overkill theory argue that the stone tools and population levels of these early people were insufficient to wipe out entire species. At present, there seems to be no clear resolution to the debate between proponents of the climate change versus overkill theories.

PART 4: EVOLUTION OF LIFE IN THE AIR

Air is a much less dense medium than water or land. Therefore, an organism must use much more energy and have lighter weight to effectively move around in air. Largely for this reason, the atmosphere was the last of the three great spheres of the physical earth to be colonized by life. However, once attained, the ability to move in the air was extremely advantageous for many obvious reasons: escaping predators, finding food, and so on. Thus, a number of groups have made the transition, as natural selection favored those individuals who could take advantage of it. Initially, this movement consisted only of leaping and then gliding through the air. (Intermediate gliding forms, such as the "flying squirrel" and gliding lizards, still exist today.) Eventually, two key adaptations made true flight possible in all flying groups: (1) light bodies, to minimize gravity, and (2) wings, to provide extra surface area to push against the air molecules and give buoyancy.

There are four main groups that have evolved these traits to achieve the power of flight: insects, reptiles, birds, and mammals (in order of appearance of flight). Because selection is limited to "tinkering" with existing forms (Chapter 2), each group solved the problem of flight somewhat differently, depending on the anatomical endowments of their ancestors. In the three vertebrate groups (reptiles, birds, and mammals), the evolution of light bodies occurred largely through the development of hollow bones and other cavities. In contrast, the insects have no bones (or even mineralized parts), and are much smaller in size. Therefore, the need to evolve lighter bodies was probably not a major hurdle in insect flight. However, the evolution of wings clearly was needed in insects, as with the other groups. As shown in Figure 3–70, the wing in each of the vertebrates is formed from limb bones, but these bones vary considerably. In flying reptiles, there is a very long finger (digit) that forms a single strut supporting a skin membrane. This weak support probably made these the weakest fliers of the four. In birds, much more of the arm, including the radius and ulna bones, is used, giving birds a stronger stroke than the reptiles had. The bat wing is different still, having all five fingers (digits) elongated, with skin in between to form a webbing. These numerous struts give bats much maneuverability. Finally, the insect wing is most different of all, being formed not of bone,

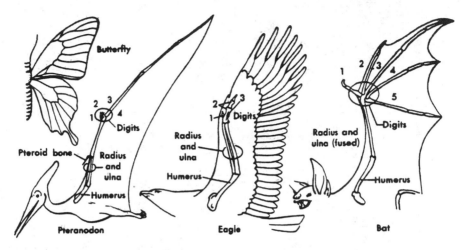

Figure 3–70. The variety of modifications used by 4 different groups to fly. Note especially differences in uses of digits on hands. Modified from G. Simpson and W. Beck, *Life*, p. 316.

but of **chitin,** which is a tough, brittle material that also forms the exoskeleton. Insect wings are chitinous outgrowths of the exoskeleton.

Unfortunately, the fossil record of all flying groups is very poor. This is not surprising given that flight requires light bodies of hollow bones or unmineralized body parts. Also, flying organisms are not likely to be buried under water where rapid sedimentation and fossilization are most likely to occur. Nevertheless, the rare chance occurrence has preserved enough evidence for us to reconstruct the major outlines of the evolution of flight. As with evolution in water and on land, the evolution of flight can be roughly correlated with the three main eras of the geological time scale:

1. Rise of primitive insects (Paleozoic)
2. Rise of advanced insects and flying reptiles (Mesozoic)
3. Rise of birds and flying mammals (Cenozoic).

These stages are shown in Figure 3–71 in the context of other evolutionary and geological events.

Rise of Primitive Insects (Paleozoic)

Recall that the millipedes and centipedes were among the first land animals. It is from these arthropods that the earliest insects arose. In 1988, the earliest known insect fossils were reported. They were dated to about 390 million years ago, or the early part of the middle Paleozoic. This is just a few million years after the first land plants and millipedes, so insects were not long in evolving. By the middle Paleozoic, insects were quite abundant in the coal-forming swamps. Imagine the early insects buzzing amid the lush humid forests of huge ferns and other spore-bearing plants, with large amphibians crawling in the moist forest floor below.

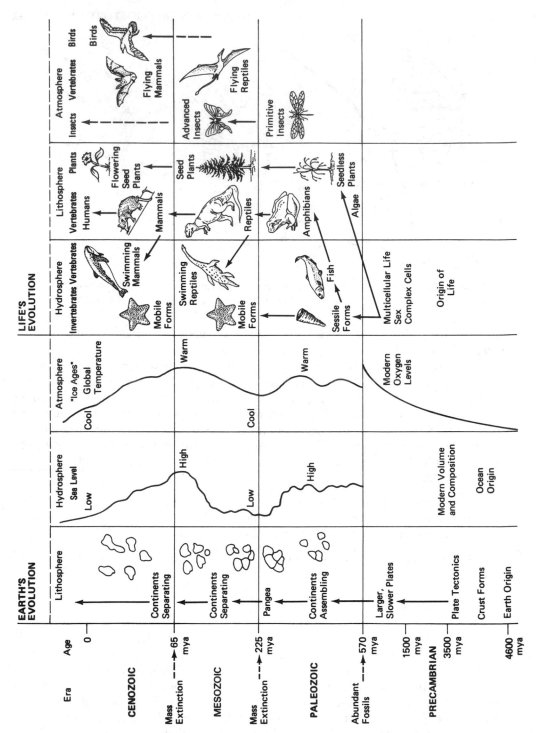

Figure 3–71. Evolution of flying animals in context of other evolutionary changes.

Insects retain the segmented body, exoskeleton, and other features of their centipede and millipede ancestors. However, they are modified in several important respects. The most characteristic of these is the possession by insects of 6 legs (3 pairs). We may divide the primitive insects into 2 basic groups: (1) the primitive wingless insects, and (2) the primitive flying insects. The primitive wingless insects were the first insects to evolve, appearing about 390 million years ago. These forms resembled the silverfish, a descendant which is still alive today. Sometime later, in the middle Paleozoic, wings evolved. Familiar members of the primitive flying insects are the dragonflies and mayflies (Figure 3–72). These and other primitive

Figure 3–72. Representatives of insects today. Modified from R. Barnes, P. Calow, and P. Olive, *The Invertebrates*, (London: Blackwell, 1988), p. 249.

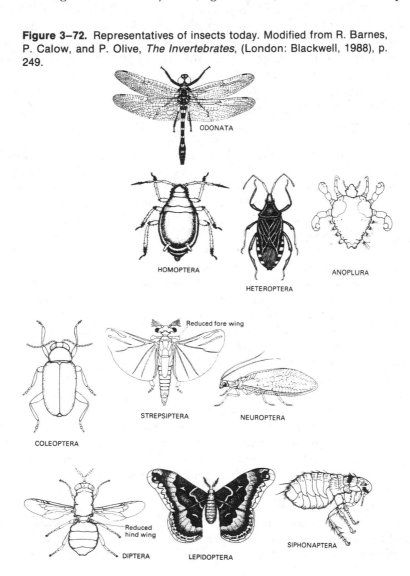

fliers became the most abundant insects by the late Paleozoic. Dragonflies are highly predatory insects, and they grew to large sizes in the late Paleozoic, some with wingspans of over 2.5 feet.

Rise of Advanced Insects and Flying Reptiles (Mesozoic)

The generally warm skies of the Mesozoic saw a much greater abundance and diversity of flying organisms than in the Paleozoic. There was a vast radiation of insects. These were not only better fliers than the primitive Paleozoic forms, but included highly social groups, such as ants, with complex behavioral patterns. Indeed, it was during this radiation that insects became the most abundant and diverse of all life forms on earth. This is still true today: Over half of all known species of life on earth are insect species.

Also appearing in the Mesozoic, for the first time, were the flying reptiles and birds (refer to Figure 3–71). However, birds did not become abundant until the Cenozoic.

ADVANCED INSECTS. "Advanced" insects include the great majority of familiar insects: roaches, flies, ants, beetles, bees, butterflies, and so on (Figure 3–72). Some of these, such as the roaches and grasshoppers, are known from the late Paleozoic. However, most advanced insects, including the social insects like ants and bees, first appear in the Mesozoic. The evolution of social insects is intimately related to the evolution of the flowering plants, which first appear in the late Mesozoic. This is one of the classic examples of **coevolution,** wherein the evolution of one group is highly dependent on the evolution of another.

We may cite four ways in which advanced insects differ from the primitive wingless and flying groups that were common in the late Paleozoic. One, advanced insects have folding wings over the back. In contrast, the primitive flying insects (such as the dragonfly) have fixed wings. Two, advanced insects can fly faster and are more maneuverable. This is because they have many more wing beats per second (up to thousands of times per second versus just a few times for dragonflies). As shown in Figure 3–73, this is because the wing muscles are attached differently, allowing more rapid action in the advanced forms. Three, advanced insects have larvae (juvenile stages) that are not restricted to water. For instance, caterpillars are larvae of butterflies and moths, which are perfectly capable of living on land. In contrast, the larvae of primitive insects must grow in water. Notice how this is similar to the early land plants and amphibians, which also retained the need for water in reproduction. They had not completely broken away from their marine ancestry. Four, advanced insect larvae are usually *specialized* and do not resemble the adult. Thus, caterpillars look nothing like adult butterflies. Instead, they feed on vegetation to grow, then spin a cocoon and undergo **metamorphosis,** which is a process of massive cellular rearrangement, before becoming an adult. In contrast, primitive insects look very similar to the adults, only they are smaller. The main change is a

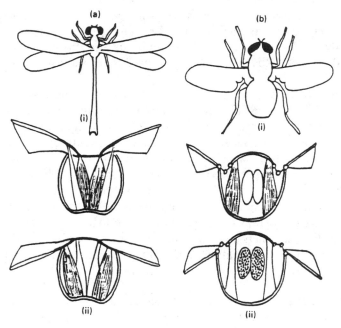

Figure 3–73. Structural and neuromuscular adaptations for flight in insects. (a) Direct flight musculature in the dragonfly. Note how flight muscles attach to wing bases. (b) Indirect flight musculature in the fly. Modified from R. Barnes, P. Calow, and P. Olive, *The Invertebrates*, p. 341.

series of **molts,** whereby the insect sheds its outer skeleton (exoskeleton) because this hard shell cannot grow.

THE INSECT THAT ATE NEW YORK? Most of us have seen science fiction movies about a giant insect of some kind. There are a number of scientific problems with such a scenario. One of the major problems is that to grow to such an extreme size, an animal must change shape to support itself; consider the thickening of legs seen in an elephant or *Brontosaurus*. Yet giant movie ants look just the same as tiny ones. Nevertheless, this does raise an interesting question: If insects and their relatives were the first animals on land, why did this highly successful group not grow to larger size? If they had done so, they likely would have excluded mammals, including ourselves, from ever evolving. Yet the largest insects, such as the goliath beetle, are only about the size of the smallest mammal, the dwarf shrew (1 to 2 inches long).

Much of the reason for the small size of insects lies in their exoskeleton, which limits them in two ways. One, while an external skeleton is effective protection and provides support against gravity at small sizes, it becomes insufficient support as an animal becomes larger. This is the result of simple engineering principles that can be observed in human structures. A builder will generally reinforce buildings, bridges, and other large structures with beams and other supports placed inside the structure where they will do the most good. Thus, if you observe the large animals that have existed, such as elephants and dinosaurs, you find that they have **endoskeletons.** This means that their bodies are supported by bones from the inside ("endo" = inner).

A second limitation associated with the exoskeleton is that insects breathe through it. The exoskeleton has many pores that admit air. This limits size for the same reason that larger animals survive in cool climates: As an animal increases in size, the ratio of surface area to body mass (volume) decreases. In the cold, this helps conserve heat, which diffuses out through the skin. In insects, this declining ratio limits growth because the animal becomes less able to breathe. The amount of air allowed to diffuse in (which is proportional to surface area) increases more slowly than the animal's need for it (total body tissue is proportional to volume). Large animals such as ourselves have solved this problem by evolving internal air sacs, or **lungs,** which contain many convolutions that maximize the surface area inside our bodies. This allows a much greater amount of air diffusion to the blood, which carries oxygen to the rest of the body.

WHY ARE THERE SO MANY INSECTS? We said that insects make up over half of all known species of life. Indeed, there are nearly one million described insect species. Furthermore, there are often many millions of individuals of any one species. In short, insects show great diversity (number of species) and abundance (number of individuals per species). If one wishes to define success in terms of sheer diversity and numbers, the insects are the greatest evolutionary success by far.

There are two basic reasons why insect diversity and abundance are so great. One is the insect's small size. Small organisms need less food than large ones so that more individuals can be supported per unit area. This accounts for the insect's greater abundance. Small size leads to more species because there is a tendency for smaller organisms to perceive the environment in a "coarse-grained" way. This means that small organisms tend to be more localized in their geographic range and habitat preference. The reason for this is intuitively obvious: A small stream may be a major obstacle for a small creature, but can be easily crossed by a larger one. This leads to more isolation of populations, and hence more speciation in small species. However, many groups, such as land snails, are small in size yet have not become as immensely diverse and abundant. Therefore, there must be some additional factor besides small size that promotes species formation. This is that insects are readily modified: Insect genes, anatomy, behavior, and development seem to be quite flexible. We see this in the development of wings, specialized larvae of all kinds, intricate social behavior, and many other ways that the basic insect design has become altered to adapt it to new ways of life. Because selection depends on variation to cause evolution, this exceptional flexibility instrinsic to insect biology is the second reason why so many insects have evolved.

FLYING REPTILES. Flying reptiles were the first vertebrates to achieve true flight. Indeed, they appear in the fossil record a few million years before birds, and it appears that they dominated the Mesozoic skies. Beginning in the early Mesozoic, and continuing for over 140 million years (over twice as long as birds have existed), flying reptiles of many shapes and sizes were filling ecological niches (roles) now occupied by birds. Only after flying reptiles died out at the end of the Mesozoic did birds truly begin their evolutionary diversification.

Flying reptiles are technically called **pterosaurs** ("ptero" = wing, "saur" = lizard). They were not dinosaurs, but were related to them. Pterosaurs appear to have evolved from tree-climbing ancestors that began as gliders. (Even today, gliding lizards exist.) Most, if not all, pterosaurs seem to have retained the tree-climbing habit. Research shows that while the fourth finger of the hand was greatly elongated as a strut for the wing membrane, the other digits (fingers) were retained on the leading edge (see Figures 3–70 and 3–74) and used as claws to climb trees. Some people have suggested that pterosaurs hung upside down like bats. While this is not certain, we do know from pelvic remains that pterosaurs were ungainly walkers, who probably could only waddle clumsily on the ground. Thus, they probably spent the large majority of time either hanging or sitting in trees or other perches, or flying. In the air, it used to be thought that because they were reptiles, pterosaurs were poor fliers, who could only glide along the air currents. However, as with the dinosaurs, they are now pictured as a diverse group of highly proficient, agile fliers that were perhaps even warm-blooded. While they lacked feathers, they had hollow bones (like birds), relying on their wings of leathery skin to give them lift. Evidence of active flight include a relatively large brain and a large sternum (breast bone) to anchor strong muscles used in flight.

There were two main groups of pterosaurs: the earlier rhamphoryncoids and the later pterodactyloids (Figure 3–74). The **rhamphoryncoids,** living mainly in the middle Mesozoic, were smaller, had tails (like a kite, for stability), teeth in their "beaks," and heavier skeletons. In contrast, the **pterodactyloids,** living in the late Mesozoic, were even more specialized for flying. They were generally larger, lacked tails (which gave them more maneuverability), lacked teeth, and had lighter skeletons. They also had a bony crest on the head; the function of this crest is not entirely clear, but it may have been a stabilizer in flight.

Within these two main groups, there were a number of diverse forms, illustrating a variety of roles in Mesozoic food webs. Agile forms, like *Rhamphorynchus* (Figure 3–74), seem to have skimmed the ocean waves, plucking fish from the water. You might visualize them as similar to pelicans at the beach: ungainly on land, but elegant in design for cruising over the wave crests, riding the air currents in search of food. At the other end of the scale from the agile, maneuverable forms were some of the huge pterodactyloids. Their massive wingspan, of at least 40 feet (the largest one known, from Texas rocks), shows that they were adapted to soaring, slowly riding the "thermals" (currents of rising hot air) to great heights. These might be visualized as similar to giant condors, which are the largest soaring fliers today. More unusual types of pterosaurs are also known. The massive feeding apparatus of one group indicates that they lived by crushing clams and other shellfish. An especially unusual group apparently lived by filter-feeding, straining plankton from the ocean with a netlike beak.

RHAMPHORHYNCHUS

PTERANODON

Figure 3–74. Common Mesozoic flying reptiles. *Rhamphorhynchus* was between two and three feet. The wing span of the *Pteranodon* was 27 feet. From W. Stokes, *Essentials of Earth History*, 4th ed., p. 361.

The pterosaurs in general seem to have lived mainly around the ocean. This is indicated not only by the fish, shellfish, and plankton diets, but from their fossil locations; pterosaurs are found mainly in marine rocks. Unfortunately, their fragile bones mean that the record is not especially good, as is also the case for birds to which we now turn.

Rise of Birds and Flying Mammals (Cenozoic)

Birds are among the most familiar creatures on earth. With nearly 9,000 living species, birds have over twice as many species today as mammals. With their mobility, they are found nearly everywhere. In spite of their generally poor fossil record, some exceptional cases of very early bird remains tell us much about the origin of birds. They clearly show that *birds evolved from reptiles by the middle Mesozoic*. But it was not until much later, beginning in the Cenozoic, that birds began to diversify into the abundant group we know today.

ORIGIN OF BIRDS. The earliest confirmed bird fossils occur in middle Mesozoic rocks of Germany, in limestone deposited in the waters of quiet lagoons. This first bird is called *Archaeopteryx*, meaning "old wing." It is one of the best cases we have of an intermediate fossil (a "missing link") showing the evolution of one group into another. Nor can it be argued that this fossil is a fluke, because there are four nearly complete skeletons of this early species (as well as other, incomplete skeletons). The skeleton of *Archaeopteryx* (Figure 3–75) is so similar to that of certain two-legged, meat-eating reptiles, that it was classified as a reptile for years. Not until a museum worker happened to notice the faint imprints of feathers did the relationship to birds become apparent. Among the more prominent reptilian features of this first bird are: teeth, tail, and pelvis for running. Birdlike traits include: feathers, long forearm, and an enlarged breastbone for wing muscle attachment.

Further evidence that birds are related to reptiles is found from observation of living birds. Most obvious to the casual observer is that both birds and reptiles lay eggs. Furthermore, bird and reptile eggs share a great many biological similarities, including similarities in the developing embryo within the egg. For instance, up to a point feathers develop in the bird embryo in the same way as reptilian scales develop. You can see this in a chicken, where the hard, scalelike covering of the leg sometimes grades into feathers higher up near the body. Finally, tissue transplant experiments on the embryos of birds have shown that the genes for growing teeth still exist in the developing bird. The bird's beak that we are familiar with occurs because these genes for teeth growth are never expressed in normal development.

There is an ongoing debate over the exact pathway that birds followed to evolve from reptiles. As shown in Figure 3–76, the *ground upwards theory* says that bird ancestors were reptiles that ran on the ground, perhaps using feathered limbs to catch insects. The *trees downwards theory* says that the reptilian bird ancestors were tree climbers that used feathers to slow descent on jumping to the ground. At present, there is no clear resolution to these, and still other, ideas.

Figure 3–75. *Archaeopteryx*, the earliest confirmed bird. From F. Racle, *Introduction to Evolution*, p. 74.

DIVERSIFICATION OF BIRDS. Aside from *Archaeopteryx*, few birds are known from the Mesozoic. The best-documented forms come from the very late Mesozoic, and seem to have lived along the ocean. One is a swimming bird with small wings and large, kicking feet. The other bird resembles modern sea gulls. Being millions of years younger, it is not surprising that both birds are anatomically much more modern than *Archaeopteryx*. They have lost the reptilian tail, among other things. However, they do retain small teeth.

The first familiar birds appear in the early Cenozoic. These are ancestors of modern gulls and ducks. However, it is not until the middle Cenozoic that the most familiar group of birds appears, the songbirds. These make up the large majority of birds that we are familiar with: blackbirds, jays, robins, sparrows, and many more. Birds of prey, such as hawks and eagles, probably evolved in the early Cenozoic, but there is no fossil record of them until the middle Cenozoic. An especially interesting

Theories of flight

"Ground upwards"
Below The image of a
scampering protobird
attempting to become air-
borne by running at high
speed has had many
doubters. One novel suggestion has been that flight
evolved as an accidental byproduct of swinging the
feathered arms in order to
catch insects.

"Trees downwards"
Above The ability to fly may have been
developed by stages which started with simple
parachuting to the ground using feathers to
help break the fall, followed by the development of feathered wings for gliding, and finally
full flight.

Figure 3–76. Two theories of flight. From D. Norman, *Dinosaur!*, p. 139.

group of birds that evolved a number of times in the Cenozoic are large, flightless birds. Most of us are familiar with ostriches, but the fossil records reveal that such birds evolved on nearly all the continents, and many islands, in the past. Some were much larger and more aggressive than ostriches. For instance, in the early Cenozoic, there was a diverse group of predatory carnivorous birds. Figure 3–77 shows an example of such an extinct bird, over 6 feet tall with a strong, flesh-eating beak and thick, muscular legs equipped with sharp talons useful for disemboweling prey. Such large carnivorous birds tended to evolve in the absence of large mammalian predators.

BATS. Bats are the most able mammalian fliers. Other mammals can glide for short distances, such as the "flying squirrels," but they are not true fliers. Bats hold a great fascination for many people, often with sinister connotations. But bats are basically very shy and are harmless, often performing essential ecological roles such as keeping down insect populations and dispersing plant seeds.

There are two distinct groups of bats: (1) the insect eaters, and (2) the fruit eaters. The insect-eating bats are the only bats found in the Americas. Insect eaters are the more diverse of the two groups, with over 700 species. As shown in Figure 3–78, they are generally smaller than the fruit eaters, flitting about, often at night, feeding on insects. They use **echolocation** when feeding. Echolocation means that they emit very high-pitched sound waves through their mouths. These waves

Figure 3–77. A large flightless predatory bird from the early Cenozoic. Modified from R. Wicander and S. Monroe, *Historical Geology*, p. 477.

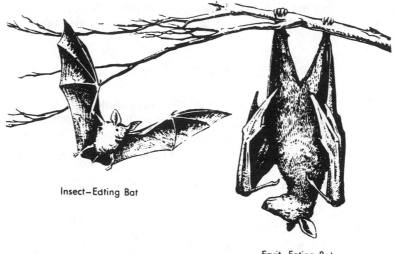

Insect—Edting Bat

Fruit—Eating Bat

Figure 3—78. Two basic kinds of bats. From E. Colbert, 1980, *Evolution of the Vertebrates* (New York: Wiley, 1980), p. 284.

bounce off objects (such as insects) and the bat's huge ears detect the "echo," or return signal. The closer an object, the faster the returned echo. (Submarines do the same with sonar.) This method is very effective and, combined with their highly maneuverable flying skills, allows these bats to be voracious feeders of insects. They need to be because, like all small mammals, they have extremely high metabolisms and need much food. This need is compounded by the flying lifestyle, which burns much more energy than walking. To conserve energy, these bats sleep during the day, upside down, often in caves, where they are generally safe from predators. An infamous kind of insect-eating bat is the vampire, of the American tropics, which makes a tiny cut in the skin of living animals and laps the blood. However, aside from carrying disease, such as rabies, these bats do little harm. The fruit-eating bats are sometimes called "flying foxes," since they are large and have a foxlike face, seen in Figure 3—78. They are not found in the Americas, but in Africa and Asia, numbering less than 200 species. They lack the ability to echolocate.

The earliest bat fossil occurs in the early Cenozoic. Like *Archaeopteryx*, it is a fortunate find near the origin of a poorly preserved group. In this case, the nearly complete skeleton is found in lake deposits. It is suprisingly modern, being very similar to some living insect-eating bats. This first known bat could apparently even echolocate, judging from the fossilized ear bones. The teeth indicate that bats arose from insectivores, a primitive group of tree-dwelling mammals that eat insects. Considering how specialized this early fossil bat is, it seems clear that bats separated from the insectivores millions of years earlier, although no one knows exactly when. The early history of fruit-eating bats is a mystery because no fossils are known before the middle Cenozoic, when they are already modern in appearance. Most workers think they separated from the insect-eating bats in the very early Cenozoic.

PRODUCTS OF HUMAN EVOLUTION

The history of human evolution involves both biological and cultural changes. Because this is a book about the evolution of life, we will focus mainly on human biological evolution. However, these changes were accompanied and strongly influenced by cultural evolution, so we will briefly summarize that aspect as well for a fuller picture of our past.

The discussion of human biological evolution will cover three basic areas:

1. Primate evolution
2. The human fossil record
3. Human races (minor biological evolution)

Our discussion of cultural evolution will very briefly summarize:

1. Technological evolution
2. Social evolution

HUMAN BIOLOGICAL EVOLUTION

Our Primate Heritage

Biologically, humans belong to the order **Primates.** This is clear from our current anatomy as well as from the fossil record that traces our separation from other members of the group. Primates are a fairly old and generalized group of mammals that probably arose in the late Mesozoic and diversified in the early and middle Cenozoic. Primates have retained many features of their insectivore ancestors at the base of the mammalian family tree. This is not surprising because primates, like these ancestors, are adapted to living in trees. Unfortunately, tree-dwelling animals do not commonly fossilize, so the primate fossil record has many gaps. Nevertheless, our overall knowledge of primate evolution is more than adequate to construct a general sequence of events.

PRIMATE TRAITS. Major primate characteristics are associated with their tree-dwelling habits. A list of basic primate traits would include the following:

1. Grasping hands that have nails instead of claws, and have an opposable thumb. Hands are, of course, useful in climbing and moving about in trees, whether it be swinging, jumping, or other movements.
2. **Stereovision,** which means that the eyes have moved to face further forward. This

gives overlapping fields of vision, which is what allows for depth perception. At the same time, the sense of smell has been reduced in many primates because there is less need for it in the trees.

3. Relatively large brains. A main reason for this is that primates are highly social animals, so that communication and complex social behavior is important.

4. Prolonged caretaking of offspring. Primates have relatively few offspring, but they are nurtured by the parents and sometimes helped by other members of the group. This allows a prolonged learning period and, ultimately, more learned behavior.

KINDS OF PRIMATES. There are four basic types of primates that have evolved since the early Cenozoic. All four basic types are alive today, and together make up 190 living primate species. In order of evolutionary appearance, the four types are: *prosimians, monkeys, apes,* and *humans.* In general, the later types are more developed in the primate characteristics just listed (such as greater brain size or more dextrous grasping hands).

Prosimians were the first to appear, and were common in the early Cenozoic (Figure 3–79). These are most similar to the Mesozoic insectivore ancestor. ("Prosimian" basically means " Before monkey.") For instance, prosimians retain a keener sense of smell and have less forward-facing eyes than monkeys, apes, and humans. Prosimians are not familiar to most people in the Americas because they are found only in Africa and Asia. The loris and lemur are examples, illustrated in Figure 3–80, along with an insectivore. Because they retain such primitive features as a

Figure 3–79. The geologic record of primates. The width of the vertical areas indicates the approximate abundance of each group. Modified from D. Eicher and A. McAlester, *History of the Earth,* p. 370.

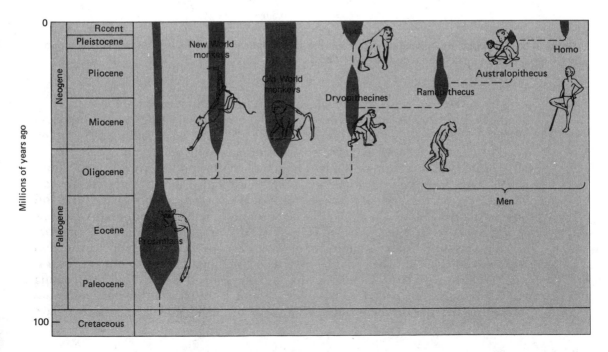

Figure 3–80. Representative primates, and insectivore ancestor. Lemur and loris are prosimians. Chimpanzee is an ape. From F. Racle, *Introduction to Evolution*, p. 126.

longer muzzle and claws, prosimians may not look "primate-like." However, close anatomical inspection shows them to be on the evolutionary path toward monkeys and later primates. They are intermediate between insectivores and monkeys in the basic primate traits, such as stereoscopic vision, grasping ability, and brain size.

Monkeys arose from prosimians sometime in the middle Cenozoic (see Figure 3–79). They have well-developed grasping hands and relatively high intelligence. There are two basic kinds of monkeys, the *New World monkeys*, found in Central and South America, and the *Old World monkeys*, found mainly in Africa and Asia. New World monkeys have been isolated from the Old World group for millions of years and have become distinct in a number of ways. Perhaps the most familiar distinction is the prehensile tail, which can be used as a "fifth hand" to grab objects. This is found only in New World monkeys. Most monkeys spend their lives in trees, eating fruits, insects, or other foods. A notable exception is the baboon, a monkey who has taken up a lifestyle on the ground. As a result of the added challenges, they have become very aggressive and have a complex social hierarchy organized in a troop society. Because early humans also coped with a ground-dwelling life on the grasslands with a cooperative, complex society, baboons have been intensely studied as an analog to how early humans evolved.

Apes are often confused with monkeys, but they are actually quite distinct. Apes tend to be larger, lack tails, live longer, and have relatively larger and more complex brains. Apes are our closest living relatives, so their fossil record is of great interest. Like monkeys, early ape fossils appear in African rocks, dating to the middle Cenozoic, about 30 million years ago (Figure 3–79). It thus appears that apes evolved from monkeys in a relatively short period of time. There are four kinds of apes living today: *gibbons, orangutans, gorillas*, and *chimpanzees*. Gibbons and orangutans live in Asia. Gorillas and chimpanzees live in the forests of Africa, spending much time on the ground. Gorillas and chimps are very closely related. Most of their physical differences result from the much greater size of the gorilla caused by a greater rate of growth in the juvenile stage. Evolutionarily, this is easy to achieve through a simple mutation in growth hormone mechanisms. However, behaviorally there are also some important differences. Chimps tend to eat more fruit than gorillas (which prefer more leaves), and chimps have a much "looser" society in that small social groups tend to form and break up often. In contrast, gorillas tend to form stable groups called harems, with a single dominant male and a number of females with offspring.

In addition to being closely related to each other, chimps and gorillas are also closely related to humans. Indeed, of all life today, gorillas and chimps are our closest living relatives. The chimpanzee, shown in Figure 3–80, in particular appears quite close to our evolutionary line. For many years, evidence from the fossil record was interpreted to indicate that the human line branched off from the chimpanzee line over 10 million years ago. However, in the last 20 years, advanced biochemical techniques have indicated that humans branched off as recently as only 5 million years ago. For example, humans and chimpanzees have identical amino acid sequences in many proteins, indicating that they have had little time to change through mutation, as occurs in distantly related animals. Further, direct analysis of the DNA in human and chimpanzee genes has shown a striking similarity: we share about 99% of the same genes. By making basic assumptions about mutation rates (the "molecular clock"), we can estimate the time at which the lineages separated. This close relation to the apes is also reflected in their anatomical and behavioral similarity to us. They are very similar in bone and tissue structure, differing mainly in rates of proportional growth.

Experiments in raising baby chimpanzees and gorillas have shown that they can learn sign language to express many concepts. They not only have a vocabulary of dozens of signs but they can combine them to form short "sentences." Sign language is used because these apes do not have the vocal apparatus to form words, as shown in Figure 3–81. Chimps also are capable of using tools, even in the wild. They have been observed using sticks and other items to extract foods such as ants from difficult places.

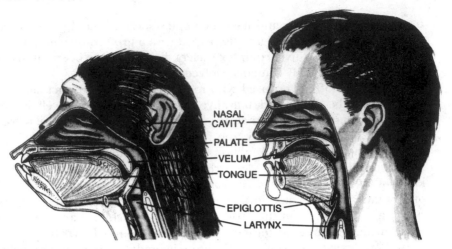

NASAL
CAVITY
PALATE
VELUM
TONGUE
EPIGLOTTIS
LARYNX

Figure 3—81. Diagram showing great differences in the vocal tract of chimpanzee and human. From *Scientific American*, April 1991, p. 147.

The Rise of Humans

If we estimate that humans and chimpanzees separated about 5 million years ago, then we should expect to find fossils of our first direct ancestors at about that time. This seems to be the case. The earliest definite human ancestors belong to the genus *Australopithecus*, which is first known from fossils that are about 5 million years old (see Figure 3—79). Our own genus, *Homo* (meaning "human"), arose from *Australopithecus* about 2.6 million years ago.

Australopithecus means "southern ape." The name was given because these fossils were first found in southern Africa. Today there is evidence that they also had lived in eastern Africa. The birthplace of humankind was clearly Africa. Technically, there may have been a number of species of *Australopithecus*. The famous "Lucy" is a specimen in this genus. As shown in Figure 3—82, the earliest species is often referred to as *Australopithecus afarensis*. (The name comes from the Afar area of Africa, where it was first found.) This was succeeded by a similar species, *Australopithecus africanus* about 2.8 million years ago. *Africanus* branched off into two lines. One line led to *Homo* and modern humans. There was also another branch, the **robust** australopithecine species, which was a dead-end side branch that existed from about 2 to 1 million years ago. This branch consisted of species larger than our direct ancestors. They appear to have been vegetarians, with little ability to use tools. There is a great amount of debate over the exact details of this family tree because early human fossils are rare and often fragmented.

Physically, *A. africanus* adults were about 4 feet tall and weighed from about 50 to 70 pounds. They had a brain that was about 500 cubic centimeters (cc.) in volume (modern humans average about 1,400 cc.). To the casual observer, these smaller australopithecines may have superficially resembled living chimpanzees,

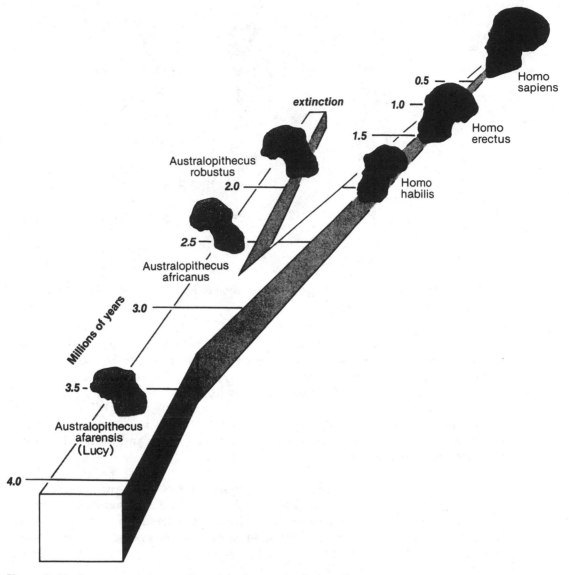

Figure 3–82. A common interpretation of the human family tree. From H. Levin, *The Earth Through Time*, p. 581.

with their small size and prominent jaws (Figure 3–83). The jutting jaw comes from having very large and numerous teeth, which modern humans have progressively lost because of increased food preparation. However, the relative brain size of the australopithecine was greater than the chimp's. Another major difference is that changes in the pelvis and legs allowed these human ancestors to walk more fully erect than the chimpanzee (Figure 3–83). This freed the hands and allowed them

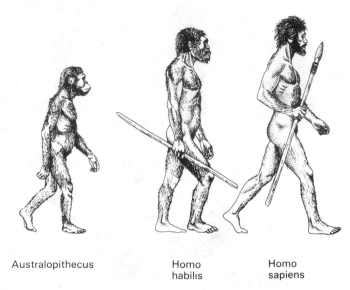

Australopithecus   Homo habilis   Homo sapiens

Figure 3–83. Depiction of fossil humans. Modified from E. Chaisson, *Universe*, p. 500.

to be used for activities besides knuckle-walking and climbing. The lifestyle of these ancestral australopithecines is of great interest because it tells us about how the human line got started. It appears that they lived not in the forests, like apes today, but in the African grasslands. Apparently they foraged a great deal for plant material (nuts, berries, fruits), and also ate some meat, mostly through hunting small game and scavaging kill of large predators. While there is no conclusive evidence of stone tool use before about 2.6 million years ago, sticks and unmodified tools were probably used earlier.

THE GENUS HOMO. By about 2.6 million years ago, members of our own genus, *Homo*, had branched off from *A. africanus* (Figure 3–82). Notice also that *Australopithecus* coexisted with our branch for some time. The larger, vegetarian form, *A. robustus*, lived on for quite some time until becoming extinct about a million years ago. Some scientists argue that *robustus* was killed off by our larger-brained, more carnivorous ancestors of the *Homo* branch.

The first *Homo* specimens are usually given the species name **Homo habilis**, meaning "handy man." This is because when they were first discovered by Louis Leakey, they were thought to have produced the first stone tools. This is now a matter of much debate, but for simplicity we will consider this to be true (see Lewin, 1989; Nelson and Jurmain, 1991). Physically, *Homo habilis* was not very different from the smaller australopithecines from which they evolved and the classification to a new genus is not well justified. Even if they were the first stone tool users, creating a new biological genus is probably not necessary. However, it is so established that the current names will probably continue. Indeed, the whole series from *afarensis* to *africanus* to *habilis* is a fairly gradational sequence showing regular trends toward increasing body and brain size. By the time *H. habilis* evolved, body size probably averaged over four feet tall and brain size had increased to about 700 cc. (Figure 3–83).

By about 1.5 million years ago, the next human species, **Homo erectus**, had evolved from *habilis* (Figure 3–82). The name "erectus" refers to the fully upright posture attained by this species. Indeed, from the neck down, they were fully modern in size and shape, with an average body size of 5 feet tall or larger. It is mainly in the skull that *H. erectus* was unlike us, having a brain size that was about 900 cc. at 1.5 million years ago. This gradually increased to over 1,100 cc. before they evolved into **Homo sapiens** (meaning "wise man") about *200,000 years ago* (Figure 3–82). Besides a smaller brain, *H. erectus* also differed from us in having a larger, jutting jaw and a heavier, more reinforced skull. The *H. erectus* stage was a threshold in human evolution. For the first time, humans began to use fire and moved out of our ancestral homeland, Africa, about one million years ago. For the first time, we became active hunters of big game. In short, we began that most human of all traits: rather than being modified by the environment, we began using our brains and tools to modify our environment to suit our needs.

While we said that modern *H. sapiens* arose about 200,000 years ago, this change was not abrupt. Once again, we see intermediates and gradational forms. The fossil record from these times is considerably better than that of earlier stages, with many remains in Africa, Europe, and Asia, so continuity of variation is even clearer. Thus, in early *H. sapiens*, brain size was roughly modern, but they retained remnants of the heavier, more elongate skulls and larger jaws seen in *H. erectus*. The most well-known of these so-called "archaic" early modern humans are the **neanderthals.** (The name is derived from the Neander Valley of Germany, where such fossils were first found.) These were a powerfully built group of humans that were centered mainly in Europe and seem to have been well adapted to the cold. In spite of the superficially "brutish" appearance (such as protruding brow ridge and jaws, elongate skulls) their cultural remains (e.g., tools and weapons) are quite advanced. In fact, these are the first humans known to have ceremonial burial of their dead (as shown by flower pollen grains on a grave). Also, they cared for their old and sick, as shown by Neanderthal individuals who bear marks of living for some time in otherwise helpless conditions. Indeed, average Neanderthal brain size was actually greater than our own.

Neanderthals disappear from the record about 40,000 years ago. Their fate is a mystery. It may be that they simply blended with more modern *H. sapiens*, as did many of the more archaic forms. Indeed, since Neanderthals were large-boned and cold-adapted, some scientists have suggested that Scandanavians may be their descendants. Others have suggested that more modern humans extinguished the Neanderthals through warfare.

By 30,000 years ago, anatomically modern humans had spread throughout Africa, Europe, and Asia. Fossils became much more abundant at this time. By "anatomically modern" we mean that (compared to archaic forms) the skull is more bulbous (not as elongated), the brow ridges and jaw bones are reduced, and the overall bone structure is less massive in general. By 40,000 years ago and perhaps sooner, humans had migrated to Australia. By at least 12,000 years ago, humans had crossed into the Americas from Siberia. At times of low sea level (as during glaciations) there is a land bridge between Alaska and Siberia.

Human Races: Minor Biological Evolution

Even though modern humans arose many thousands of years ago, biological evolution in our species did not stop. As populations spread throughout the world, gene flow among the groups decreased as populations became more distant and isolated from one another. Genetic drift and local selection pressures caused populations to differentiate into races. Given enough time and isolation, some of these may have become separate species (Chapter 2). However, a powerful force worked against such separation: the extreme mobility and adaptability of the human species. Human populations moved very rapidly and easily breached geographic barriers that would have isolated many animal populations. For instance, it only took the mongoloid race (American Indians) a few thousand years to completely cover the entire Western Hemisphere. As a result, all human groups can interbreed with other humans to produce fertile offspring, proving that they all belong to one species (Chapter 2).

We need not rely only on the ease of interbreeding to show that human racial variation is minor. Direct genetic analysis of races is now possible with a number of techniques. One major study examined 150 protein and blood group genes and found that all races were the same for 75% of these. Similar studies have led geneticists to estimate that overall, when all 100,000 human genes are considered, about *85% of these genes are the same* in all races. The geneticist Richard Lewontin has said it this way: "If everyone on earth became extinct except for the Kikuyu (tribe) of East Africa, about 85% of all human variability would still be present in the reconstituted species."

THE FOUR BASIC RACES. In spite of our overall similarity, it is obviously true that regional differences in human populations do exist. While the race concept is useful to help classify this variation, there has been much confusion over just how many human races there are. The fact is that such attempts, like any classification process, involve arbitrary choices. For example, do we want to use skin color? If so, we find that groups with the same skin color vary greatly in other respects, such as blood factors, having more in common with populations of another color. However, by using multivariate techniques, which are statistical techniques that consider many traits at once, most scientists agree that four basic races can be usefully identified: negroid, australoid, mongoloid, and caucasoid. There is nothing rigid about this classification; it is simply a convenient way to generalize characteristics of human diversity.

Figure 3–84 shows the four basic races, with the branches being determined

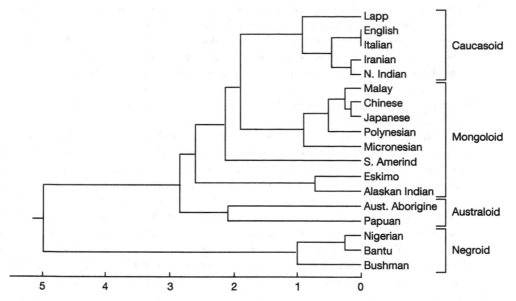

Figure 3–84. A tree diagram of human races based on genetic distances. This is only an estimate of racial separation. Modified from D. Futuyma, *Evolutionary Biology*, p. 523.

by multivariate analysis of genetic traits. If we consider each branching event to be the separation of one race from the rest, we see that the negroid line was the earliest race to branch off. As we see below, some workers theorize that they form the basic stock from which all humans arose. Australoids (such as the Australian "aborigine") are not negroid in spite of their dark skin. Rather, they are more closely related to the mongoloids. This is not surprising given that australoids apparently once inhabited the Asian continent and were pushed out by advancing mongoloid populations. In addition to dark skin, australoid traits include much body and facial hair, large teeth and jaws, and a bonier skull with a heavier brow ridge than other races. There are a number of carbon-14 dates of archaeological sites on Australia about 40,000 years old—this appears to be about the time that australoids reached that and nearby islands, migrating southward from Asia.

Mongoloids were so named because the race was believed to have originated in Mongolia, just as the australoids were thought to have originated in Australia. Prominent mongoloid traits include: thick, straight black hair; sparse body and facial hair; a broad face and the "epicanthic fold" (tissue around the eye giving it the so-called "slanted" appearance). Caucasoids were so named because they were thought to have originated in the Caucasus Mountains of Russia. Genetic analyses estimate that the caucasoids separated from their most closely related race, the mongoloid, about 40,000 years ago.

Aside from the four basic races, a number of subraces within them may be familiar to you. As shown in Figure 3–84, the Polynesians, the seafarers of the

Pacific (including Hawaii), are a branch of the mongoloid race. The American Indian (including the Eskimo) is likewise a mongoloid branch. The resemblance to Asian mongoloids is best seen in Indian children who sometimes have an epicanthic fold. Of course, straight black hair is also the rule in American Indians and mongoloids. This relationship is not surprising since, as noted above, the American Indian crossed into the Americas from Asia via the Bering Land Bridge, which connects Siberia to Alaska during low sea level. Such low sea levels were common during the glacial advances of the recent Ice Ages.

The earliest confirmed dates of archaeological sites in the Americas cluster around 11,000 to 12,000 years B.P. (before present). Thus, many researchers believe that this is roughly when the Indians arrived, but some argue for an earlier date. If the 11,000 to 12,000 year dates indicate the earliest occupations, it tells us a great deal about human migration and race formation because, surprisingly, archaeological sites in the southern part of South America have also been dated to roughly that time. This indicates that human populations are capable of very rapid movement into unpopulated areas. It has been estimated that it took less than 2,000 years to colonize the entire Western Hemisphere (based on traveling at a rate of 10 miles per year). Such rates are common in hunting and gathering societies today.

CAUSES OF RACIAL DIFFERENCES. Racial differences have two basic causes: (1) different selection pressures on the separated populations (races), and (2) random changes in the population gene pool, such as genetic drift. Regarding selection, it is often very difficult to determine the exact selection pressures that caused specific traits to become common in certain races. However, we can make reasonable inferences based on such evidence as medical data, which show how different traits confer advantages or disadvantages under given environmental conditions. For example, dark skin occurs because the person's skin cells produce a large amount of melanin. This pigment blocks solar radiation, which has a number of beneficial effects in tropical conditions. One, it reduces sunburn. Two, it reduces the incidence of skin cancer. Three, it prevents production of too much vitamin D. This vitamin is rare in most foods, and humans must rely on sunlight-driven chemical reactions in the skin to produce it. Without it, children develop rickets and other diseases because it is needed for calcium metabolism. However, production of too much can be harmful. This seems to explain why the negroid and australoid races, both tropical in much of their range, have dark skin. In contrast, caucasoids and mongoloids are less pigmented in skin (and sometimes hair) color. They inhabit colder, cloudier climates where there is less sunlight. Here, overexposure to sun is not a major danger. Instead, in order to synthesize enough vitamin D, it is good to block as little sunlight as possible. Medical data thus show that caucasoids suffer much more from sunburn and skin cancer in sunny hot climates while negroids have a much better chance of getting rickets in cold, cloudier climates.

Other racial traits have no demonstrable adaptive value and may be due to random processes such as genetic drift. Recall (Chapter 2) that genetic drift occurs when small populations become isolated. This can cause traits to become disproportionately common in the small population simply via random sampling processes.

Most important, although these traits may have little adaptive value, their high proportional genetic abundance can cause them to become widespread as the population grows. For example, blue eyes are common in some caucasoid populations even though they have little or no significant adaptative value. They are thought to have originated as a mutation in a population of northern Europe a few thousand years ago. The mutation causes de-pigmentation of the iris in the eye, causing reflected light to have a blue color for the same reason that reflected light makes the sky blue. Because this mutation is not significantly adaptive, it is thought to have spread throughout northern Europe through largely random processes.

THE EVE HYPOTHESIS: ALL RACES FROM ONE MOTHER? A major question that has been debated for many years is whether the living human races evolved (1) from the same parent stock of modern humans, or (2) from separate *H. erectus* populations. Evolution from the same parent stock is called the **out of Africa model** because this model states that modern humans evolved in Africa and migrated out of Africa about 100,000 years ago. As they migrated into Europe, Asia, and the Americas, they replaced archaic humans such as the Neanderthals. It was only after this migration that races formed. In contrast, the **multiregional model** states that modern human races evolved "in place" from separate populations of premodern humans, such as *H. erectus*. Using this model, human races would be much older than the age postulated by the out of Africa model.

Until recently, the main evidence for both models was from the fossil record. Because this record is highly fragmented, there was no promise of a clear resolution to the debate. However, new developments in DNA analysis have added another line of evidence. Analysis of DNA in modern races indicates that all humans may be related to the same African woman, who existed about 200,000 years ago. This so-called **Eve hypothesis** thus supports the out of Africa model. The analysis shows that DNA in the cell mitochondria of all humans, from mongoloids to negroids to caucasoids and others, is much more similar than would be expected if the races separated more than about 200,000 years ago. Mitochondrial DNA is used because it passes unchanged directly from the mother to the baby, unlike the DNA in sperm and egg cells, which is combined upon fertilzation. This makes it much easier to estimate mutation rates for the DNA. These act like a "clock" because DNA often mutates at a fairly constant rate. Thus, the differences in mitochondrial DNA among living humans can be used to estimate how long ago the races separated by indicating how many mutations have occurred. These results have been criticized and the DNA evidence (along with the fossil evidence) for races will likely be debated for years to come. However, at present, the general trend seems to be toward favoring the out of Africa model.

HUMAN CULTURAL EVOLUTION

Whereas **paleontology** is the study of fossils to reveal biological evolution, **archaeology** is the study of humanmade objects ("artifacts") to reveal cultural evolution. No one will ever know exactly when humans first began to modify sticks and many other available objects because such objects do not preserve well. Archaeologists have been able to more reliably study the durable stone artifacts. It is useful to distinguish such technological evolution from social evolution. Technological evolution involves changes in relation to our physical and biological environment. On the other hand, social evolution involves changes in social organization, affecting our relations with other humans. In both cases, evolution has been toward increasing complexity, toward more complex tools, and more complex societies. First, we briefly examine the evolution of technology.

Evolution of Technology

For many years, archeologists have divided human technological evolution into four basic stages: Paleolithic, Mesolithic, Neolithic, and Metal Ages. The first three stages refer to the "stone ages" (lithos = "stone" in Greek). As shown in Figure 3–85, the **Paleolithic** ("old Stone Age") is by far the longest, lasting from about 2.6 million years ago until about 10,000 years ago. During this time humans were hunters and gatherers, living off the land without domestication of plants and animals. Nevertheless, many cultural advances were made during this time: biface (two-sided) tools, fire, dwellings, and other achievements are shown in Figure 3–85. Notice that a number of changes in human biological evolution also occurred in the Paleolithic. It begins with *H. habilis* and goes up through early modern humans.

The **Neolithic** ("new Stone Age") was a time not only of more advanced stone tools but marks the beginning of plant and animal domestication, roughly 10,000 years ago. This time corresponds with the approximate end of the Ice Ages. Indeed, the climatic and ecological upheaval from this was probably a main factor leading to widespread adoption of domestication. This was truly a revolution in human evolution for two main reasons. One, the land can support many more people with agriculture and animal husbandry. Two, people no longer had to be mobile to gather and hunt. Both of these led to the formation of cities and thus civilization. The **Megolithic** ("middle" Stone Age") marks the short transition from hunting and gathering to domestication. The **Metal Ages** include the Bronze Age and Iron Age, when humans first began to work metals.

Social Evolution

From archaeological data and observation of other societies, anthropologists have created a classification scheme that roughly summarizes four stages of social evolution: band, tribe, chiefdom, and state. These are shown in Figure 3–86, along with the appearance of major social institutions. It is worth emphasizing that these stages, like all evolutionary progressions, are *not* meant to imply a progression toward "better" or "worse." This is simply a very approximate scheme showing changes in how societies became larger and organized in more complex ways.

The **band** was the basic social unit throughout the vast majority of human evolution. Hunters and gatherers move often in such bands and have little sense of

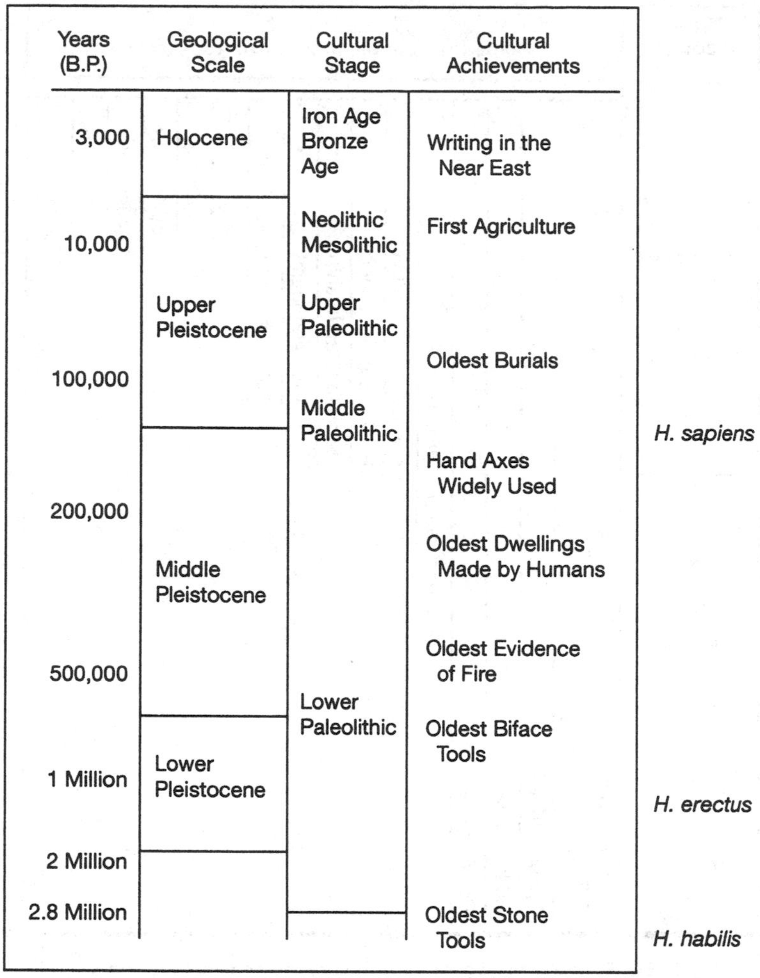

Years (B.P.)	Geological Scale	Cultural Stage	Cultural Achievements	
3,000	Holocene	Iron Age Bronze Age	Writing in the Near East	
10,000		Neolithic Mesolithic	First Agriculture	
100,000	Upper Pleistocene	Upper Paleolithic	Oldest Burials	
		Middle Paleolithic		*H. sapiens*
200,000			Hand Axes Widely Used	
	Middle Pleistocene		Oldest Dwellings Made by Humans	
500,000		Lower Paleolithic	Oldest Evidence of Fire	
1 Million	Lower Pleistocene		Oldest Biface Tools	*H. erectus*
2 Million				
2.8 Million			Oldest Stone Tools	*H. habilis*

Figure 3–85. Some major technological events in human prehistory. (B.P. = Before Present).

territoriality and private property. Usually there are no more than a few score people in bands, belonging to a number of families. The organization is egalitarian in that there is no elite or inherited differences in social status. Leadership is provided by force of personality and personal skills rather than being inherited.

Tribes form when population densities increase beyond that of hunter-gathering societies. Usually, such densities are associated with plant and animal domestication. Thus, the first tribes are known from the Neolithic of the Near East, not long after the origin of food production. Thereafter they appear in other food-producing centers. However, not all tribes formed in agricultural societies. Where

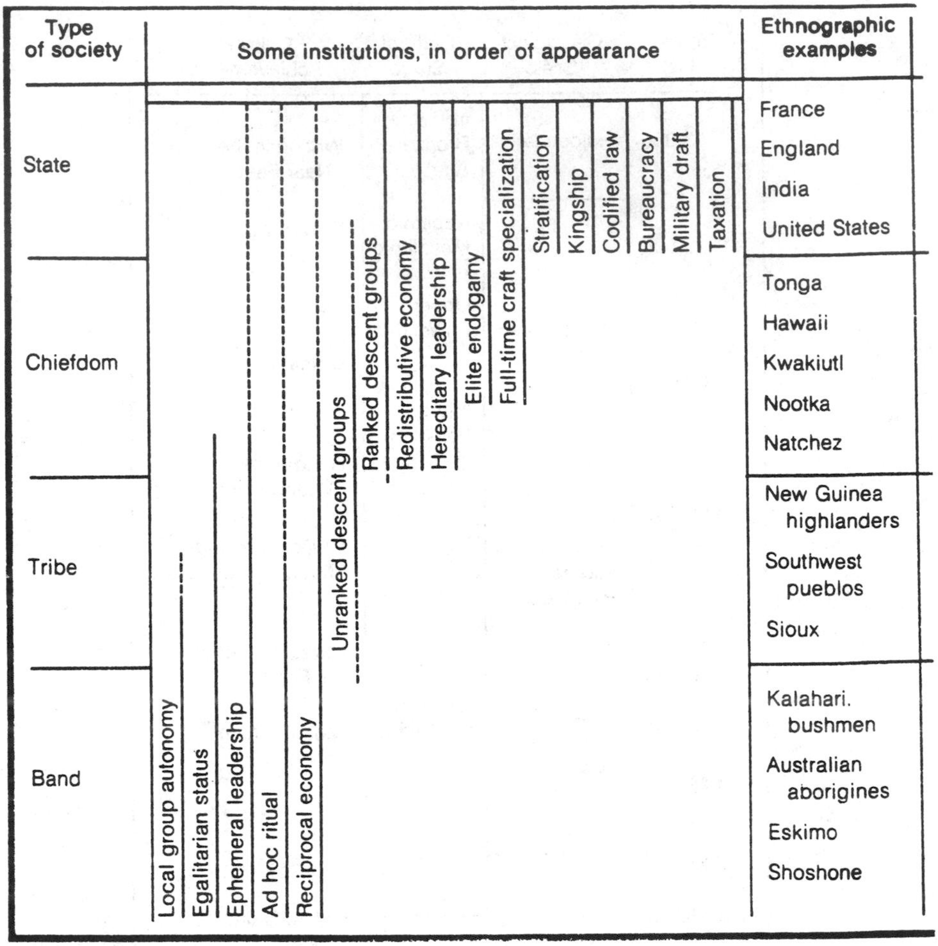

Type of society	Some institutions, in order of appearance	Ethnographic examples
State		France / England / India / United States
Chiefdom		Tonga / Hawaii / Kwakiutl / Nootka / Natchez
Tribe		New Guinea highlanders / Southwest pueblos / Sioux
Band		Kalahari bushmen / Australian aborigines / Eskimo / Shoshone

Institutions in order of appearance (Band → State): Local group autonomy, Egalitarian status, Ephemeral leadership, Ad hoc ritual, Reciprocal economy, Unranked descent groups, Ranked descent groups, Redistributive economy, Hereditary leadership, Elite endogamy, Full-time craft specialization, Stratification, Kingship, Codified law, Bureaucracy, Military draft, Taxation.

Figure 3–86. Four kinds of social organizations and institutions.

humans learned to intensely exploit a very rich natural resource, tribes also formed. Many Indians of the American Great Plains, such as the Sioux, are an example. Tribes are similar to bands in many ways, but are different in the tribe's larger size. Another difference is that group decisions tend to be made collectively by family leaders instead of a single leader.

Chiefdoms occur at still higher population densities, generally found only in agricultural societies. The earliest ones known evolved from some of the tribal villages of the Neolithic Near East. This process seems to have taken about 1,500 years. Chiefdoms show a number of major organizational changes from tribes, shown in Figure 3–86. Leadership becomes an inherited position. Generally, this position is religious in nature. That is, chiefdoms are often theocracies, which are societies governed by religious leaders. One of the most important changes from an economical point of view is craft specialization. For example, pottery specialists, or stone-working artisans, appear. However, they do this only on a part-time basis, with farming duties still being required.

States represent the largest social unit of modern societies. People live in cities instead of villages as population density increases to a maximum. Many of the changes seen in Figure 3–86 result from increased food production. For instance, this allows full-time specialization of artisans, soldiers, and other workers. It also leads to surplus accumulation of wealth, which becomes inequitably distributed. Leadership becomes less theocratic as practical knowledge accumulates and the mystique of the priesthood erodes. Laws, taxes, bureaucracy, and other mechanisms needed to sustain the state become instituted. The first known states occur in Egypt and the Near East (Mesopotamian area) about 5000 B.P.. In MesoAmerica they emerged about 2000 B.P..

SUMMARY

The **red shift** indicates an **expanding universe** in which all other galaxies are moving away from our own at a rate correlating to their distance as described by **Hubble's law.** This expansion is a result of the **Big Bang,** an explosion which created all of the matter and energy of the universe. Two eras occurred after the Big Bang: The **radiation era** describes the first hundred seconds of the universe and the **matter era** describes all time until present.

The **Milky Way** galaxy is about 12 billion years old, spiral in shape, and contains some hundred billion stars. These stars have three stages of life: contraction, main sequence, and fuel exhaustion. **Nuclear fusion** is the joining of hydrogen atoms to form helium, which occurs during the **main sequence** of a star. Small stars simply die out, while larger stars go through a red giant phase as they die, burning off by-products of fusion reactions. Novas and supernovas are the result of violent explosions, which occur after a large star collapses on itself, implodes, and ejects matter and energy into space. The remaining compressed matter may form a **neutron star.** The dense remains of supermassive stars are **black holes.**

The solar system originated from contractions of gas and dust clouds and elements created by explosive remnants of other stars. There are four inner planets—Mercury, Venus, Earth, and Mars—which are smaller and rockier than the remaining outer planets.

The earth originated 4.6 billion years ago. This age was partly determined by using three laws that allow ordering of geologic events. These laws are **superposition, original horizontality,** and **lateral continuity. Stratigraphy** is the study of rock layers. **Correlation** of these layers over great distances allows reconstruction of past geologic events. **Biostratigraphy** is the study of fossils in relation to these layers. There are two types of fossils, trace fossils and body fossils. The study of fossils is called **paleontology.**

The ages of geologic events were placed in chronological order (**relative age**) using these methods, but an exact age of events could not be determined until the discovery of radioactivity. Using radioactive dating techniques, an exact age in years (**absolute age**) could be determined. This allowed a geologic timetable to be compiled, which is called the Geologic Time Scale. The time scale is divided into four major eras—the **Precambrian, the Mesozoic,** the **Paleozoic,** and the **Cenozoic.** Each of these eras is subdivided into periods.

The earth can be divided into four spheres: the lithosphere, or solid portion of the earth; the hydrosphere, or the water of the earth; the atmosphere, or the air surrounding the earth; and the biosphere, or the life on earth. The lithosphere is composed of three layers—the **crust, mantle,** and **core,** which were formed by **differentiation.** The process by which convection causes movement of plates on top of the "fluid" mantle is called **plate tectonics.** Plate tectonics has created continents and changed the shape of continents and seas over time. Two types of plate boundaries are **divergent** and **convergent** boundaries.

The **hydrosphere** was formed by **outgassing,** or the expulsion of water from volcanoes. This led to the formation of **epicontinental seas,** which were warm, shallow seas covering much of the continents in the early Paleozoic.

The **atmosphere** was also created by outgassing and is held to the earth by gravitational forces. There have been three distinct atmospheres in earth's history. The initial atmosphere was hydrogen rich and the second nitrogen rich. The third is oxygen rich and was created by the release of oxygen by plants from photosynthesis.

There are three steps in the evolution of increasing complexity: the rise of **prokaryotes,** or simple, single-celled organisms (such as bacteria and blue-green algae); the rise of **eukaryotes,** or complex, single-celled organisms containing a nucleus and specialized organelles; and the rise of multicellular organisms (such as metaphytes and metazoans).

Eukaryotes are thought to have formed as a result of symbiosis between prokaryotic organisms. Multicellular organisms formed as a result of symbiosis between single-celled organisms (**symbiotic hypothesis**) or by reproduction of cells that stayed together in a colony (**colonial hypothesis**).

Evolution of life in the hydrosphere occurred in three stages: the rise of sessile invertebrates (such as stalked echinoderms, corals, bryozoa, and brachiopods) and fish; the rise of mobile invertebrates (such as sea urchins, clams, crustaceans, and cephalopods) and marine reptiles; and the rise of marine mammals (such as whales and porpoises). The transition from sessile to mobile organisms is known as the mobile marine revolution.

By the mid-Paleozoic four groups of fish were common: **jawless fish, placoderms, sharks,** and **bony fish.** Fish were the first vertebrates and originated as a result of heterochrony, or the change in timing of development. There were also four groups of marine reptiles: **ichthyosaurs, plesiosaurs, mosasaurs,** and **turtles.**

Evolution on land required organisms to overcome the problems of drying out, higher gravity, and the need to breathe oxygen. There were three stages of the evolution of life on land: the rise of seedless plants and amphibians, the rise of seed plants and reptiles, and the rise of flowering seed plants and mammals.

There were four types of seedless plants in the Paleozoic **psilopsids, ferns, sphenopsids,** and **lycopsids.** These plants exhibited alternating generations and required moisture to survive. The first invertebrates on land were arthropods such as millipedes and scorpions. The first vertebrates on land were the amphibians, the **labyrinthodonts.** However, amphibians never fully adapted to land because they required water to reproduce.

The second radiation of life on land was the nonflowering seed plants, or **gymnosperms,** which had a single reproductive generation and seeds instead of spores. Four types were common in the Mesozoic: **seed ferns, cycads, ginkoes,** and **conifers.** The first reptiles were archaic reptiles that had many reptilian features (for example, a hard-shelled egg, which allowed them to fully adapt to land). These primitive reptiles gave rise to three groups: **mammal-like reptiles, nondinosaurs** (such as turtles, lizards and snakes, and crocodiles), and **dinosaurs.**

Dinosaurs arose from reptiles called **thecodonts.** There were two major groups of dinosaurs: the **lizard-hipped,** which include **sauropods** and **theropods,** and the **bird-hipped,** which include the **ornithopods, stegosaur, ceratopsians,** and **ankylosaurs.** Although many dinosaurs were quite large due to warm, stable climate and low metabolism, some were small, fast, agile, and moderately intelligent.

There are several theories regarding the extinction of dinosaurs. The most accepted theory is that they were killed in a mass extinction resulting from a meteorite impact. The discovery of the **iridium spike** and **shocked minerals** supports this theory. The extinction of the dinosaur caused an **ecological release,** which enabled mammals to thrive.

The third radiation of life on land was the flowering seed plants, or **angiosperms.** These plants were more reproductively efficient, relying on animals to carry pollen from the stamen to the pistil. Mammals arose from mammallike reptiles and had mammalian features such as specialized teeth, hair, and constant body temperatures. Three groups of mammals arose in the Mesozoic: **monotremes, marsupials,** and **placentals.**

Evolution to air required two key adaptations: light bodies and wings. There are four groups that have evolved flight: insects, reptiles, birds, and mammals. The evolution of flight occurred in three stages: the rise of primitive insects in the Paleozic, the dominance of advanced insects in the Mesozoic, and the dominance of birds and mammals in the Cenozoic.

Insects are the most abundant and diverse organisms due to their small size and ability to modify development easily. Their size is limited due to their **exoskeleton.** The first reptiles capable of flight were the **pterosaurs.** Birds evolved from reptiles by the mid-Mesozoic. Evidence of this is the first bird seen in the fossil record, the **Archaeopteryx,** which is an example of a "missing link." Bats are the most able mammalian fliers; they have developed echolocation to aid them in flying and catching food.

There are four types of primates: **prosimians, monkeys, apes,** and **humans.** Apes are the closest human relatives with chimpanzees and humans having 99% of their genes in common.

The earliest human ancestor was *Australopithecus.* The species *Australopithecus africanus* eventually led to modern humans. The next human ancestor was *Homo habilis; H. habilis* used stone tools. The next to evolve was *Homo erectus,* known for standing upright. This species was followed by *Homo sapiens.* Neanderthals were early modern humans.

There are four primary races: **negroid, australoid, mongoloid,** and **caucasoid,** with racial differences resulting from different selective pressures on the separate populations and random changes in the gene pool.

Did modern humans evolve from a single parent ancestor or from different populations of *H. erectus*? There are two theories regarding this point of human evolution. First is the **out of Africa model,** which argues for the first case and is supported by the **Eve hypothesis.** Second is the **multiregional model,** which argues that modern humans arose from different populations of *H. erectus.*

There are four stages of cultural evolution: **paleolithic,** in which was a society of hunters and gatherers; **mesolithic,** in which was a transitional period; **neolithic,** in which began agriculture; and **metal ages,** in which humans worked metals.

There are four stages of social evolution: the **band, tribe, chiefdom,** and **state.** These do not necessarily represent a trend toward "improvement," but rather toward more organization and social institutions.

KEY TERMS

red shift	fossils
expanding universe	biostratigraphy
Hubble's law	relative time
radiation era	absolute time
matter era	half-life
Milky Way	radiometric dating
nuclear fusion	geologic time scale
main sequence	era
red giant stage	biosphere
nova	lithosphere
supernova	differentiation
neutron stars	crust
black holes	mantle
inner planets	core
outer planets	convection
law of superposition	plate tectonics
law of original horizontality	divergent boundaries
law of original lateral continuity	subduction boundaries
stratigraphy	Pangea
correlation	hydrosphere

outgassing
epicontinental seas
atmosphere
photosynthesis
Milankovich cycle
sedimentary rocks
paleontologist
body fossil
trace fossil
pull of the recent
prokaryotes
eukaryotes
stromatolites
ecosystem
primary producer
consumer
symbiosis theory
symbiotic hypothesis
colonial hypothesis
metaphytes
metazoans
Ediacara Formation
phyla
invertebrates
sessile
stalked echinoderm
brachiopod
coral
bryozoans
trilobite
vertebrates
jawless fishes
placoderm
shark
bony fishes
ray-finned fishes
lobe-finned fishes
heterochrony
sea urchin
clam
snail
crustacean
cephalopod
mollusk
Mesozoic marine revolution
arms race
ichthyosaur
convergent evolution
plesiosaur

mosasaur
whale
porpoise
green algae
vascular system
spore
gametophyte
sporophyte
psilopsids
fern
sphenopsids
lycopsids
coal
millipedes
scorpions
exoskeleton
book gills
labyrinthodont
gymnosperms
seed
seed fern
cycads
ginkoes
conifers
archaic reptiles
hard-shelled egg
mammal-like reptiles
nondinosaurs
dinosaur
thecodont
lizard-hipped dinosaur
bird-hipped dinosaur
theropod
sauropod
Ceratopsians
stegosaur
ankylosaur
ornithopod
ecological release
iridium
shocked minerals
nuclear winter
extinction selection
angiosperms
stamen
pistil
monotreme
marsupial
placental

insectivores	apes
archaic ungulates	*Australopithecus*
odd-toed ungulates	*robustus*
even-toed ungulates	*Homo habilis*
megafauna	*Homo erectus*
Bergmann's rule	*Homo sapiens*
chitin	neanderthals
coevolution	out of Africa model
metamorphosis	multiregional model
molts	Eve hypothesis
endoskeleton	paleontology
lungs	archaeology
pterosaurs	Paleolithic
rhamphoryncoids	Neolithic
pterodactyloids	Megolithic
echolocation	Metal Ages
primates	band
stereovision	tribe
prosimians	chiefdom
monkeys	state

REVIEW QUESTIONS

Objective Questions

1. What is red shift? How does it indicate an expanding universe? What is Hubble's law?
2. Briefly describe the three stages in the life of a star.
3. Describe the following: nova, supernova, neutron star, black hole.
4. What are the four inner planets? How do they differ in composition from the outer planets?
5. What are the three laws of stratigraphy? How are they used? (Refer to Figure 3–11.)
6. What is the difference between absolute time and relative time?
7. What are the four spheres that categorize the world?
8. What three layers make up the earth? How was this layering created?
9. Why is convection a possible mechanism for plate tectonics?
10. What is Pangea? How was it formed? During what time period did it exist? (Refer to Figure 3–21.)
11. What are fossils? Describe the two types of fossils.
12. What are some prokaryotic features? What are some eukaryotic features? (Refer to Figure 3–30.)
13. Compare and contrast the symbiotic and colonial hypotheses.

14. What is unique about the Ediacara Formation?
15. What are the three stages of the evolution of life in the oceans?
16. Briefly describe each of the four types of fish. (Refer to Figures 3–41 and 3–42.)
17. Define heterochrony.
18. What three problems were faced by life evolving on land?
19. Describe the life cycle of seedless plants. In what two ways does it differ from that of seed plants? (Refer to Figure 3–50.)
20. How is coal formed?
21. Why did amphibians never completely adapt to land?
22. What three modestly successful groups make up the nondinosaurs?
23. Briefly describe the following groups: theropods, sauropods, ornithopods, horned dinosaurs, stegasaurs, ankylsaurs.
24. Compare and contrast the reproductive strategies of gymnosperms and angiosperms. How do their seeds differ? (Refer to Figure 3–65.)
25. Describe several mammalian traits that allowed mammals to flourish and diversify.
26. Give an example of an odd-toed and an even-toed ungulate. Which group is most successful?
27. How do advanced insects and flowering plants represent an example of coevolution?
28. Give two reasons why the exoskeleton is the limiting factor of insect size.
29. What are four primate traits?
30. What are the four stages in the evolution of human technology? Which stage lasted the longest?

Discussion Questions

1. How is the Eve hypothesis related to the out of Africa model?
2. If you start with 10 grams of a radioactive substance that has a half-life of 20 years, how much would be left after 10 years? after 20 years? after 40 years?
3. What problems led to the incompleteness of the fossil record? What two factors promote fossilization? How does this lead to biased preservation? How does the pull of the recent distort our perception of the fossil record?
4. How does the "arms race" analogy reflect coevolution between predator and prey species? Explain and give examples.
5. There are five prominent theories regarding the extinction of the dinosaurs. Which theory does the discovery of the iridium spike support? Explain.
6. How does the earth today compare with the past? Explain in terms of sea level, global temperature, and life forms.

TABLE 3–4. Origin of Names of Eras and Periods in the Geologic Time Scale

GEOLOGIC TIME SCALE		
Life Eras	Geologic Periods	Origin of Names
Cenozoic (Recent Life)	Quaternary	Fourth group of strata
	Holocene	"Whole recent"
	Pleistocene	"Most recent"
	Tertiary	Third group of strata
	Pliocene	"More recent"
	Miocene	"Lesser recent"
	Oligocene	"Few recent"
	Eocene	"Dawn recent"
	Paleocene	"Ancient recent"
Mesozoic (Middle Life)	Cretaceous	Chalk deposits, England and France
	Jurassic	Jura Mountains, Switzerland
	Triassic	Threefold division of rocks, Germany
Paleozoic (Ancient Life)	Permian	Perm province, USSR
	Pennsylvanian	Pennsylvania, USA
	Mississippian	Mississippi Valley, USA
	Devonian	Devonshire, Britain
	Silurian	Silures, Celtic tribe in Wales
	Ordovician	Ordovices, Celtic tribe in North Wales
	Cambrian	Cambria, Roman name for Wales
Precambrian		Before the Cambrian
Proterozoic		First life
Archeozoic		Archaic life

7. In Table 3–4, how does the origin of names for Cenozoic periods differ from the origins of Paleozoic and Mesozoic periods? What do you suppose "-cene" means? Why might periods be named after geographic locations?

SUGGESTED READINGS

On the history of the universe and solar system

CHAISSON, E. 1988. *Universe: An Evolutionary Approach to Astronomy.* Prentice-Hall, Englewood Cliffs, NJ.

On the history of earth

CLOUD, P. 1988. *Oasis in Space: Earth History from the Beginning.* W.W. Norton, NY.
LEVIN, H. 1988. *The Earth Through Time.* Saunders, NY.

On the history of life

COWAN, R. 1990. *History of Life*. Blackwell, Cambridge, MA.
LANE, N. 1986. *Life of the Past*. Merrill, Columbus, OH.
MCALESTER, A. 1977. *The History of Life*. Prentice-Hall, Englewood Cliffs, NJ.

On dinosaurs

NORMAN, D. 1991. *Dinosaur!* Prentice-Hall, Englewood Cliffs, NJ.

On the history of fossil humans

LEWIN, R. 1989. *Human Evolution*. Blackwell, Cambridge, MA.
NELSON, H., AND JURMAIN, R. 1991. *Introduction to Physical Anthropology*. West, St. Paul, Minn.

4

Evolution Past: Patterns in the History of Life

OVERVIEW: RATES AND DIRECTIONS OF EVOLUTION

Having outlined the history of earth, life, and humans, let us now see what patterns have been uncovered by scientists, who have analyzed this history more closely. This search for consistencies is extremely important in rendering the past as more than just a sequence of contingent events. It is only by searching for regularities ("patterns") that we can understand the operation of evolutionary processes (Chapter 2) over long periods of time.

Because evolution is change through time, it is convenient to subdivide its patterns into two parts: rate and direction of that change. While most of us intuitively know these concepts well, it is worth being more precise. As shown in Figure 4–1, **rate** may be defined as the amount of change (Δ) in some phenomenon ("variable") over some period of time. In graphical terms, this means that rate is the *slope* of the line tracing the phenomenon through time. If the line is straight, the rate is constant. If the line is curved, then rate is changing. For instance, in Figure 4–1, rate is steadily increasing in curve A and decreasing in B. In curve C, the rate varies. **Direction** may be defined as the tendency of a phenomenon to persistently increase or decrease over the long term. For instance, curve C (Figure 4–1) has no long-term direction in spite of undergoing considerable change because the change has no persistent tendencies. In contrast, curves A and B both show such directional tendencies. A common word for such a persistent tendency is *trend*.

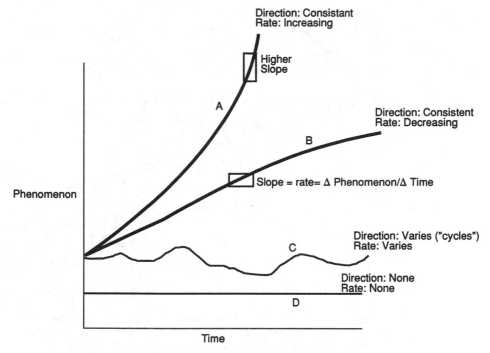

Figure 4–1. Graphical view of direction and rate of change.

Both rate and direction are usually *scale dependent*: Our view of rate of change and direction of change depends on the scale (interval) over which we observe the phenomenon. Figure 4–2 shows how a trend that appears over a long scale may not be visible when we view only a short time span. The main reason for scale dependence is that change is rarely constant at all scales. In other words, reversals and fluctuations occur at different scales. For instance, the stock market (Dow Jones Average) may show an upward trend that lasts for years, followed by a downward trend of similar length. Superimposed within those trends are many upward and downward trends, each lasting for only a few weeks.

PATTERNS OF PHYSICAL EVOLUTION

DIRECTIONS AND RATES IN EARTH'S EVOLUTION

Lithosphere

Recall from Chapter 3 that the various plates on the earth's surface brought the continents together in the late Paleozoic. However, this was only the most recent unification of plates; this has happened a number of times in the past. Thus, the

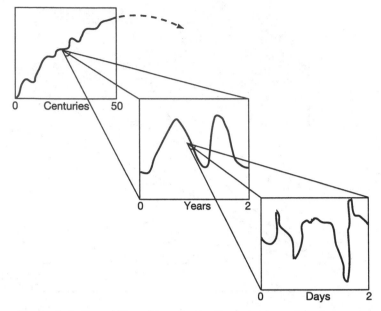

Figure 4–2. Perception of trends usually depends on how long we observe change. Modified from R. O'Neill et al., *A Hierarchical Concept of Ecosystems* (Princeton: Princeton University, 1986), p. 84.

motion of plates, to and fro, across the face of the earth has been effectively "random" and without persistent direction when geological time scales of hundreds of millions of years are considered. On a smaller time scale of a few million years, this plate motion would appear to have nonrandom trends, such as Europe persistently moving away from North America since the Mesozoic. Ultimately, this movement will change, illustrating the scale dependence of trends noted above (Figure 4–2).

In addition, the rate of plate motion has decreased, as the internal heat of the earth has decreased. As shown in Figure 4–3, the rate of heat flow decrease has been exponential. This is because the heat is generated by radioactive decay of minerals within the earth and many of the shorter-lived isotopes have lost most of their radioactivity.

Atmosphere and Hydrosphere

We saw in Chapter 3 how global temperature and sea level have fluctuated. There have been trends, such as cooling temperatures during the Cenozoic, for instance. But in the very long run, a cyclical return toward warmer temperatures can be expected, illustrating again the scale-dependence of many trends. However, unlike the lithosphere, there is no slowing of the rate at which these cycles occur. This is because most atmospheric and oceanic processes are driven by the "external engine" of the sun, which has not yet begun to cool down because it has much more fuel

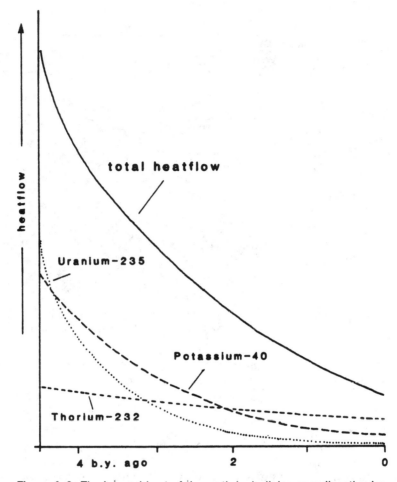

Figure 4–3. The internal heat of the earth is declining as radioactive isotopes decay. From T. van Andel *New Views on an Old Planet* (London: Cambridge University, 1985), p. 200.

than the internal earth. Indeed, most astronomers believe that the sun has become hotter over the last 5 billion years.

Summary: the Cyclical Earth

The cyclical nature of the earth is illustrated in Figure 4–4. Matter moves in all of the spheres, driven by external energies from the sun and internal heat from within the earth. In addition to slowing rates of cycles driven by the internal heat in the earth, the rotation of the earth on its axis is slowing as well. The proximity of our relatively large moon has caused friction from oceanic tides to slow our rotation so

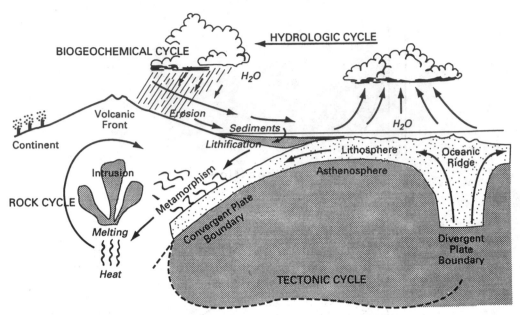

Figure 4–4. Some of the many cycles on earth, driven by the earth's internal heat and solar heat. From D. Botkin and E. Keller, *Environmental Studies* (Columbus, OH: Merrill, 1987), p. 39.

that the days are becoming longer. For instance, in the Paleozoic, a day was a few hours shorter than now. In contrast, the hydrologic cycle, being driven largely by the increasingly hot sun, may have increased in rate of cycling.

THE SUN AND UNIVERSE

The sun is an average star with enough hydrogen fuel to last about another 5 billion years. Thus, it will be a long time before it begins to die. The current status of the universe itself however is much less certain. Figure 4–5 shows that since the Big Bang, the universe has grown in size, but that this expansion will begin to slow—if it has not already begun to do so. (Astronomers are not sure at this time.) Some astronomers believe that the universe will eventually contract, with all matter being drawn to a common center. Others think it will expand forever (Chapter 5).

PATTERNS OF BIOLOGICAL EVOLUTION

Patterns of biological evolution can be conveniently subdivided into patterns of origination and patterns of extinction. **Origination** involves the creation of new species or other groups, while **extinction** is the opposite, the termination of species or groups. Between them, these represent the only two evolutionary alternatives to

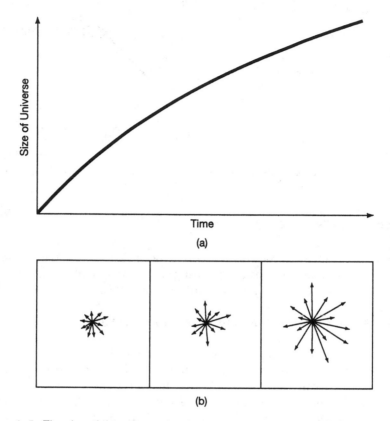

Figure 4–5. The size of the universe has been expanding since the Big Bang.

environmental change: Either a species adapts by evolving, or it becomes extinct. In terms of our equation used in Chapter 2:

Variation + Environmental Change = Evolution or Extinction

As shown in Figure 4–6, we would say that evolution (origination) occurs when existing variation is sufficient to adapt to environmental change. Extinction occurs where it is not.

Both origination and extinction show patterns of rate and direction. This section has three parts:

Part 1: Origination Rates
Part 2: Origination Directions
Part 3: Extinction Rates and Directions

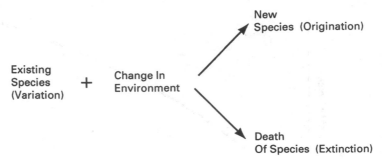

Figure 4–6. Environmental change eventually results in either origination or extinction.

PART 1: ORIGINATION RATES

First, we discuss the measurement of evolutionary rates. We will see that evolution can be either fast or slow, depending on a number of variables. Then we look at what causes these variations in evolutionary rates.

Measurement of Origination Rates from Fossils

In measuring evolutionary rates, remember that evolutionary change involves change on at least two main levels: the individual organism and the gene. In other words, if we want to measure change through time in a species, we can look either directly at change in the anatomy of the organisms themselves, or changes in the gene pool. Because the gene is the actual "blueprint," it seems most desirable to measure rates of change in the gene pool. Unfortunately, this is almost never possible because genes are not preserved as fossils. Instead, we must usually content ourselves with measuring changes in those anatomical aspects of fossils that are preserved. Such direct measurement of an anatomical trait through time is called, appropriately enough, an **anatomical rate.** A second kind of rate is the **group origination rate.** Here, instead of measuring anatomical change in one or a few traits, evolutionists measure the rate at which new groups (such as species) appeared in the fossil record. This still involves changes in anatomy, since new groups are defined by such changes, but a larger scale of change, involving many species, is the focus. First, we turn to a discussion of anatomical rates.

ANATOMICAL RATES. Figure 4–7 shows an example of such anatomical rates in a shelled invertebrate (brachiopod) from the early Paleozoic. The measured trait in this case are the "ribs," which are the ridges strengthening the shell. At about 415 million years ago (as determined by radiometric dating), the ratio of rib height to rib width was quite high (nearly 50%), indicating that the ribs were prominent. However, as time passed (in other words, as we go "upsection" in the rocks where these fossils are found), the ratio decreased. By about 10 million years later (405 million years ago) the rib height was very small and the shell was nearly smooth.

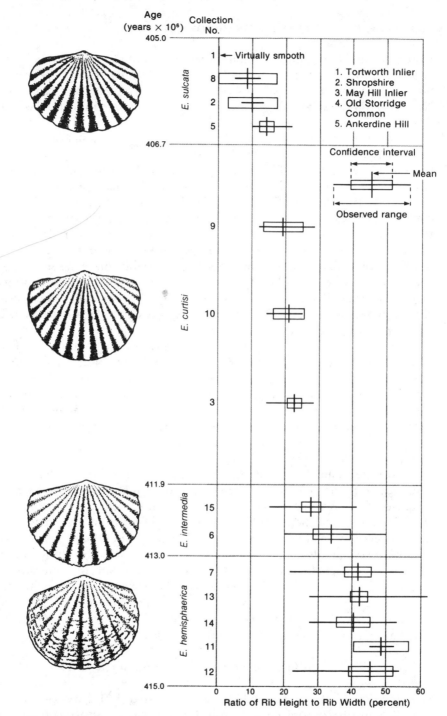

Figure 4–7. Populations of these fossil brachiopods tend to have smoother shells through time. Vertical bar is population average and horizontal bar is range. Modified from T. Dobzhansky et al., *Evolution* (New York: Freeman, 1977), p. 329.

Note that this is an example of nonbranching evolution (Chapter 2). Paleontologists classifying this lineage have defined four species, using anatomical changes because we cannot know if these species were reproductively isolated from one another. Note also that these measurements are not taken on one individual of each species. The main interest is what the species' average for the trait was. This gives a much better idea of changes in the gene pool than measurements of just one or a few individuals.

The main point of Figure 4–7 is that, for this trait, the *anatomical rate varies through time.* During the first 2 million years, the earliest species shows only minor, fluctuating change. This fluctuation could truly reflect the species' evolution, or it could be simple sampling error (statistical "noise"). On the other hand, within the few hundred thousand years following 413 million years ago, there is a relatively fast rate of change, as the second species comes into being. There are huge gaps here, on the order of tens or hundreds of thousands of years, so we cannot really specify if the change took any less time than that. Similar observations could be made for other changes later on in the lineage. Note the very long period of little change between about 412 and 406 million years ago. Such periods of little or no evolutionary change are called **stasis.**

A second example, shown in Figure 4–8, is the tooth area of a primitive mammal (early Cenozoic) from Wyoming. Teeth are often used because they are the most common mammalian fossil and tooth area is a good indicator of body size (and diet) in mammals. This example differs from the one above because it shows branching evolution, but the same method of measuring anatomical rate can be used. For instance, once again we see varying rates of change as we go upsection toward younger fossils: periods of little change in average tooth size interspersed with periods of more rapid change.

Another important difference between Figures 4–7 and 4–8 is that the investigator studying the mammals in Figure 4–8 was unable to date the rocks as precisely as those of the brachiopods of Figure 4–7. This is not uncommon, since marine rocks have more of the microfossils that are useful for dating. In addition, marine deposition tends to be more continuous. This illustrates the key distinction between relative versus absolute rates of evolution. *Absolute rates* of evolution are seen when we have adequate absolute dating of rocks and can therefore measure the rate of anatomical change per unit time. Looking at Figure 4–7 we can say that the absolute rate of change in the rib ratio was, for instance, about 2.5% per million years between 415 and 413 million years ago. This is because the ratio went from about 45% to 40% in that time period. Thus, we calculate:

$$(5\%)/2 \text{ million years} = 2.5\%/\text{million years.}$$

Similar calculations can be made for any time span in Figure 4–7.

In contrast, *relative rates* of evolution are seen when we compare anatomical change at different times, but we cannot measure the time intervals involved. For instance, in Figure 4–8 we can observe that evolution was very slow for the species *H. loomisi* that occurs very low in the rock unit (200 feet). However, the relative rate increases rapidly at about the 400 foot mark, where splitting occurs and there is

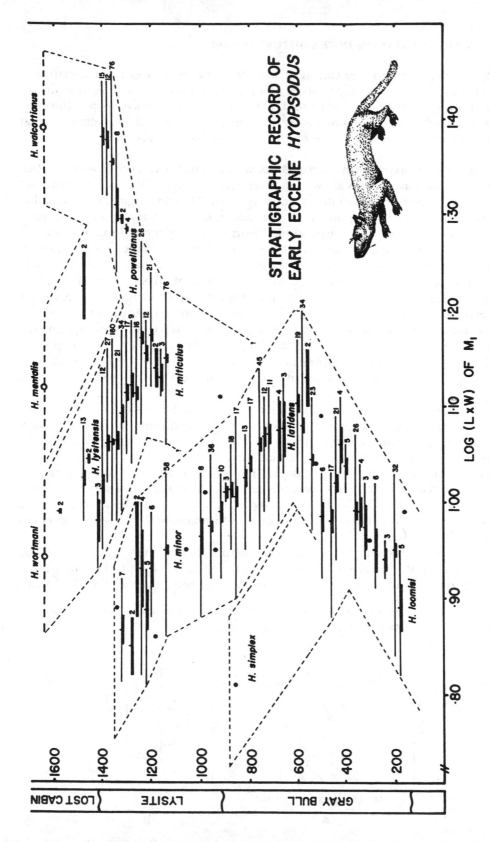

Figure 4–8. Change in tooth (first lower molar) shape (length × width) in the branching evolution of a fossil mammal of Wyoming. Heavy bar = most specimens in sample, light bar = total range in sample. Rocks not precisely dated, but total thickness of rocks shown spans about five million years.

evidence of rapid tooth area change. Time intervals cannot be estimated accurately from rock thickness (as in Figure 4–8). It is indeed possible to use rates of sediment deposition to estimate how long it took a certain rock thickness to form. However, the rate of sediment deposition varies so greatly over time and between areas that such estimates are regarded with much skepticism by geologists.

THE GREAT RATE DEBATE: PUNCTUATED OR GRADUAL EVOLUTION? Evidence from many fossil studies (such as the two above) has given rise to a major debate over whether evolutionary rates are generally rapid or gradual. Darwin originally argued that evolution was usually a gradual process and most evolutionary theorists have agreed. However, in recent years, some theorists have suggested that evolution is more often a very rapid process. This theory is called **punctuated equilibrium.** It states that during most of a species' existence, there is virtually no evolutionary change (stasis). Instead, most change occurs in very brief bursts, "punctuating" the long-term stasis (equilibrium). As shown in Figure 4–9B, change between species is therefore concentrated in the speciation events. In contrast, the theory of **gradualistic evolution** holds that evolutionary change is not limited to any specific punctuations but can occur at any time and in small steps. As shown in Figure 4–9A, gradual evolution gives the overall impression that evolution is a slowly branching tree, compared to the punctuated view that evolution is more like an upwardly sprouting bush.

The debate between these two schools continues. Sometimes it is mistakenly implied that this debate means that biological evolution is being questioned. This is *not true.* Virtually without exception, all evolutionary biologists agree that Darwin's basic ideas of natural selection as the cause of evolution are correct. Further, they agree that evolutionary rates vary, depending on a number of circumstances. The real debate is over which rates are more common in creating new species: rates of rapid or gradual change. This involves more than rates alone because, as we will discuss, punctuated patterns may be caused by fundamentally different processes than gradual ones.

Resolution of such a debate may seem simple enough. Why not simply add up the many fossil studies already done (such as those cited above) and see how many show gradual versus punctuated patterns? This has been done and we find that a

Figure 4–9. Branching evolution in (A) gradual and (B) punctuated patterns. Modified from D. Futuyma, *Evolutionary Biology* (Sunderland, MA: Sinauer, 1986), p. 402.

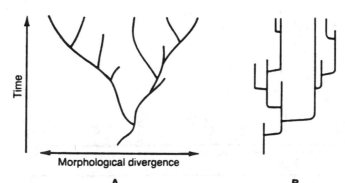

Time

Morphological divergence

A   B

great many (perhaps the large majority of) fossil examples show the punctuated pattern. There is little change, often for millions of years, and then a relatively rapid change. Unfortunately, there have been two basic problems impeding such a direct (and therefore desirable) interpretation of the data: (1) gaps in the fossil record, and (2) disagreements over definitions of the term "rapid." The first problem, as Darwin himself noted, is that the fossil record is very incomplete. There are always long periods of time, often hundreds of thousands or even millions of years, when no deposition occurred or was eroded away. This means that we only get "snapshots" of any trait at limited periods of time. As shown in Figure 4–10, gradual evolution in such cases may look punctuated: If intermediate forms are not preserved, then the trait may appear to "jump" quite rapidly. Complicating this greatly

Figure 4–10. Resolving power of the fossil record has great influence on whether we perceive the same event to be rapid (top) or slow (bottom). From W. Berggren and J. van Couvering, Eds., *Catastrophes and Earth History* (Princeton: Princeton University, 1984), p. 84.

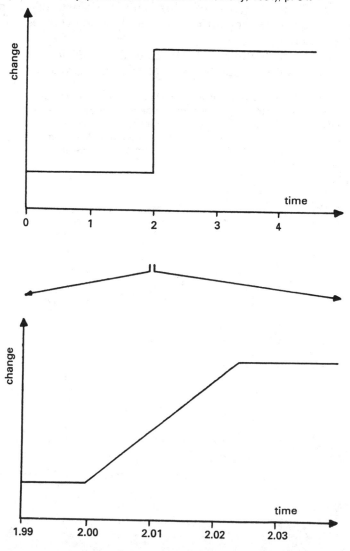

is the difficulty of dating sedimentary rocks. If dating were easy, then we could at least date the "snapshots" and know how much time had passed between them. As it is, we usually cannot say exactly how long the gaps were. If they were short, perhaps the lack of intermediates was real, and evolution actually was rapid.

The second problem is defining what is meant by "rapid." Evolutionary scientists often study evolution at different scales. For instance, to paleontologists, who see little change in fossil lineages for millions of years, a change that occurs in a few thousand years is rapid. Yet geneticists working with fruit flies under intense selection in the laboratory often see major anatomical changes in populations in a few years or even months. Therefore, geneticists tend to argue that a few thousand years is actually very gradual and that Darwin was thus correct. However, this is more than just a problem of perception. Figure 4–11 shows that differences in evolutionary rates do exist, depending on the interval of time studied. Short intervals, studied by geneticists, yield much higher rates than long intervals of fossil change, studied by paleontologists. This is because longer time intervals, as noted above, allow more time for reversals and periods of stasis that decrease the average rate of change.

At this time, the rate debate continues with no sign of a resolution. The only sure way of solving both the gap and definition problems is to obtain precise dating

Figure 4–11. Inverse correlation between anatomical evolutionary rate versus time interval over which rate is measured. I = lab experiments, II = historical events, III = recent fossils, IV = older fossils. Modified from D. Briggs, and P. Crowther, *Palaeobiology* (London: Blackwell Science, 1990), p. 156.

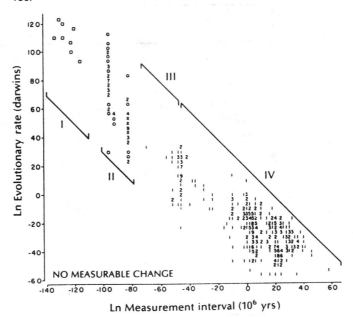

on a large number of fossil lineages. Only then can we say with certainty whether punctuations result from long gaps in the record and specify just how rapid the evolutionary changes were (if they are not from gaps). Unfortunately, such dating is not easy to obtain, although a number of promising studies have been reported recently that trace fossil lineages of microscopic plankton in deep-sea sediments, where deposition is much less disturbed, continuous, and more easily dated. Yet even these studies raise many questions because many of them show both rapid and slow rates of change. Furthermore, even if deep-sea plankton clearly showed mainly punctuated or gradual patterns, it is not at all certain that the same evolutionary patterns would hold true for other organisms, such as mammals or clams, who live in much different environments and who follow different patterns of living and growing.

CORRELATED TRAITS VERUS MOSAIC EVOLUTION. So far we have considered only evolutionary rates in one or two anatomical traits. Shell ribs and tooth area are only examples of the huge number of isolated traits that could be measured on a fossil sequence. However, when a species evolves, many traits are usually altered together. These are called **correlated traits,** defined as groups of traits that tend to evolve together. Correlation is especially common among traits that must function in unison. For instance, if chewing teeth in the lower jaw increase in size, then chewing teeth in the upper jaw must increase as well. The correlation of brain size with body size in human evolution is a good example. This occurs because larger bodies require larger nervous systems and more sensory information. In general, scientists have found that most traits are correlated to at least some degree: Increasing the size of some anatomical part usually corresponds with increased size of most others. This is because most anatomical parts interact with all others, at least in some indirect way. Where such traits are correlated, anatomical rates of evolution will tend to be correlated as well. Thus, rapid evolution of tooth size in the mammal example above would tend to accompany rapid evolution of body size and other parts. (Larger tooth size permits more food to be chewed for the larger bodies).

However, there are important exceptions to the positive correlation of traits in that some traits are uncorrelated, or even inversely correlated. Usually these are of great interest to evolutionists because they indicate the direct action of natural selection on the specific parts that "go against the grain" of the changes in other parts. To take another example from human evolution, we will see later that, while human brain size increased along with body size, average tooth size did just the opposite. Thus, as human body size increased over the last 5 million years, tooth size tended to decrease because our ancestors began to prepare their food more, cooking it and using other means of softening it. This lessened the need for large teeth. This phenomenon, whereby some traits are uncorrelated and therefore evolve at different rates, is called **mosaic evolution.** This term aptly describes the "mosaic" produced when anatomical parts change at rates independent of one another.

In summary, anatomical rates vary in time and space. That is, the evolutionary rate of a single trait will vary through time. Whether most such change is rapid is a major point of debate. Evolutionary rates can vary through space in that different traits can have different rates of change, leading to mosaic evolution.

GROUP ORIGINATION RATES. These are simply rates which measure the number of groups that originate during some period of time. For example, in Figure 4–8, there are nine species of this primitive mammal that originate from a single ancestral species (*H. loomisi*). As noted, we do not have precise dates for the 1,600 feet of rock shown, but for illustrative purposes, let us say that the 1,600 feet of rock represents 1 million years. In such a case, the group origination rate would be 9 species/million years. As with anatomical rates, we can then compare this rate to rates at other times. For instance, we might find that during earlier times, the species' origination rate of a particular mammal was much lower for some reason. This example also shows the connection between anatomical rates and group origination rates: New species (and other groups) are classified on the basis of the anatomical changes directly measured by anatomical rates (see Figure 4–8).

Group origination rates are coarser than anatomical rates in that they do not focus on specific trait changes. However, this coarser view gives group origination rates three advantages over anatomical rates. One advantage is that group rates represent a measurement of changes in many traits at once. This is because new groups (such as the species of Figure 4–8) are identified and named on the basis of many traits, not just one or a few. In using many traits, a much more complete view of evolution is seen. The second advantage is that group rates can be used to characterize the large-scale patterns of major types of organisms. For example, Figure 4–12 shows the origination rates for genera and families of the trilobites. These sea-dwelling relatives of insects became extinct at the end of the Paleozoic but, as seen in the graph, in the early Paleozoic, trilobites showed very high rates of origination which tapered off before extinction. (This is a common pattern in fossil groups.)

The third advantage of group rates is that they allow rate comparison among different kinds of organisms. This can lead to very powerful evolutionary insights, as shown in Figure 4–13, where we see that trilobites and mammals are among those groups with very high rates of evolution. In contrast, other groups, such as corals and clams (bivalves), have very low rates of evolution. (We discuss possible causes of this interesting pattern later.) *Only the use of group origination rates permits this comparison.* If we were limited to anatomical rates, how could we meaningfully compare the rate of evolution of a brachiopod shell to that of a mammal tooth? Are we comparing apples and oranges? When we compare species or family origination rates between brachiopods and mammals, for example, at least we are comparing species to species or families to families. Unfortunately, even here there is no guarantee that families of trilobites, brachiopods, and mammals are directly comparable. Ultimately, classifiers must rely on anatomical traits.

In summary, group origination rates, like anatomical rates, vary in time. That is, the origination rate for any one group will vary through evolutionary time. For

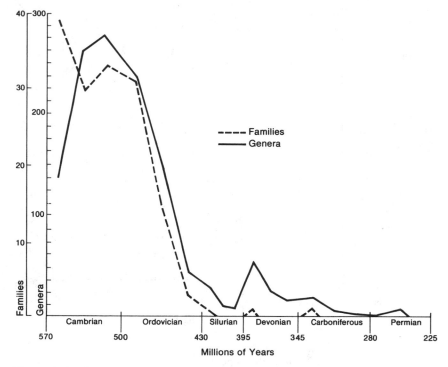

Figure 4–12. Rate of appearance of trilobite groups are highest in the early phases of trilobite evolution. Cambrian and Ordovician = early Paleozoic. From T. Dobzhansky et al., *Evolution*, p. 334.

example, the trilobites showed high rates early in their existence and much lower rates later. Group origination rates also vary among groups. Some groups, such as mammals, generally have much higher rates than other groups, such as clams.

Measurement of Evolutionary Rates Using Living Organisms

Living organisms do not suffer from the partial preservation of the fossil record. We can observe the entire organism: soft tissue, biochemistry, genes, everything. Therefore, if evolution were often very rapid (say, over a few years), we could easily describe the evolutionary process by observing living organisms that are evolving. There would be little debate about evolution in general. However, evolution in nature usually takes at least many hundreds and probably thousands of years to produce new species. Therefore living organisms are most useful when we compare them to their recent ancestors. Only then can we see enough change to measure the evolution of new species. This is an advantage over just the fossil record alone because the more recent fossils are generally better preserved and better dated.

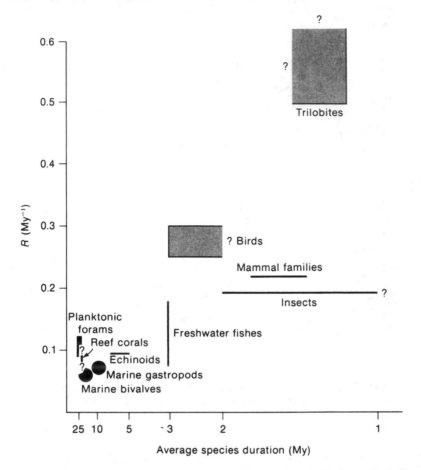

Figure 4–13. Origination rate (R) seems correlated with "complexity" and species duration. Modified from S. Stanley, *Macroevolution* (New York: Freeman, 1977), p. 231.

Also we can observe the life habits and complete anatomy of the living descendants today.

Such observations using living organisms confirm the fossil evidence that evolutionary rates vary greatly. For example, American plant populations of sycamores have been separated for over 20 million years from those of Europe and Asia because of continued continental drift. Yet these populations can still interbreed after over 20 million years to create fertile offspring, indicating that they can be classified into the same species. This is also true for other American and European organisms, such as plantains and some birds (creepers and ravens, for example). These indicate that rates of evolution can be extremely slow since these groups show little sign of diverging even after 20 million years.

This is accompanied by similar evidence for very slow anatomical rates of evolution when comparing living groups to their fossilized ancestors. Paleontologists have documented many examples where the fossil ancestors of living species are anatomically indistinguishable from their living counterparts. The most extreme cases of this are the so-called **living fossils,** which have not significantly changed in many millions of years. Examples of these are as follows (with approximate time since they have not changed): oppossum (70 million years ago), crocodiles (70 million years ago), sequoia trees (70 million years ago), and the horseshoe "crab" (200 million years ago). The best-known record setter is a type of brachiopod, *Lingula*, that has changed very little in over 350 million years.

On the other hand, living evidence also indicates that evolutionary rates can be rapid. Many lakes in the Death Valley region have only been separated for 20,000 to 30,000 years, when the Ice Ages drew to a close, drying up the region. Yet some of these lakes contain different species of fish (pupfish), unable to interbreed, that must have evolved since the lakes separated. Similarly, Lake Nabugabo, in Africa, was once a part of Lake Victoria, until it was cut off by a strip of land. Radiocarbon dating shows that this separation occurred only about 4,000 years ago. Yet there are a number of fish species unique to Lake Nabugabo that must have evolved since the separation. Most rapid of all have been experiments with fruit flies, which have created very different populations in a few years. Indeed, there is at least one report of a new species, incapable of reproducing with others. (However, the extremely intense selection experienced in the lab by these flies is vastly greater than the force of selection in nature. That new species have been only rarely created after decades of such intense artificial selection illustrates how slowly natural evolution usually is when measured on the scale of a human lifetime.)

MOLECULAR EVOLUTION. Aside from a more complete anatomy and interbreeding ability, living species also offer the opportunity to directly examine their genes. This is done with a variety of biochemical techniques. For example, **DNA hybridization** involves the mixing of single-stranded fragments of DNA of two species. By measuring the proportion and rate at which single strands relink to form double strands, we can measure how similar the DNA strands (and therefore the genes) of the two species are. Another, very common, method is to compare molecules that compose tissues in the two species. Most useful to compare are amino acid molecules, which make up the many proteins. **Amino acid sequencing** compares the sequence of amino acids that compose the protein molecules. Examples of such molecules would be: hemoglobin (in the blood), myoglobin (in muscle), and immunoglobin (in the immune system). This helps us reconstruct evolutionary rates because organisms differ in the specific amino acid sequence that composes each protein. Through time, mutations occur after species separate so that their sequences begin to vary. For instance, part of the human hemoglobin protein has 141 amino acids linked together in a chain. If, as in Table 4–1, we compare hemoglobin in humans with the same protein in other animals, we see that there is a greater difference as time

TABLE 4–1. Amino Acid Differences in Hemoglobin Alpha Chains Between Man, Dog, and Other Vertebrates, Compared With the Age of Their Divergence (million years)

SPECIES PAIR	% AMINO ACID DIFFERENCE	TIME SINCE DIVERGENCE
Human–dog	16.3	90
" –kangaroo	19.1	140
" –echidna	26.2	225
" –chicken	24.8	300
" –newt	44.0	360
" –carp	48.6	410
" –shark	53.2	450
Dog–kangaroo	23.4	140
" –echidna	29.8	225
" –chicken	31.2	300
" –newt	46.1	360
" –carp	47.9	410
" –shark	56.8	450

Source: V. Grant, *The Evolutionary Process* (New York: Columbia Univ. Press, 1985).

of separation increases. About 16% of the sequence differs with that of a dog, while over half differs when compared to a shark's hemoglobin. In contrast, our hemoglobin has exactly the same amino acid sequence as a chimpanzee (0% difference), and only about 2% difference when compared to a rhesus monkey.

We can consider these biochemical traits as a kind of anatomical rate of evolution in that we are comparing parts of the organisms. Only it is at a much finer scale than with bones or other large organs or traits. These biochemical rates, or more accurately, **molecular rates,** of evolution have shown that, like larger traits, different molecules also evolve at different rates. For example, hemoglobin molecules evolve quite rapidly. In contrast, molecules of insulin (used in metabolizing sugar) evolve much more slowly. Some workers have argued that many molecules do not experience strong natural selection and that some molecular rates are therefore often fairly constant, being controlled mainly by biochemically determined rates of mutation. This **neutral theory** of molecular rates is still being hotly debated. However, it appears that at least some molecules do have constant rates.

Molecular evolutionary rates are very helpful because, unlike organs and other large anatomical traits that are controlled by many genes, molecular traits directly mirror the genes that produced them. This is because molecular traits are based on amino acids, which link up in direct accordance to the DNA sequence in the gene (Chapter 2). Hence, the amino acid sequence directly reflects the DNA sequence. Currently, techniques are available to directly "map" human (and other organisms') genes (see Chapter 5). These will provide an even fuller picture of how genes have changed through time.

Causes of Evolutionary Rates

There are a number of factors that directly affect evolutionary rates. It is convenient to divide these factors into two basic categories: intrinsic and extrinsic factors. **Intrinsic factors** are simply those of the organism itself, whereas **extrinsic factors** refer to environmental processes affecting the organism from the "outside." This is easy to visualize if you recall our equation:

$$\text{Variation} + \text{Environmental Change} = \text{Evolution}$$

Evolution is the interplay of the extrinsic factors of the environment and the intrinsic factors of the organism. Therefore, the rate of evolution will be affected by both kinds of factors.

By far, the most important extrinsic factor influencing evolutionary rates is the rate of environmental change. If the environment is undergoing rapid change, the rate of selection on a species will be greater than if environmental change was slow. Thus, an extreme drop in temperature will more rapidly select cold-tolerant individuals (and genes) than a more gradual one. It is important to note that a rapid change in environment is often initiated by the organism itself. It is not always a matter of a species helplessly watching its environment change. The most obvious example is that of migration into a new environment, which will often vary in many ways (either subtle or drastic) from the former one. A very different environment can lead to rapid evolution if the organism can adapt to it. A classic example of this is **adaptive radiation,** defined as the rapid diversification ("radiation") of a group while adapting to unoccupied environmental zones. An example would be the rapid evolution of the first plant and animal groups to inhabit land.

Such unoccupied environments are often made available by the evolution of some **key innovation.** For instance, the lung allowed animals to inhabit land (Chapter 3). For this reason, the early evolution of many major groups is characterized by relatively high rates of evolution compared to later periods of the groups' evolution. Thus, as we saw in Figure 4–12, early trilobite evolution showed high rates of evolution as the group expanded into the relatively unoccupied early Paleozoic marine environment. In other cases, the unoccupied environments may be created when some catastrophic event eliminates competing groups that formerly dominated the environments. We have called this ecological release when discussing the example of the radiation of mammals in the early Cenozoic (Chapter 3). They radiated into the air, sea, and many land environments after the extinction of dinosaurs.

INTRINSIC FACTORS. Many traits of a group can affect its rate of evolution, but we will focus on two of the most important: *mobility* and *complexity*. Generally speaking, less mobile organisms tend to have higher rates of evolution. This is because smaller populations will generally evolve faster: the more individuals there are in a group, the longer it will take the group as a whole to change. Or, in terms of the gene pool, the more genes in a pool, the longer it will take the proportion of gene types to change (Chapter 2). For example, some types of marine invertebrates have larvae that can float in the water column for many months, widely dispersing the gene pool. We would expect groups with this kind of mobility to evolve more slowly than other groups that have larvae that last for only a few days. In this latter case, small isolated populations would form much more often, and be susceptible to rapid change.

Combining what we have said so far about environmental change and mobility, we would expect species that live in stable, unchanging environments and are widespread, with large, mobile populations, to have slow rates of evolution. This also often seems to be the case. For example, many of the "living fossils" listed above are widespread in stable environments on land (oppossums), rivers (crocodiles) and the sea (horseshoe "crab").

The second intrinsic factor to influence evolutionary rates is complexity. In general, greater complexity tends to increase evolutionary rate. This was shown in Figure 4–13, where mammals, birds, and insects (and insect-relatives, the trilobites), have higher rates than clams, foraminifera (single-celled animals), snails (gastropods), and corals. Mammals, birds, and insects are generally considered to be more complex than the others in two important ways: anatomical and behavioral complexity. Greater anatomical complexity refers to the fact that mammals, birds, and insects have more cell types, organs, tissues, and body parts in general than simpler organisms. This can lead to more frequent evolution because there are more things to change. For example, complex anatomies require a more complicated growth and developmental sequence to "build" the animal from a single cell. This means that there are more steps that can be altered by mutation. In contrast, an organism with fewer parts has fewer things to change. Since evolution is change through time, there will be less evolution.

Behavior complexity leads to more rapid evolution for a number of reasons, but one of the most obvious is that animals with complex behavior often have complex mating "rituals" that must be followed very precisely. It is very easy for new species to evolve in such circumstances because even a minor change in mating behavior can cause some individuals to stop interbreeding with others, leading to reproductive isolation. Another reason is that changes in behavior often precede changes in anatomical evolution: When the giraffe ancestor began to shift its diet to leaves higher on the tree, natural selection began to favor longer necks and other anatomical changes. Organisms with complex behaviors tend to change such behaviors more often, leading to more anatomical change.

In addition to mobility and complexity, there are many other intrinsic factors affecting rates of evolution. One that may have occurred to you is generation length, the amount of time it takes for an organism to reach maturity and reproduce. Microbes can reproduce in less than a day. In contrast, a large organism like a whale, elephant, or human can take over 15 years to reproduce a new generation. Since evolution depends on the accumulated differential production of successful offspring, it would seem that organisms with shorter generation times could evolve faster. However, most data collected on evolution in fossils have not supported this. The reason seems to be that environmental change usually occurs too slowly for generation time to be a major factor. Climatic and sea level changes, for instance, usually take many thousands of years (Chapter 3). It is only when human activities create very rapid and extreme environmental selection (such as through pollution or in the lab) that rapid reproduction is needed to keep up with the environmental change.

Another intrinsic factor that you may have thought of is a group's tendency to undergo mutations. Since evolution is the interplay of variation and environment, it would seem that evolution is more likely the more mutations that are produced. However, as with generation length, this is another intrinsic factor that does not seem to be clearly supported by evidence. It is true that different groups have differing tendencies to undergo mutations in their genes, but the resulting effect on rates of evolution is not consistent. There are many reasons for this, but one of the most important is that mutation rate is itself governed by environmental selection.

PUNCTUATIONISM VERSUS GRADUALISM: CAUSES. We discussed above the debate over whether most change in the fossil record consists of long periods of stasis punctuated by rapid evolution (punctuated equilibrium) or whether gradual, transitional changes are seen (gradualistic evolution). Proponents of both schools realize that evolutionary rates vary, but punctuationists argue that rapid change is more common. However, this is *not* merely a question of which rates are more common. A widely omitted point in the popular press is that punctuationists have a fundamentally different view of evolutionary processes. They view the stasis commonly observed in the fossil record as a result of the genes in any species' gene pool being so highly integrated that small changes in just a few genes cannot occur without the whole gene pool being rapidly rearranged. An analogy would be a house of cards: they are so tightly interdependent that either no change (stasis) or a rapid collapse (rapid evolution) occurs, after which a new interdependence (and period of stasis) is achieved.

On the other hand, gradualists do not deny that stasis is common. However, they believe it is simply due to the fact that a species will follow its environment and do whatever it can to live in conditions that it is adapted to. As long as a species can do this, there will be little change. To the gradualist, genetic integration is not rigid and even large populations can gradually become altered over long periods of time: The gene pool is much more "plastic."

A huge number of technical papers have argued on this subject since punctationism became prominent in 1972. However, the issue is still not clearly resolved. As we have noted, while much of the fossil record does indeed show a punctuated pattern of stasis and rapid change, we cannot say that this is not due to gaps and other artifacts. Nevertheless, there is no conclusive evidence that gene pools are as integrated as punctuationists have often argued. For example, many species show significant geographic variation in the form of gradients (clines) of subspecies (Chapter 2). This obviously indicates that minor, gradual changes in gene pools can occur. In addition, there are some well-documented examples of gradual change in the fossil record. An excellent example is shown in Figure 4–14, where we see the change in body shape of a fossil sea urchin from the late Mesozoic of England. Note the gradually diverging change in the variation of the populations under selection.

It seems likely, then, that the punctuated patterns often seen in the fossil record can be explained by traditional Darwinian views. Even if the rapid changes represent truly rapid evolution and are not due to gaps, the most likely cause is that some of the factors listed above acted to increase rates of change in the populations under selection. For instance, we said that small populations and rapidly changing environments increase rates of evolution. These changes might well appear as "punctuations" in the record, even though they took many years to occur. Even so, most evolutionary scientists remain open-minded to the possibility that evidence from genetics or other fields may yet challenge this view.

PART 2: ORIGINATION DIRECTIONS

Many directions, or trends, are found in the history of life. Nearly all of them can be explained by directional tendencies inherent in the interplay between mutations and environmental selection. The large majority of these trends are *statistical tendencies* with fluctuating patterns of change that are often temporarily reversed. For example, we will see that many groups of organisms show size increase through time. Yet, closer inspection reveals that while the maximum size of the group became larger, many species within the group remained the same size and some were smaller.

Kinds of Trends

Just the fact of existing usually entails many kinds of trends. For instance, there are many directional patterns, or trends, in our daily lives, society, and the physical environment. To name a few of a huge number of examples: graphs of your metabolism taken during your lifetime will show a trend downward; the U.S. gross national product shows a long-term trend upward; and the global temperature during the Cenozoic has shown a general trend downward (Chapter 3). Therefore, it is no surprise that the evolution of life should also show many different kinds of trends. For simplicity, most of these can be placed into two basic categories. **Lineage trends** are those trends seen in an evolving lineage. Such a lineage can range from a single evolving species to a large, multispecies branching lineage, often called a "clade," such as horses or humans and our ancestors. **Biosphere trends** are those trends that we see in the evolution of life as a whole. For example, the upper limits of both

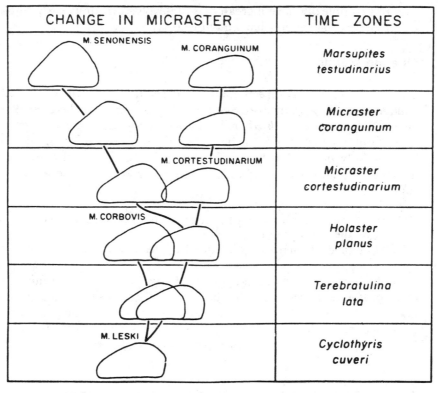

CHANGE IN MICRASTER	TIME ZONES
M. SENONENSIS M. CORANGUINUM	Marsupites testudinarius
	Micraster coranguinum
M. CORTESTUDINARIUM	Micraster cortestudinarium
M. CORBOVIS	Holaster planus
	Terebratulina lata
M. LESKI	Cyclothyris cuveri

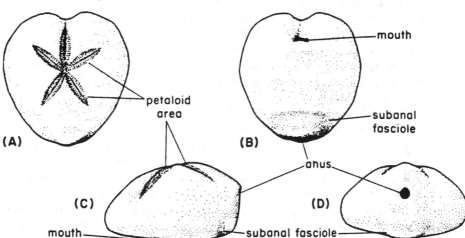

Figure 4–14. Branching gradual evolution of a Mesozoic sea urchin ("sea biscuit") in the fossil record of England.

size and complexity have increased as life evolved from its early single-celled state to the diverse biosphere that we know today.

Lineage Trends

Lineage trends reveal directional tendencies in the various traits and organs that change as the lineage evolves. Such trends typically occur because these traits have been under *consistent directional selection*. They thus provide crucial information about how an organism has adapted to its environment. For example, we have already noted that human evolution is characterized by a general increase in brain size and body size, with a decreasing trend in tooth size. The selective causes were:

1. Intelligence was consistently favored as cultural adaptation grew ever more important (thus, the larger brain).
2. More food was prepared, therefore large teeth were no longer needed.

We have also said that traits often change in unison. This means that the same trends often occur among a number of correlated traits (such as human brain and body sizes) rather than in isolation. Let us begin our discussion with trends of single traits in order to illustrate some basic ideas about evolutionary trends. Most of these ideas will also apply to trends in correlated traits, but they are easier to visualize for one trait.

TRENDS IN SINGLE TRAITS. An excellent example of a trend in a single trait occurred in the brachiopod lineage, illustrated in Figure 4–7. The trend was for the shell to become smoother with time. Yet, as in most such trends, the change showed some temporary reversals. Such "zig-zags" illustrate the statistical nature of directionality. It often means that the directional selection was not entirely consistent. This is not surprising because most things do fluctuate, even where a trend is involved. The reason they fluctuate is that there is usually a complex interplay of *many contributing factors* which vary in duration and strength.

Nevertheless, it is often possible for the evolutionist to isolate key factors that were among the most important in the environmental selection that caused directional changes in traits. For example, in Figure 4–15 you see variations in average tooth size of the European brown bear over the last half million years of the Ice Ages. During cold times (glacial advances), the brown bear increased in size (as judged from tooth size, which is proportional to body size). Warmer periods (interglacials) saw body size decreasing. This trend correlates well with modern biological knowledge. Among living mammals, larger species or races of a group tend to be found in cooler climates. For instance, in North America, the largest bear is the polar bear, while the next largest, the grizzly, is found to the south. Still farther south, in still warmer climates, is the relatively small black bear. This generality, that cooler climates tend to favor larger warm-blooded animals, is called **Berg-**

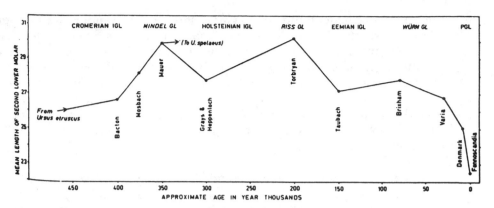

Figure 4–15. Tooth size evolution in the Ice Age brown bears of Europe. Peaks of largest tooth size correspond to times of maximum glaciation. From J. Bonner, *The Evolution of Complexity*, p. 31.

mann's rule. It occurs because larger mammals and birds conserve body heat better than small ones.

The key point in the above example is that one environmental factor played a dominant role in selecting the size of individuals, causing the species as a whole to vary in size through time. In the example of the Ice Age bear there was no persistent directional trend because temperature was cyclical. However, if temperature had steadily decreased, a directional trend to larger body size would probably have occurred. In such cases, we say that **orthoselection** occurs. This literally means *straight-line selection*, but we define it here as simply selection in one direction, acknowledging that it is not really "straight-line," but, like all directional trends, will fluctuate. Orthoselection then is the *cause* of directional trends: It is the environmental selection that drives the changes that occur. (Be careful not to confuse it with orthogenesis, which attributes straight-line evolution to unproven spiritual or internal drives.) Other examples of orthoselection are selection for increased brain size in humans or smaller tooth size, as we have noted.

In many cases, the orthoselection is not as easily identified as is temperature on body size, or food softness on human tooth size. For instance, body size changes very often in many lineages. Yet, many of these show no evidence of temperature change. Instead, we know from studies of living organisms that body size is affected by many environmental factors. The appearance of larger predators can cause smaller body size to become an advantage in prey species because larger predators often select larger prey. In other cases, a drop in food availability will select for smaller size since smaller individuals need less food. Isolating such factors is often very difficult, if not impossible, in many fossil lineages. For example, the large predators causing size change may not fossilize and food availability is almost impossible to estimate.

This identification of cause is often further complicated because even a seemingly simple directional trend in a trait is not necessarily caused by a single factor, pushing in one general direction. This is illustrated by **random walk** experiments. These are statistical trials carried out using coin flips or other random processes. As shown in Figure 4–16, these experiments show that a series of random events can often generate a trend by chance. In the figure, 20,000 coin flips produce a pattern that looks highly deterministic (as if a single, underlying cause were determining it). This means that some evolutionary trends that appear to have a single underlying environmental cause may actually result from a number of conflicting (randomly acting) causes that just happen to produce a directional trend that looks like a single cause.

TRENDS IN TWO OR MORE LINEAGES. Trends in two or more lineages can be usefully compared. We may identify two basic types: divergent and convergent trends. As shown in Figure 4–17, **divergent** trends represent the splitting of a lineage into descendant species as their gene pools become progressively separated through time. **Convergent** trends do not arise from the same ancestral stock. Instead, unrelated groups follow trends that cause them to become similar (to converge) anatomically (Figure 4–17). This occurs because the converging groups have independently taken up similar life habits, putting them under similar selective pressures. This convergence is very common since the environment imposes so many constraints on ways of surviving. For example, the wings of bats, birds, and flying reptiles of

Figure 4–16. A "trend" produced by 20,000 flips of a fair coin ("random walk"): heads = up 1 unit, tails = down 1 unit from location of preceding flip. Modified from *Paleobiology* Magazine, 1987, p. 451, Fig. 1.

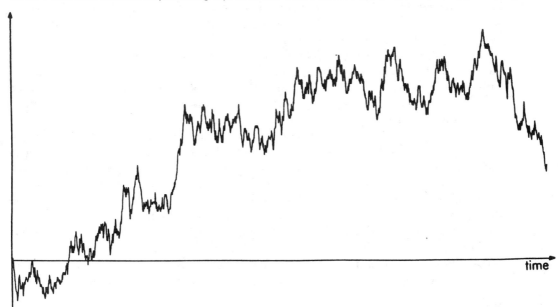

time

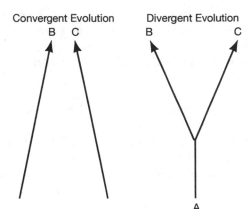

Convergent Evolution   Divergent Evolution

Figure 4–17. Divergent evolution from a common ancestor versus convergent evolution from different ancestors. Modified from R. Wicander and S. Monroe, *Historical Geology*, (St. Paul, MN: West Publishing, 1989), p. 126.

the Mesozoic are superficially similar because the constraints of flying in air demand certain basic characteristics in a wing. Such convergence is not restricted to single traits. Many traits may converge in different groups, from similar ways of life.

Convergent and divergent evolutionary trends both illustrate what is perhaps the single most important principle of evolution: that evolution proceeds only by tinkering with preexisting traits. We discussed this in Chapter 2, but it bears repeating in the context of directionality. Tinkering may seem obvious, but the fact that natural selection can operate only on what exists at any given time greatly restricts evolution. For instance, it means that organs are often not designed with maximum effectiveness, as an engineer would design them if he could "start from scratch." Rather, organs are designed with whatever materials are available. Because of such limitations, evolution has often been called *opportunistic*, meaning that it must use whatever raw materials are at its disposal. In the case of divergent evolution, tinkering is shown by the presence of **homologous organs.** These are organs (or traits) that have different functions in the diverging groups, but they clearly arose from modification of the same original structures. The example illustrated in Chapter 2 is the limb bones of many mammals. Humans, dogs, whales, and birds are obviously very different in having arms, legs, fins, and wings, respectively. Yet the major bones composing each are the same. They are simply modified into different sizes and shapes to perform different tasks.

Convergent evolution also illustrates the tinkering process, only in the opposite way. Instead of finding underlying similarities in superficially different (divergent) groups, we often find underlying differences in superficially similar (convergent) groups. For example, clams and brachiopods have an apparent similarity of form, but a detailed dissection of the internal organs will show many major differences. The feeding mechanisms are completely different, with clams using gills and the brachiopods using a ropelike organ. A more familiar example of such organ convergence was already noted: the superficial similarity of the wings of birds, bats, and flying reptiles. By tinkering with organs that differed to begin with, the resultant anatomy is superficially similar but different in underlying detail. Insect wings are composed of a hardened protein material, while bat wings are composed of five fingers. The flying reptiles had wings based on only one finger (Chapter 3). Such convergent organs are called **analogous organs,** and reflect similarity of adaptation rather than the similarity of ancestry seen in homologous organs.

TRENDS IN CORRELATED TRAITS. The same patterns discussed above for single organs or traits can occur in many traits. In fact, this is usually the case because traits seldom evolve in isolation: An organism is a functioning whole, but the parts making up the anatomy of each of us must work together in an integrated fashion. Therefore, evolutionary change in one body part is often accompanied by compensatory changes in others. We discussed such "correlated traits" above.

The presence of correlated traits means that a trend in one trait will often be accompanied by a corresponding trend in the traits correlated with it. A classic example is the evolution of the horse. Horses are one of the most well-documented lineages in paleontology and their history shows a clear trend among correlated traits. As shown in Figure 4–18, horse evolution exhibits *three concurrent trends*:

1. Increase in tooth size
2. Increase in body size
3. The loss of toes from 4 toes to 1

The modern horse (*Equus*) is relatively large in body and tooth size, with only one toe (the hoof).

These trends all occurred as part of the horse's adaptation to the spreading grasslands of the middle Cenozoic. Horses originated as small forest dwellers, browsing on buds and leaves. As they moved into the expanding grasslands, their tooth size increased because grasses are much tougher than buds and leaves and destroy the enamel faster. The toes became reduced because running on an enlarged central toe allowed the animal to run faster. This was necessary because open exposure to predators is much greater on grasslands. Finally, body size increased because larger animals have longer digestive tracts and can more efficiently process grass in the diet because it is in the tract longer.

The study of evolutionary change among correlated traits is called **allometry.** A very effective way of analyzing such change is to measure the traits (such as tooth size) and plot the measurements against one another. This allows us to directly measure the amount of correlation, instead of just visually comparing the traits as we did in Figure 4–18. For example, there was a lineage of rhinolike animals, called titanotheres, that coexisted with horses during the early and middle Cenozoic. Like horses, these creatures also made the transition to grasslands, becoming larger, too, in the process. In their case, larger body size was correlated with larger horn, perhaps used for defense on the open grasslands. As shown in Figure 4–19, we can plot skull length (representing body size) versus horn length. This **allometric plot** shows that there was a high degree of correlation between body size and horn size:

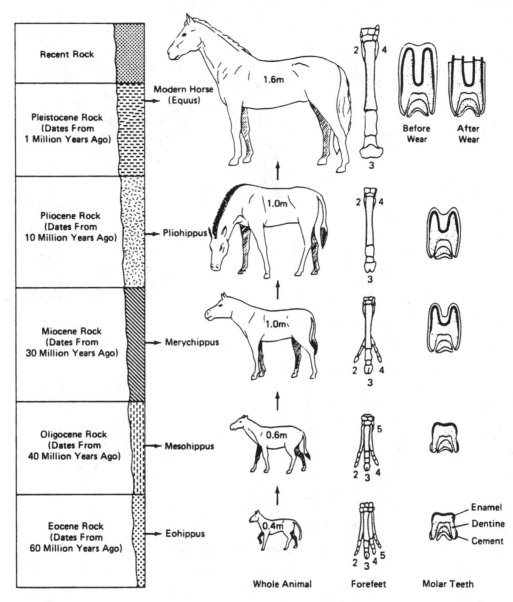

Figure 4–18. A very simplified version of three trends in horse evolution: size increase, toe loss, and tooth size increase. It omits many horse lineages that did not show these trends.

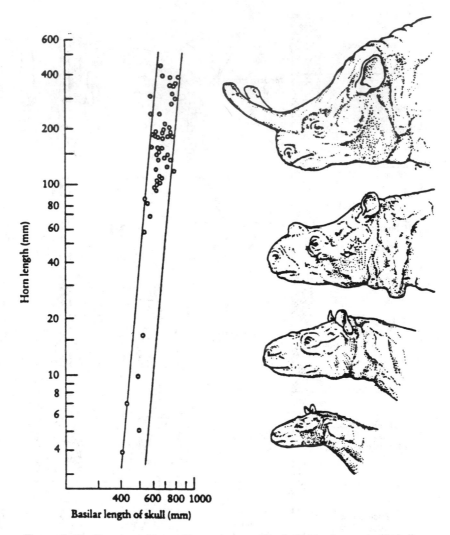

Figure 4–19. Regular pattern of horn change with skull size change (called allometry) in an extinct rhino relative.

A trend in one was thus accompanied by a trend in the other. A similar plot could be constructed for body size versus tooth size in horses.

A crucial point about all correlated traits is that selection on one trait may cause a change in a trait correlated with it. Consider the example in Figure 4–19. Suppose the environment favored larger body size in the titanotheres. If horn growth was strongly correlated with body growth, then horn size would increase whether selection favored larger horns or not. Larger horns would be a byproduct of selection on body size. Because correlation among traits is extremely common, it may be that many, perhaps even most, evolutionary changes in traits may be *byproducts of selection on other traits*. The implications of this are profound for evolutionary theory. For instance, it means that many traits are not as precisely tuned to their environment as we might otherwise think. This is very difficult to test on fossils, but artificial selection experiments with living organisms confirm that when one trait is selected for, many correlated traits are also indirectly "carried along."

Horses and titanotheres are but two of many fossil lineages that show a trend in body size increase and its correlated traits through time. Indeed, trends of increasing body size are so common that this tendency has been given a name, **Cope's rule,** which states that many lineages tend to increase in body size through time. Aside from many hoofed mammals, other lineages that increased in size over time include many dinosaur lineages, marine invertebrates, and even our own (the *Australopithecus* was about 4 feet tall, while today our average height is about 5 feet). Indeed, *size change is the most common evolutionary change* observed in fossils. Very often, many other traits change along with body size because they are correlated with it.

Why is this increasing size trend so common? In part, it is because large body size is favored under many circumstances. In horses and titanotheres, larger size was favored due to a grazing lifestyle. In other cases, larger animals are at an advantage when competing for food or mates, defending against predators, and, as we saw with the Ice Age bear, they retain body heat better. However, this does not fully explain the trend. If large size is so beneficial, why don't lineages simply begin at a large size, with big ancestors? If they did, then no trend to large size would occur. The answer is found in a basic observation of modern ecology: *Most species are small.* No matter what group we observe (reptiles, mammals, invertebrates, and many others), we find that there are fewer large species than small. An example is shown in Figure 4–20, for birds. The point is that the ancestor of any lineage is much more likely to be small instead of large just because there are so many of them.

Finally, we must note that most fossil trends toward increasing size and its correlated traits are *oversimplified*. Like all trends, size increase in a branching lineage has many reversals and branches that do not increase. For instance, the true evolutionary history of horses, shown in Figure 4–21, is considerably more complex than the grossly oversimplified "straight-line" evolution illustrated in Figure 4–18. Thus, Figure 4–21 shows that only three of the many kinds of horses have been picked out for depiction in Figure 4–18 (*Hyracotherium, Merychippus*, and *Equus*). In truth, the transition to grazing from browsing had many side branches, and not all of these showed the same trend toward large size, teeth, and one toe. In addition, you can see that browsing horses coexisted with grazing horses for a long time.

CONVERGENCE AND CORRELATED TRAITS. We discussed earlier how single traits could converge between different organisms that shared similar ways of life. Such convergence also extends to correlated traits. For example, Figure 4–22 shows how organ-

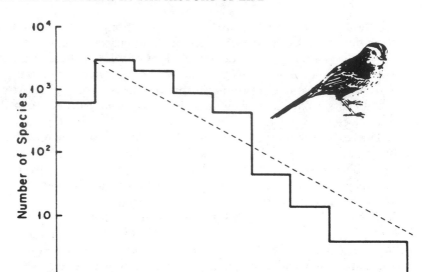

Figure 4–20. Number of bird species declines rapidly with increasing size. Log scale for size means that decline is exponential: more rapid than shown.

isms with many different backgrounds, from fish to birds to mammals to reptiles, have anatomically converged in many traits designed to adapt them to a swimming life in the sea. The streamlined shape of the body, the tail fins for propulsion, the pectoral (front) fins for steering, and other correlated traits operate together to allow the animal to move efficiently in the water.

Biosphere Trends

Aside from trends in individual lineages, we can also identify "megatrends" in the history of life. Two such trends are of central importance: (1) a diversity trend, and (2) an ecological trend. The diversity trend refers to the general increase in the kinds of organisms that have occurred since the origin of life. The ecological trend refers to the general increase in the kinds of interactions among these organisms. Both of these trends are probably intuitively obvious to you. If life began from a single ancestral microbe (Chapter 2), then it is clear that the 5 million or more species now on earth represent a considerable increase in diversity. It is also logical that with so many kinds of species, the ways of interacting are bound to increase as well. However, the details of these trends are of considerable interest and reveal many not-so-obvious lessons about evolution. We begin with biosphere diversity.

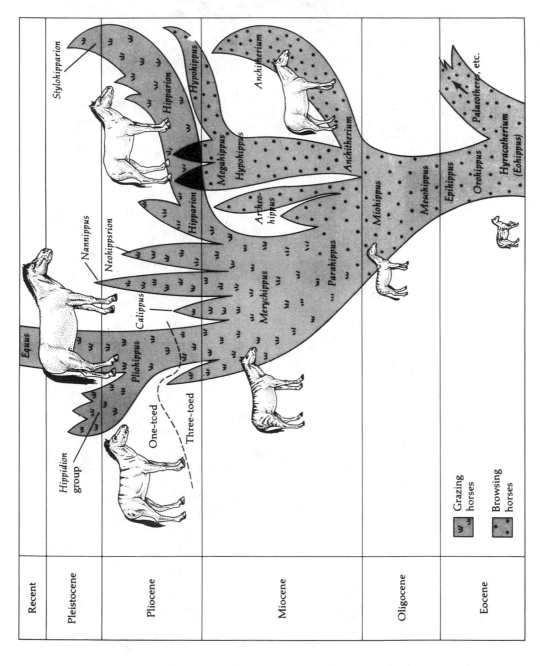

Figure 4–21. The complex evolutionary "bush" of horses over the last 50 million years. From S. Luria et al., *A View of Life* (Menlo Park, CA: Benjamin/Cummings, 1981), p. 640.

Figure 4–22. Convergent evolution among aquatic life. From W. Stokes, *Essentials of Earth History* (Englewood Cliffs: Prentice-Hall, 1982), p. 476.

BIOSPHERE DIVERSITY. Figure 4–23 shows the diversity (number of families) of fossilized marine invertebrates that have been described for the last 600 million years. You can see that after a rapid rise in the early Paleozoic (the "explosion of life" described in Chapter 3), there has been a more gradual increase in the number of families, except for a major interruption caused by the mass extinction at the end of the Paleozoic. Since then, the Mesozoic and Cenozoic have shown an especially clear trend toward increasing numbers. Thus, if we take this graph at face value, we would conclude that the marine biosphere has increased in diversity according to this general pattern.

However, we cannot take the fossil record at face value. We discussed in Chapter 3 how the record is not only incomplete but can contain systematic biases that

Figure 4–23. Number of families of shallow-water marine life over the last 600 million years. From S. Luria, et al., *A View of Life*, p. 649.

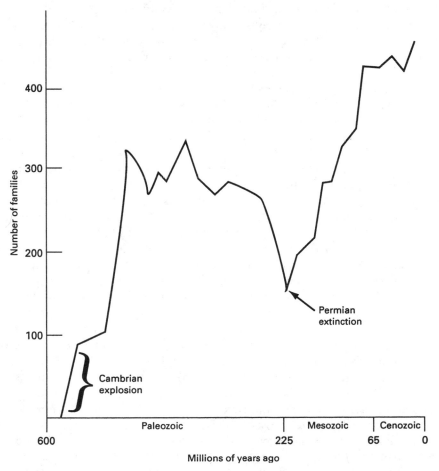

can actively mislead us. Most prominent here is the bias of the pull of the recent, wherein younger sediments (and fossils in them) are better preserved. (This is because more recent rocks have had less time to become eroded, destroyed, or covered up.) Thus, many paleontologists have argued that the increasing diversity seen in Figure 4–23 is a reflection of better preservation in younger rocks. They point out that we would see an increase in diversity even if diversity was constant because earlier rocks would preserve fewer species. Since the marine fossil record is the best there is, the same could be said for diversity on land. Furthermore, there is evidence from modern ecology that ecosystems can become saturated with species. Because the biosphere is an ecosystem (the largest of all), it is possible that it became saturated at some point, perhaps as soon as the early Paleozoic, shortly after the "explosion of life." This is the **equilibrium diversity hypothesis:** biosphere diversity reached an equilibrium long ago and the diversity increase seen in Figure 4–23 is from a biased fossil record.

While a case can still be made for the above hypothesis, the current consensus among most experts is that a second hypothesis is the correct one. This is the **increasing diversity hypothesis,** which states that biosphere diversity actually has been increasing through time. This does not deny that fossil preservation acts to increase the apparent diversity toward the recent. Instead, the increasing diversity hypothesis says that this distortion merely acts to *magnify* an increase that actually occurred. Various types of evidence have been put forth to support the increase in diversity, including sophisticated statistical methods that attempt to subtract out the pull of the recent and other biases of the fossil record. However, the most convincing evidence for many people has been the evidence for **niche packing** in ecosystems. A **niche** is simply an organism's way of "making a living." Therefore, niche packing refers to the process wherein evolution invents new ways of making a living in any given ecosystem. For example, detailed comparisons of Paleozoic marine ecosystems to Mesozoic and Cenozoic ecosystems in similar environments (such as tropical shallow waters) show that more species occur in the more modern ecosystems. In other words, more species have been added to the same environment as they have evolved new ways of fitting in (new niches). The driving force behind this creation of new ways of life is natural selection that favors individuals who can exploit new resources.

The invention of "new ways of doing things" is dramatically visible when we look at the appearance of major groups in the biosphere. For instance, Figure 4–24 shows that the diversity increase of Figure 4–23 is actually composed of the rise of three different faunas. The two largest are the so-called **Paleozoic fauna** and the **modern fauna.** The modern fauna, composed of modern marine organisms such as sea urchins, clams, and snails, became much more diverse after the Paleozoic. A main reason is that they invented new ways of doing things—they became deeper burrowers, their armor improved, and they experienced other changes not common in the Paleozoic. Most of the Paleozoic organisms (fauna) were stationary filter feeders. This changeover to modern ways of life has been called the "Mesozoic marine revolution" (Chapter 3). The crucial point is that this invention of "new ways" does not result in total replacement of the older groups. If it did, there might

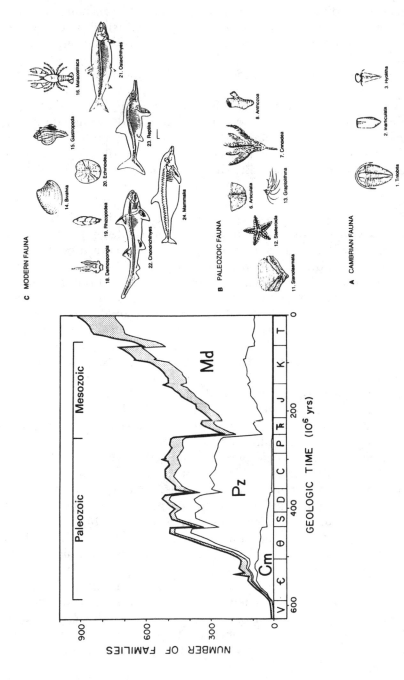

Figure 4–24. Marine animal diversity increase is comprised of three separate, sequential "suites" of organisms. Cm = Cambrian (early Paleozoic), Pz = Paleozoic, Md = Modern. Note effects of huge end-Paleozoic mass extinction. Stippled area = poorly fossilized groups. Modified from various sources.

not be an increase in diversity since one group would just be exchanged for another. Instead, the process is cumulative: the newer groups are usually added to the older ones. Even though the older ones decline, they still contribute significantly to diversity.

More examples of this principle are seen in the appearance of land organisms. Figure 4–25 shows the appearance and diversification of land vertebrates and land plants. The amphibians, which never fully made the transition to land, were followed by the rise of reptiles, and later, mammals and birds. While the later groups have become more recently prominent, the earlier groups (amphibians and reptiles) still contribute significantly to the kinds of organisms in the biosphere. A similar scenario occurred with the three basic kinds of plants, listed here in order of their appearance: the spore-bearing group, which still requires moisture to reproduce: the gymnosperms (such as conifers); and finally, the flowering plants. Each successive "wave" of land organisms invented new ways of doing things: flowers for easier pollination (flowering plants), flying (birds), the hard-shell egg (reptiles), and so on.

While niche packing can account for much of the pattern of increasing biosphere diversity, two other factors probably influenced this general trend. These factors were: (1) competitive replacement, and (2) ecological release. **Competitive replacement** is the replacement of one group by a competitively superior group. The extent to which this occurs in evolution is a hotly debated and unresolved issue. It is virtually impossible to prove if one group directly caused another group's loss of diversity, or whether other causes, such as differential susceptibility to environmental changes, were involved (see Chapter 5). Nevertheless, many paleontologists believe that the reduction of diversity of early groups was at least partly caused by competition from later groups. Probably the best case for this has been made for plant evolution. A number of prominent paleobotanists have cogently argued that each phase of plant evolution has involved the competitive reduction of the preceding group (Figure 4–25). For example, the replacement of many species of gymnosperms by flowering plants (angiosperms) is thought to reflect the advantages of rapid growth and more efficient pollination of flowering plants.

The role of competition in replacement is less clear in animal evolution. It has been suggested that competition is more important in plant evolution because all plants must compete for more similar resources (soil and light) than animals, and plants must compete while being stationary as adults. Thus, while many workers believe that amphibian diversity losses were largely caused by reptiles, others argue that factors other than competition were at play. Perhaps the best example of this is the replacement of reptiles by mammals as the most diverse group in the Cenozoic (Figure 4–25). We have already noted that competition alone cannot account for this because mammals were dominated by reptiles for over 100 million years in the Mesozoic. It was not until a catastrophe wiped out many of the reptiles that the "ecological release" occurred, allowing mammals to radiate into the niches formerly occupied by reptiles.

INCREASING BODY SIZE AND COMPLEXITY. Within the overall trend of increasing diversity, we can also observe two other important trends in biosphere evolution: (1) increas-

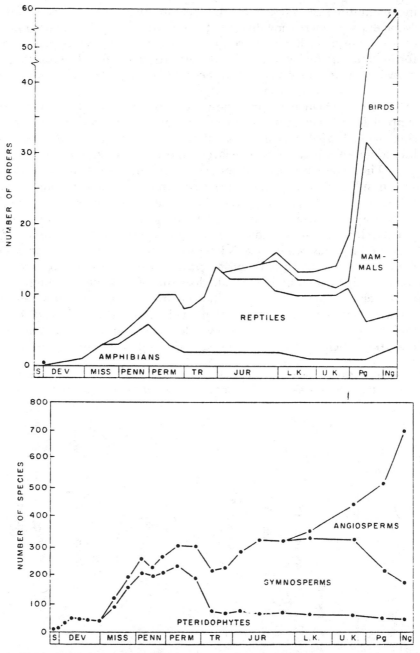

Figure 4–25. Increasing diversity of vertebrates (top) and plants occurs when earlier groups are superseded but not entirely replaced by later ones. Pteridophytes = spore plants. Modified from J. Bonner, *The Evolution of Complexity*, p. 230.

ing maximum size, and (2) increasing maximum complexity. Both of these are really "subtrends" of increasing diversity because the very act of creating many new kinds of life from a small, simple cell will inevitably lead to larger and more complex forms. Put another way, as life evolves into new ways of doing things, two of the most important "new ways" have been to get bigger and more complex. Both of these allow the organism to do things that smaller or simpler forms cannot do and therefore help avoid predation and, especially, competition with those smaller, simpler forms. For example, a worm is able to make a living doing things that no single-celled organism can do. Similarly, a human is adapted to yet another way of life, much of it because we are larger and more complex.

The evolutionary trend toward maximum size increase in the biosphere is shown in Figure 4–26. This trend occurs in both plants and animals. It has leveled off in recent years because physical constraints limit how large organisms can get. Whales are the largest animals because they have water to help buoy them against gravity. The largest land mammal (an extinct relative of the rhino) probably never exceeded 20 tons, while today's blue whale is over 100 tons. Today's elephant is a mere 8 to 10 tons. In addition to gravity, acquiring enough food for large sizes could be difficult. The blue whale does this by scooping up huge amounts of the extremely abundant, highly nutritious plankton found in the water. Similarly, it is unlikely that plants much larger than the huge sequoia could evolve: Gravity would pull them down and soil nutrients would have to be very abundant and rich. Notice that the principle of diversity increase through accumulation is very much in evi-

Figure 4–26. Increase in maximum size of plants and animals over the last 3.5 billion years. Note logarithmic scale. From J. Bonner, *The Evolution of Complexity*, p. 27.

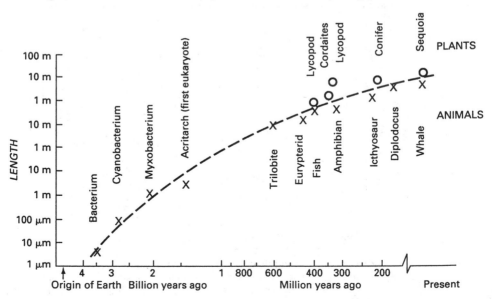

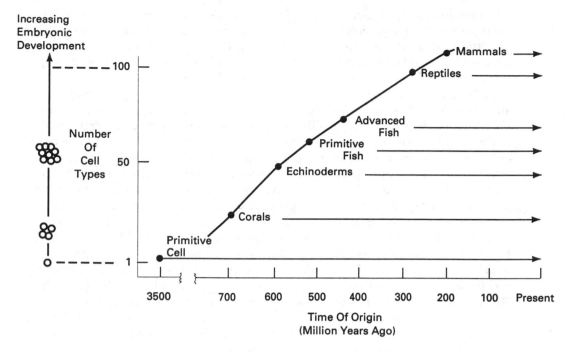

Figure 4–27. Increasing number of cell types and increasing development has evolved through geologic time. For example, each individual mammal has over 100 cell types and these cells organize themselves in complex ways during development. In contrast, the development of each individual fish involves fewer cell types. Based on data from J. Bonner, *The Evolution of Complexity*.

dence here: Figure 4–26 shows the maximum size of organisms in the biosphere at a given point in geological time. Thus, smaller sizes, which had evolved earlier, are still very much alive at all times. As we saw above, in nearly all ecosystems there are many more species of smaller organisms than larger organisms. This is because there are many more ways of doing things (niches) for smaller organisms as each small individual requires fewer ecosystem resources than large organisms. For example, a whole colony of ants can be supported on a small patch of land, while a single elephant must forage in a much larger area to feed itself.

Evolutionary increase in maximum complexity can be seen in Figure 4–27. Be aware that "complexity" is a very greatly abused term because it is so hard to measure and define. Nevertheless, it is of great importance and as long as one specifies what is meant, much insight into evolution can be gained from it. Recall that something that is more complex usually has more kinds of parts (Chapter 2). In Figure 4–27, we thus estimate organismic complexity as the number of kinds of cells that an organism has. Beginning with a single cell, the maximum number of cell types in an organism has increased, with mammals currently being the organism with the most kinds of cells. We have over 100 cell types: liver cells, brain cells, muscle cells, blood cells, and so on. Once again the process is additive, with simpler organisms continuing to exist so that total diversity increases.

Figure 4–27 also shows the relationship between embryological development and evolution. *More complex organisms have more complex development.* All organisms start life as a single cell, the fertilized egg. As the cell splits into 2 and continues to multiply, it will eventually multiply into trillions of cells and there will be over 100 cell types in the individual mammal. Organisms of intermediate complexity have intermediate levels of development between mammals and the single cell. Thus, once again we see that the creative force of evolution is that it can build upon preexisting forms. As evolution has progressed, mutations have occurred that created new cell types during development. Where natural selection favored such mutations, they were passed on, such as in the origin of mammals from reptiles (see Figure 4–27). A major result of this process is *recapitulation*: more complex organisms "repeat," in a very general way, the development of their older, simpler ancestors (Chapter 2). For instance, early human embryos not only have fewer cell types but have similar anatomies to simpler organisms, such as fishlike gill slits.

It is worth reemphasizing that graphs such as Figure 4–27 should not be interpreted as a "scale of nature" (such as the Chain of Being), with evolution striving to create "better" organisms (Chapter 1). There is often a tendency for us to think of increased complexity as necessarily "better." However, "better" is defined in purely subjective, and therefore human, terms. As far as nature is concerned, simpler organisms are just as well adapted for what they do as humans and other complex organisms are for their lifestyles. Increased complexity is just one more new way of doing things that has been favored by natural selection.

ECOLOGICAL TRENDS. The second major kind of biosphere trend involves ecological changes. **Ecology** is the study of interactions among organisms and their environment. We have seen how diversity (kinds of organisms) has increased through time by finding new ways of doing things (such as size and complexity change). We should also expect an increase in the kinds of interactions. Organisms can interact in many ways, but ecologists consider two main types to be the most important: competition (for food, space, mates, and other resources) and predation (who eats whom). Evolutionary trends have occurred in both. For instance, the maximum size of plants has increased because plants compete for sunlight and taller plants are thus at an advantage in many situations. However, for a variety of reasons, trends in predation are more thoroughly documented in the fossil record, so we will focus on those.

The most general trend in predation is that of the **food chain.** The food chain is the sum of all predation in an ecosystem. This is shown in Figure 4–28, where

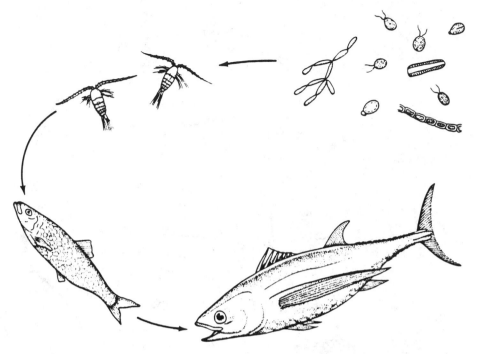

Figure 4–28. A marine food chain, with plant plankton at the base. From J. Barrett et al., *Biology* (Englewood Cliffs: Prentice-Hall, 1986), p. 938 (top).

you can see that **producers** form the base of the chain. These are nearly always photosynthetic organisms that produce food from sunlight. In the ocean plant plankton are the producers. Above producers are the **consumers,** which feed on them. Primary consumers (plant eaters or herbivores) feed directly on the plants. Second-level consumers feed on the plant eaters. Notice that our use of the term "predation" includes not only big fish eating little fish, but also plant eaters eating plants. Indeed, predation also includes the activities of parasites (which consume other organisms, but may not kill them) and decomposers, such as bacteria (which digest dead organisms). Decomposers are not shown in Figure 4–28, but they are a crucial part of any ecosystem.

The food chain in Figure 4–28 is much simpler than what exists in most real ecosystems. A more realistic representation is the **food web,** shown in Figure 4–29. It shows that the eating relationships among the organisms are more complex, as more organisms are added. The key point for evolutionary trends is this: As diversity increased, the complexity of food webs increased as well. This is easily visualized if you recall that complexity is defined by the number of parts that something has. Since organisms are the parts in a food chain, then complexity increases with the

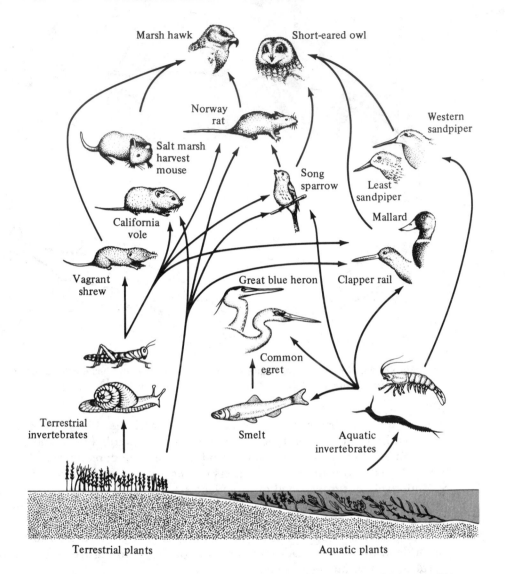

Figure 4–29. A nearshore-land food web. From J. Barrett et al., *Biology*, p. 938.

kinds of organisms because the number of interactions (labelled with arrows in Figure 4–29) increases.

A way to visually simplify the predation relationships in an ecosystem is to classify organisms according to the level at which they feed. This is a **food pyramid,** shown in Figure 4–30. All producers are at the bottom and consumers that feed at the same level are grouped together. Thus, all plant-eating species are first-order consumers above the producers. This stacking of food levels creates a pyramid because there is a substantial *"leakage" of energy between each level.* On average, about

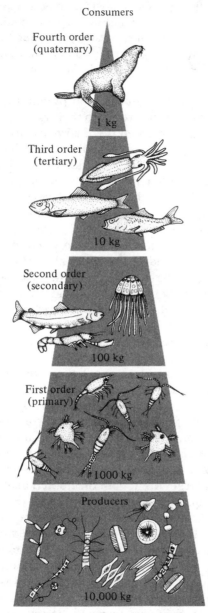

Consumers

Fourth order
(quaternary)

1 kg

Third order
(tertiary)

10 kg

Second order
(secondary)

100 kg

First order
(primary)

1000 kg

Producers

10,000 kg

Figure 4–30. A marine food pyramid; plant plankton are the base. From J. Barrett, et al., *Biology*, p. 939.

90% of the energy is lost in each transition from a lower level to an upper level. (The exact percentage lost varies from about 50% to over 95%, depending on the ecosystem. For instance, pyramids with many insects tend to have a much lower percentage lost because they conserve energy very effectively.) As an example, if all of the 10,000 kg. of plant plankton in Figure 4–30 is eaten by first-order consumers, about 90% of the calories gained will be lost as "wasted" heat energy. Only about 10% will be converted into living tissue that will then be available to consumers on the level above it. It thus takes about 10,000 kg. of plant plankton to support 1,000 kg. of plant eaters, which in turn will support only about 100 kg. of second-order customers (Figure 4–30). At the top of the food chain, a mere 1 kg. of seal tissue can be supported by all the plant plankton because so much is lost in food links in between.

Such energy leakages between levels in the food pyramid explain why large animals near the top of food chains are so rare. Both their rarity and their need for a large food base contribute greatly to the demise of large carnivores such as large cats (e.g., panthers, lions, tigers), wolves, eagles, and many others. The leakages also explain why ecologists have found that most pyramids only have three to four levels in natural ecosystems: So much energy is lost between predators and prey that there is usually little energy available beyond four levels.

The increasing diversity of the biosphere has affected the evolution of food pyramids on earth in two important ways. One, the height of food pyramids on earth has increased. The original founders of any ecosystem must be producers because they form the base of the food chain. Only after that can higher-level consumers be added. Even these must be added in order because each level depends on organisms below it. Two, the number of food pyramids on earth has increased. For example, the evolution of new kinds of plants has created new producers as the basis of new food chains. This is illustrated in Figure 4–31 with the origin of land plants, flowering plants, and other kinds of producers, and the origin of consumers that they support.

Aside from food chains and pyramids, more specific trends in predation are seen among specific groups. **Coevolutionary trends** are those in which interacting groups influence one another so that they "push" each other in a certain direction. A good example is the evolution of brain size in mammals. From the early mammals of the Mesozoic, average relative brain size has increased dramatically. This trend is coevolutionary because as one group becomes smarter, another group, such as one feeding upon it, must become smarter as well. Thus, as seen in Figure 4–32, carnivores must generally have a larger brain than the hoofed mammals (ungulates) in order to successfully prey upon them. At each step, from the archaic, through the mammals of the Paleogene (early Cenozoic) and Neogene (late Cenozoic), to the Recent, meat eaters have had larger brains. This kind of predator-prey coevolution has been called an **arms race** and is very much analogous to the arms races of modern nations. Each group must become continuously enhanced (in this case, smarter) in order to survive. Natural selection in each group is affected by the presence of the other group. This arms race concept is also apparent in the Mesozoic marine revolution (Chapter 3), wherein prey marine invertebrates (clams, urchins, and others) became more heavily armored and mobile over time as predators (snails, starfish, and others) became more proficient in killing.

There are other examples of coevolutionary trends. This is not really surprising considering that other organisms form a major selective force in any group's adaptations. Nor are all coevolutionary trends caused by predator-prey arms races.

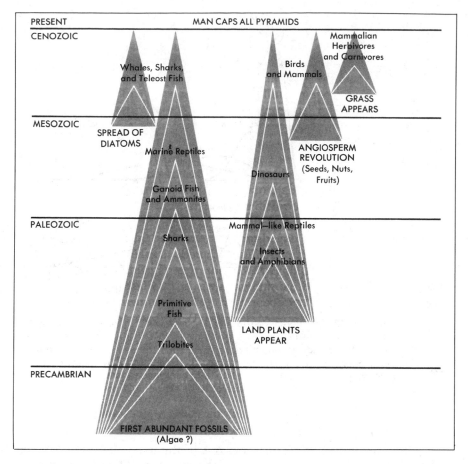

Figure 4–31. Building food pyramids through time. Each successively larger pyramid subsumes all surviving life below. From W. Stokes, *Essentials of Earth History*, p. 471.

There are other kinds of interactions among groups that can drive them along, such as cooperation and other kinds of symbioses. These are not as well documented, although there is the classic example of flowering plants and insects. They began their mutually beneficial interaction (called "mutualism") in the late Mesozoic, wherein plants are pollinated by insects and insects derive food from the plants. In this case, the existence of pollinating insects caused a progressive trend toward more flowering plants, often designed to attract specific insects. Similarly, the increasing number of such plants led to the evolution of more insects specialized for certain plants.

HOW FAR HAVE ECOLOGICAL TRENDS GONE? We have seen how trends toward increasing interactions, such as predation and competition, have built up connections among

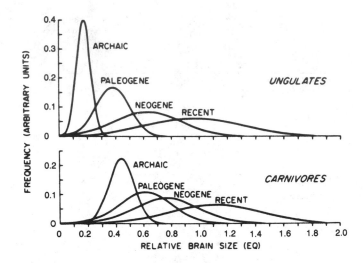

Figure 4–32. Frequency distributions of relative brain sizes of ungulates (hoofed mammals) and carnivores. Paleogene = early Cenozoic, Neogene = late Cenozoic. Both average and maximum brain size increase through time in both groups, with carnivores being generally smarter. Modified from K. McNamara, *Evolutionary Trends* (Belhaven, 1990), p. 39.

life in the biosphere. This raises one of the major (and most intriguing) questions in ecology: How interconnected is the biosphere? This has crucial practical implications as well. If biosphere food webs and other connections are very tightly interconnected, then just a minor disturbance (such as a human-caused event) may cascade through the system, killing off many other species, and collapsing the system. On the other hand, if there are relatively few connections, the system may be much more resistant to disturbances, which will remain localized.

There is currently much debate over this question. At one extreme is the **Gaia hypothesis,** which states that the biosphere is highly interconnected not only among its life forms, but also to the earth's atmosphere, oceans, and land. (Gaia was the ancient Greek goddess of earth.) This idea goes back over 200 years, but has been recently revived by James Lovelock in many books and articles. The basic point of the hypothesis is that life has become so interconnected with other life forms and its physical environment that the whole earth-biosphere system forms a living *superorganism.* The evidence proposed for this is that as life evolved, it did not just passively adapt to the physical world; it actively modified it. For example, we saw in Chapter 3 how the evolution of plants eventually created the present 21% oxygen atmosphere, which can support so much life. If the oxygen level were much more or much less, life (at least as we know it) would not exist. For example, a 25% oxygen atmosphere would lead to widespread, uncontrolled wildfires. Similarly, as humans pump tons of carbon dioxide into the atmosphere, much of the excess is being absorbed by the oceans. This presence of mechanisms to produce and maintain optimal conditions for life is called **homeostasis.**

Critics of the Gaia hypothesis argue that the biosphere is indeed a working system with lots of interconnections (and therefore negative and positive feedback mechanisms to sustain optimal conditions). However, they point out that these connections and feedbacks are not so "tight" and numerous that the whole system actually forms a superorganism. The components of the biosphere are not nearly as highly interconnected as cells in a human body. While they agree that life has modified the earth, the critics argue that connectedness and homeostasis alone do not comprise a superorganism. For example, a refrigerator is composed of highly connected parts and is homeostatic (it maintains a constant internal temperature against a changing outside temperature). But few people would be tempted to call a refrigerator a superorganism, or even an organism.

While most scientists do not believe that the Gaia hypothesis is correct for these and other reasons, it is also clear that the other extreme is false: The earth-biosphere system is not just a loosely organized set of disconnected components. As with most ideas, the truth probably lies somewhere in the middle. Ecological trends toward increasing interactions have created many interdependent relationships, but they are not so interdependent that the entire biosphere can easily collapse in domino-like cascades from minor (localized) disturbances. The reason for this intermediate level of interconnection is probably that natural disturbances (hurricanes, varying solar output, catastrophic volcanoes, and so on) are relatively common throughout geologic time scales. These intermittently disturb the biosphere (and the local ecosystems that comprise it) so that large networks of extremely high interdependence are not allowed to build up.

PART 3: EXTINCTION RATES AND DIRECTIONS

The evolutionary patterns discussed so far have involved originations: the creation of new species. However, extinction is also a major part of the evolutionary process. Recall once again, the Darwinian equation:

$$\text{Variation} + \text{Environmental Change} = \text{Evolution (or Extinction)}$$

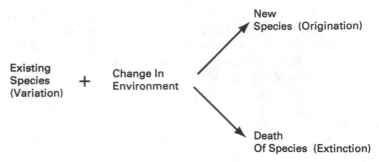

Figure 4–33. Environmental change eventually results in either origination or extinction.

As shown in Figure 4–33, if there is enough anatomical (and genetic) variation, then environmental change will result in evolution, as the most adaptive variations are favored. Extinction occurs when environmental change is too great for the amount of variation existing at that time. For example, if temperature change is slow enough, the genes that permit tolerance of the change may have time to appear and accumulate. However, if change is too fast, there will not be enough time for this adaptation.

Ultimately, all species become extinct. This is perhaps not surprising from your experience: All things must end. In the case of extinctions, this means that the everchanging world in which all life exists eventually undergoes some change that cannot be coped with. It often surprises students to learn that over *99% of all species to ever exist have become extinct.* Estimates from the number of known fossil species indicate that about 1 to 3 billion species have existed since life began. Yet only about 3 to 30 million species exist today, while the rest have died out. The average fossil species lasted a few million years. Put another way, the rate of origination has barely exceeded the rate of extinction (by less than 1%). It is this slight surplus of origination that accounts for the existence of life through time.

Defining and Classifying Extinctions

Extinction is the permanent disappearance of a group, when all individuals of that group die out. Most of us think only of species going extinct, but the "group" in our definition can refer to the disappearance of any group from a local population to entire families of species.

There are three basic categories of extinction: pseudoextinctions, mass extinctions, and background extinctions. **Pseudoextinction** occurs when an entire species disappears by slowly evolving into another species. This is simply nonbranching evolution, discussed elsewhere and shown in Figure 4–7. **Mass extinction** is the disappearance of many species of different kinds of organisms in a relatively short period of time. In addition, they occur over a wide geographic area. There have been five major mass extinctions in the last 570 million years (since the beginning of the Paleozoic era). These five mass extinctions killed off over 50% of the species alive at the time, from many groups and on many parts of the earth. **Background extinctions** are extinctions that occur during that vast time between the brief calamities of mass extinctions. These are caused by ongoing environmental changes of local and even regional scales: the eruption of a volcano, the drying up of a lake or small ocean basin, and so on. While these seem less impressive to the average person, about *95% of extinctions occurred during these background times*, going out with a whimper instead of a bang.

RATES AND DIRECTIONS OF PAST EXTINCTIONS

Measuring Extinction Rates in Fossils

Using compiled data and sophisticated statistical methods, paleontologists have been able to analyze mass and background extinctions. Figure 4–34 shows the extinction rate over the last 600 million years for marine organisms. The **extinction rate** is the number of fossilized groups (in this case, families) that go extinct per million years. It is based upon massive compilations of all known fossil groups and when they first and last appear in the fossil record. This represents the work of thousands of paleontologists over the last hundred years or so. Even so, this rate is still a crude estimate of what really happened in the history of life because our ability to date such ancient events is limited. Thus, it is a "per million year" average even though some extinction events may have occurred in a much shorter period. For instance, mass extinction from a meteorite impact, which may have killed the dinosaurs, could have occurred in only a few months or years (Chapter 3). Yet our calculated rate will be much lower because we are spreading those losses over millions of years. Ongoing studies that seek to date events with higher resolution (hundreds of thousands of years) may help reduce some of the "fuzziness," but it seems unlikely that we will ever be able to pinpoint the time boundaries of very rapid catastrophes like meteorite impacts. An additional problem is that not all species fossilize well, so we cannot say how they were affected. While these and other problems with the data continue to challenge researchers, many interesting and testable patterns have emerged.

Figure 4–34 shows the five extinction peaks representing the five major mass extinctions since the beginning of the Paleozoic. These five peaks of extinction stand significantly above the background rates, almost all of which fall below the dashed line. Two of these major events, the end-Paleozoic and the end-Mesozoic mass extinctions, were discussed in Chapter 3. These are the best known of the five since they separate the three major eras and also because of their general notoriety: the **end-Paleozoic mass extinction** was the largest extinction and the **end-Mesozoic mass extinction** was the one that destroyed the most famous of all fossils, the dinosaurs. The other three major mass extinctions were also important, however. Two of them occur in the Paleozoic era, before the end-Paleozoic event: the **early**

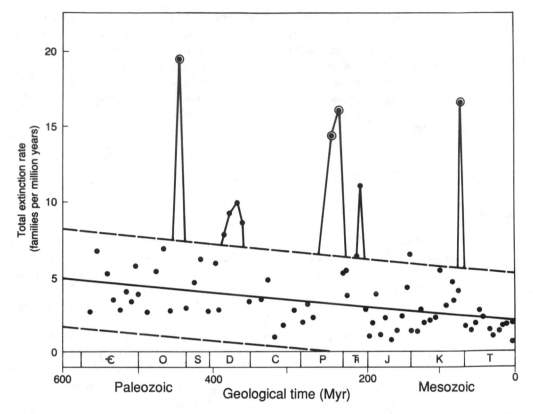

Figure 4–34. Extinction rate of families of marine organisms has generally declined in the last 600 million years, shown by solid trend line. Dashed lines represent statistical variation around the trend line. Peaks are "mass extinctions" of exceptionally high rate, well above that expected from statistical variation. Modified from D. Futuyma, *Evolutionary Biology*, p. 361.

Paleozoic mass extinction and **middle-Paleozoic mass extinction.** The third is the **early-Mesozoic mass extinction.** The more technical names for these events, used by geologists, are given in the caption of Figure 4–34.

While the five extinction peaks are calculated in terms of family extinction rates, it is possible to estimate how many species were killed off to obtain those rates. Figure 4–35 shows a **reverse rarefaction curve,** which converts the number of species killed in a group into the number of families or genera killed. For instance, killing off 70% of the species will eliminate only about 40% of the genera and less than 20% of the families. This is because each family usually contains more species than each genus, so more species must be killed to eliminate the whole family. Since Figure 4–34 was calculated with family extinction rates, we can use Figure 4–35 to estimate how many species must have died to yield those family rates. For instance, the massive end-Paleozoic family extinction rate (Figure 4–34) represents about

Reverse Rarefaction

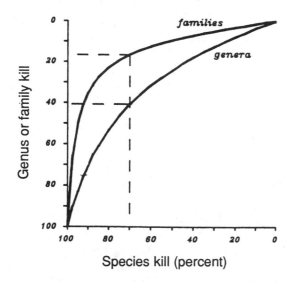

Figure 4–35. Reverse rarefaction estimates percentage of species killed from known percentages of families or genera killed. Thus, 40% genera killed is about 70% of existing species killed. From D. Raup, *Extinction: Bad Genes or Bad Luck?* (New York: Norton, 1991), p. 73.

52% of the families existing at the time. Reverse rarefaction estimates an astounding 96% species killed to achieve this. Unfortunately, this oft-cited species rate is probably wrong. Reverse rarefaction uses a statistical sampling method that assumes that species are killed randomly, but there is growing evidence that this is not the case. Because species killed were probably concentrated in various families (and genera) this curve probably overestimates the percentage of species killed. Nevertheless, the end-Paleozoic and other mass extinctions were still truly profound in their impact, killing over 50% of the species. Notice that the large-mammal extinctions at the end of the Ice Age (about 10,000 years ago) are not included on Figure 4–34. Compared to the truly massive extinctions of earlier times, this was a small perturbation, affecting relatively few organisms. The Ice Age extinctions are notable because they occurred a relatively short time ago and because of the grand size of the animals affected (see the discussion in Chapter 3).

The difference between mass and background extinctions in Figure 4–34 is not as clear-cut as one might wish. For example, one background point is above the line. Also, some of the mass extinction events (such as the middle-Paleozoic event) are "smeared": They have a number of points, indicating that they occurred over millions of years. This has led to a continuing (and unresolved) debate among scientists over whether mass extinctions are *qualitatively different* from background extinctions, or just "more of the same." Present evidence seems to favor the latter view. For example, Figure 4–36 shows a *continuous gradation* when the extinction rates of Figure 4–34 are examined on a finer scale by using genus extinction rates. If mass and background extinctions were qualitatively different, then we might expect to see extinction rates from the five mass extinctions to be separated (dis-

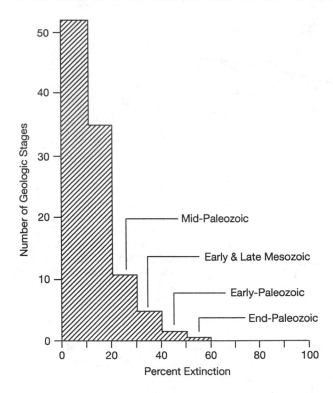

Figure 4–36. Frequency bar graph of extinction rate (percent genera killed) for 106 geologic intervals (stages). Mass extinctions are not clearly distinct. From D. Raup, *Extinction: Bad Genes or Bad Luck?*, p. 81.

junct) from the others. Instead, they intergrade with background rates. On the other hand, it may be that the coarse nature of the data noted above account for this. Averaging rates over long periods (each stage in Figure 4–36 is millions of years long) may be generating so much "noise" that the true extinction rates (perhaps as high as millions of species *per year*) are lost. In such a case, the random mixing of rates by noise alone could cause the gradation. (Consider what would happen if you took a nonrandom set of numbers, such as 1 through 10, and divided through by random numbers; the pattern would disappear and the distinction between the numbers would blur.) However, there are theoretical reasons for suspecting that the gradation between background and mass extinctions is real, as we discuss in the following section.

Extinction Directions: Trends and Possible Cycles

An important directionality seen in Figure 4–34 is the apparent decline in background extinctions through time. This is seen in the dark line drawn through the center of the point cloud and approximating the average background rate. The reason for this decline is under debate. Some suggest that it occurs because organisms have become better adapted to their changing environments through time so that fewer become extinct through random processes and competition. Others point out that families tend to gain more species through time so that there is less of a chance that a family will die out by losing all of its species.

At first glance, the timing of the five mass extinctions illustrated in Figure 4–34 appears to be what one would intuitively expect: They generally occurred at random. Three occur in the long Paleozoic, two in the shorter Mesozoic, and none in the shortest of all, the Cenozoic. However, a number of highly sophisticated statistical studies of the pattern have recently indicated that the timing is not random, but *cyclical*. This is visible in Figure 4–37, which is a more refined view of the data presented in Figure 4–34. You can see the "cycles" of extinction: The extinction peaks tend to occur on the vertical lines, which mark 26-million-year intervals. (Figure 4–37 shows mainly post-Paleozoic extinctions, but there is evidence for a similar pattern in the Paleozoic.)

Many researchers have hypothesized possible causes for such a cyclical pattern. Some theories suggest *extra-terrestrial* sources, especially periodic bombardment of the earth. Figure 4–38 shows an apparent pattern of large meteorite impact craters with peak frequencies of roughly 26 million years. What could cause such periodic bombardment? One theory notes that earth passes through the galactic plane about every 26 million years and might be bombarded with cosmic debris at that time.

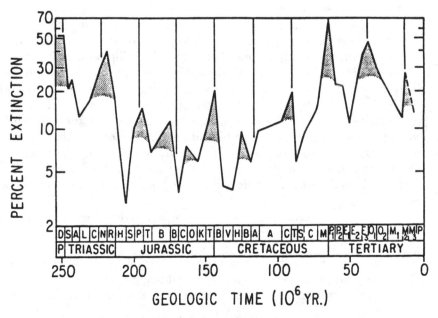

Figure 4–37. Extinction rate distribution after the late Paleozoic seems to show peaks that tend to be periodic at 26-million year intervals (solid vertical lines = 26 m.y.) From V. Grant, *The Evolutionary Process* (New York: Columbia University, 1985), p. 403, Fig. 36.4.

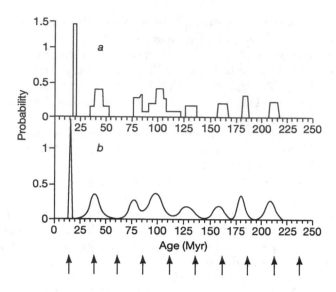

Figure 4–38. Impact craters on earth show apparent periodicity at 28 million year intervals (arrows). From J. Levinton, *Genetics, Paleontology, and Macroevolution* (London: Cambridge University Press, 1988), p. 457.

However, most astronomers doubt that the debris is dense enough to affect the earth. The most accepted extra-terrestrial theory is that there is an unseen, dark star or some other large body in orbit beyond the farthest planet, Pluto. This body, sometimes called Nemesis, is thought to have an orbit that takes it near the large cloud of comets existing beyond Pluto. Every 26 million years, it comes close enough to the cometary cloud to disturb the orbits of many comets. Having been disturbed from their own orbits, these comets then spiral in toward the sun. Some of these would then collide with the earth, causing extinctions. Other theories suggest terrestrial (earth-bound) causes for the extinction cycles. The most feasible seems to be that internal processes, deep within the earth, periodically affect plate tectonic motion, causing volcanic eruptions. This would also cause atmospheric (climatic) disturbances and other changes as gases are released by the volcanoes. Changes in the earth's magnetic field could also be affected because the field is generated by internal processes.

Many other researchers believe that past extinctions were not cyclical. These researchers point out that:

1. The peaks do not always exactly occur at 26 million year intervals (see Figure 4–37).
2. The peaks are sometimes very small, indicating small extinctions.
3. The dating of past events is often very crude so that errors in extinction timing are likely.
4. The statistical tests that indicate cycles are open to question on many methodological grounds (for example, by defining extinctions in terms of "peaks," there is a guaranteed interval that must occur between events, introducing nonrandom spacing).

5. The data are suspect in that the fossil record is incomplete and biased in ways that might produce deceptively systematic patterns.
6. There is evidence that causation differed among the five major mass extinctions.

As we shall discuss shortly, the end-Mesozoic extinction seems to have been caused by a meteorite, while the other four major extinctions seem to have other causes, mainly climatic. Nevertheless, in science all theories are given a fair trial and the search for evidence on cyclicity continues. If nothing else, this serves to illustrate that science is a continuously changing field, and that not all statements should be considered as undeniable fact. A stimulating personal account of this is *The Nemesis Affair* (1986) by David Raup, who is one of the major scientists working on past extinctions.

Causes of Past Extinctions

We said that extinction is caused by a change in the environment that is too great for a group to adapt to, even through evolution. This change can either be subtle, such as a minor change in water chemistry, or catastrophic, involving wholesale destruction of habitat, such as a meteorite impact. Reconstructing the exact cause of extinction for each species is impossible. The rock record does not record many environmental changes, such as temperature, and of course many species are not recorded at all. In addition, causes of extinction are surprisingly complex when population dynamics and organismic interactions are included so that even modern extinctions are often poorly understood (see Chapter 5). However, we can make educated inferences about past extinction causes.

ENVIRONMENTAL CHANGE AND BACKGROUND EXTINCTIONS. To start, we might infer that background extinctions were caused by relatively minor changes in the local or regional environments, affecting fewer numbers of species than the more widespread changes of mass extinctions. We would therefore expect that background extinctions tend to record those species (and any group) that have relatively restricted geographic distribution. Generally speaking, the more locally restricted the distribution, the more likely the group is to go extinct because it is more likely to have all of its habitat disturbed. We can test this by comparing patterns of distribution with the patterns of extinction. For instance, Figure 4–39 shows a *hollow curve*, illustrating that most groups (genera in this case) tend to be geographically restricted, with more widespread groups being progressively fewer. Similarly, Figure 4–39 also shows a hollow curve for the duration of fossil genera: Most genera last the shortest time (less than 10 million years), with longer durations becoming progressively fewer. The agreement of these two curves is what we would expect from our inference. If most groups (species and genera) are restricted, then we would expect most groups to have relatively short durations before some localized environmental change killed them off. These we would call background extinctions

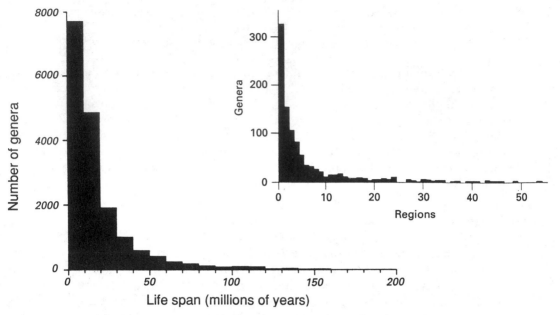

Figure 4–39. Frequency bar graphs. Top: Regions occupied by living bivalve (clam) genera. Bottom: Life spans of fossil genera. The vast majority of organisms occupy small geographic ranges and have the shortest span of existence. Top: From D. Briggs and P. Crowther, *Palaeobiology*, p. 457; bottom: From D. Raup, *Extinction: Bad Genes or Bad Luck?*, p. 55.

and they would be the most abundant extinctions of all. (Recall that about 95% of species died out as background extinctions.)

We can also infer from the above that as the amount of area (habitat) altered increases, more species will become extinct. This is indeed seen in what ecologists call the **species area effect.** As shown in Figure 4–40, there is an increase in the number of species supported as the size of the area surveyed increases. This is mainly because larger areas contain more habitats. Conversely, if a large area is reduced in size by environmental changes, some of the existing species will become extinct as habitable area is reduced. The more habitat that is altered, the more species go extinct.

ENVIRONMENTAL CHANGES THAT CAUSED MASS EXTINCTIONS. The fossil record is replete with specific examples of the kinds of local and regional environmental changes that could cause background extinctions: small seas drying up (the Mediterranean), islands forced underground in bad storms, and so on. However, to explain the five mass extinctions, we must seek the specific environmental changes that occurred on a global scale. This is one of the most active areas of research in paleontology and there is considerable debate about the causes of each mass extinction. The pro and con arguments for these are clearly discussed in two popular books, *Extinction:*

Species—Area Effect

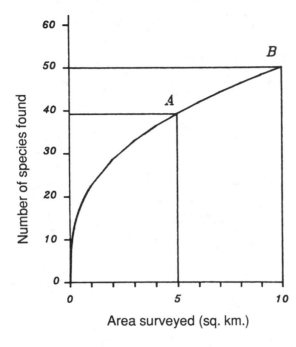

Figure 4–40. Species-area effect: as area increases, the number of species found increases rapidly at first, then slows. From D. Raup, *Extinction: Bad Genes or Bad Luck?*, p. 130.

Bad Genes or Bad Luck? (1991) by David Raup, and *The Miner's Canary* (1991) by Niles Eldredge.

Most of the debate over mass extinctions focuses on three main phenomena as causes of the global environmental change: *sea level change, climate change*, and *meteorites* (or comets). The evidence supporting change in sea level as a contributing factor is summarized in Figure 4–41, which shows that each of the five mass extinctions were associated with sea level lowering. Such drastic decreases would cause massive environmental change (habitat loss) for marine species because many shallow water areas would become land. Since about 90% of marine life resides in shallow waters (especially the continental shelves), this could cause a drastic reduction of species as described by the species-area effect. However, this apparently simple explanation is disputed by a number of paleontologists. For instance, calculations of modern marine species show that drastic sea level reduction may not cause a mass extinction because of **refugia,** such as coral atolls, which can provide refuge for shallow water groups. Another major complication is that sea level reduction is often correlated with cooling climate. As we saw in Chapter 3, water locked up in glaciers causes lower sea levels. Therefore, some paleontologists (in particular, see Steven Stanley, *Extinction* (1987)), argue that global climatic cooling is the ultimate cause of mass extinctions. A particularly cogent argument to support this theory is that tropical species tend to be the most highly affected during mass extinctions.

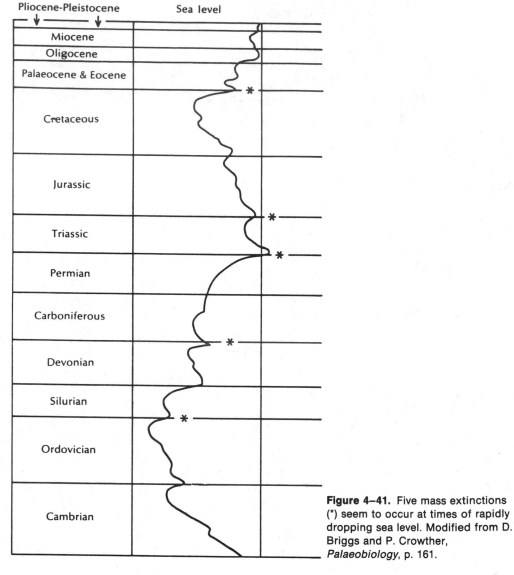

Figure 4–41. Five mass extinctions (*) seem to occur at times of rapidly dropping sea level. Modified from D. Briggs and P. Crowther, *Palaeobiology*, p. 161.

The evidence for extra-terrestrial (meteor or comet) impacts is most prominently argued by David Raup. We have already seen one line of evidence for this in Figures 4–37 and 4–38 where periodic extinction peaks seems to align with bombardment peaks. Yet, as we have also seen, others debate such evidence.

Keeping in mind these contentions about causes of mass extinctions, let us try to summarize current thinking. Table 4–2 lists the causes that most recently have been put forth. In addition, two minor extinction "events" are added because they

TABLE 4–2. Suggested Causes of Past Mass Extinctions and Two Extinction "Events."

EXTINCTION EVENT	PROBABLE CAUSE
Major Mass Extinctions:	
end-Mesozoic	Meteorite impact
early-Mesozoic	Global cooling, low sea level
end-Paleozoic	Global cooling, low sea level due to Pangea forming
middle-Paleozoic	Global cooling, low sea level
early-Paleozoic	Global cooling, glaciation
Smaller Extinctions Events:	
late-Cenozoic (end-Ice Age)	Global warming plus human overkill
early-Cenozoic	Global cooling

Source: Stephen K. Donovan, *Mass Extinctions* (New York: Columbia Univ. Press, 1989).

have been exceptionally well studied. Of the five major mass extinctions, four are associated with global cooling that can be seen in the evidence for glaciation, falling sea level, and other indications of cooling from the rock record. (Such methods as isotope ratios can be used to directly deduce paleotemperatures because isotopes of minerals in the rocks are affected by temperature.) Furthermore, both of the best-studied minor mass extinctions also involved temperature change. Thus, while the debate over what caused these past extinctions continues, global cooling is a prime candidate as a major factor in most mass extinctions. Indeed, there is evidence that even the sole exception, the end-Mesozoic extinction event, was influenced by long-term temperature decreases. The meteorite impact simply delivered the coup de grâce to many groups already in decline from temperature change. Moreover, one of the main results of the meteorite impact itself is global cooling from the huge suspended dust cloud (refer back to Chapter 3).

CATASTROPHES: THE INEVITABILITY OF THE IMPROBABLE. The basic lessons learned from the fossil record tell us about both background and mass extinctions:

1. All species die out from environmental change.
2. Such environmental change is inevitable.

The only way that life has survived is by that small surplus of originations over extinctions. An analogy is that of the Red Queen from *Alice in Wonderland*: life must keep going (adapt) just to stay in place (stay alive).

Most environmental changes are relatively small and common. You know this from your own experience: Large earthquakes, storms, floods, and so on, are rare compared to quick rain showers, minor snowstorms, and other common changes. The logical conclusion is that if one waits long enough, a highly improbable catastrophe will eventually happen. For instance, Table 4–3 shows that astronomers have found that meteorite and comet impact craters of 150 km. (about 100 miles) in diameter are made about once per 100 million years, with smaller craters being

TABLE 4–3. Average Time Interval Between Impacts of Space Debris With Earth

CRATER DIAMETER	AVERAGE TIME INTERVAL
> 10 km	110,000 years
> 20 km	400,000 years
> 30 km	1.2 million years
> 50 km	12.5 million years
> 100 km	50 million years
> 150 km	100 million years

Source: D. Raup, *Extinction: Bad Genes or Bad Luck?* (New York: Norton, 1991).

made progressively more often. If plotted, these data would form the same kind of "hollow curve" shown in Figure 4–39, with small craters being the most common. It takes a meteorite about 10 km. (7 miles) in diameter to create a crater of 150 km. in diameter. This is the estimated size of the meteorite that is thought to have killed the dinosaurs at the end-Mesozoic mass extinction 65 million years ago. On average, then, we are due for another major strike in about 35 million years or so. This average interval between environmental events of specified size is called the **return time** and can be used as a valuable predictor of change.

David Raup has quantified the return times of extinction rates in the fossil record, based on 20,000 fossil genera. As shown in Figure 4–42, this so-called **kill curve** shows that an extinction rate of 5% of global species occurred once every 1 million years. As the kill curve is derived from the hollow curve distribution characterizing all environmental changes, where larger extinctions are progressively less common, we see that a mass extinction of 65% of global species occurs only about once per 100 million years.

Does the kill curve provide any direct clues to causes of past extinctions? Superficially, you might think that it does. For instance, the very idea that events of certain sizes have an average time interval of occurrence seems reminiscent of cycles of meteorite bombardment. However, because nearly all natural phenomena (sea level change, temperature extremes, volcanic activity, and so on) tend to follow the hollow curve, the kill curve of Figure 4–42 would be generated even where many environmental changes co-occurred to cause extinction. For example, Figure 4–43 shows how extreme changes in volcanic activity, cooling, sea level, and other factors may act like a dice roll. In most rolls, just one or two of the dice would show sixes. But eventually a roll with all sixes would occur. To be exact, $(1/6)^8 = 1$ in 1,679,616 rolls. Not all the factors shown in Figure 4–43 were likely involved, and, unlike dice, volcanic activity, sea level, and climate change are not independent. Climate change can affect sea level, and volcanoes can affect climate, for example. However, some factors are largely independent, such as meteorite impacts and volcanoes, and the general idea that mass extinctions occurred from a confluence of rare events is a valid one. We have already noted this as a possibility in the end-

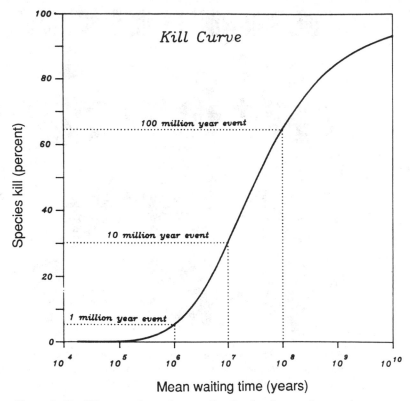

Figure 4–42. Kill curve shows that small extinction "events" are very common (about every 1 million years). About once per 100 million years, events that kill 65% of existing species occur. From D. Raup, *Extinction: Bad Genes or Bad Luck?*, p. 85.

Mesozoic mass extinction: a meteorite strike as the coup de grâce after a decline caused by long-term decreasing temperatures.

Extinctions and the Direction of Evolution

We have seen how patterns of extinction show directionality (trends), such as the decline in extinction rates through time, and the possibly nonrandom placement (cyclicity) of peak extinction events. However, extinctions also have an important influence on the direction of evolution in general. During most of earth's history, extinctions have reflected evolution according to the basic ecological rules of natural selection: those organisms that can best compete, acquire food, care for young, and are generally best adapted to their environment, tend to have more offspring. But during rare, catastrophic events, the environment is radically altered and a new set

THE "RARE EVENT" OR THE 8-DICE GAME

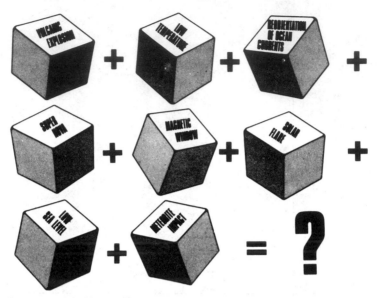

Figure 4–43. The inevitability of the improbable: eventually even rare events can occur simultaneously. From W. Berggren and J. van Couvering, Eds., *Catastrophes and Earth History*, p. 79.

of rules is temporarily created. Survivors are those who are favored during these new, temporary rules. In this way, mass extinctions act as *bottlenecks*, which can have major effects on the evolution of life. Even if a group is highly fit and can thrive during normal times, it may be eliminated during a catastrophe.

MASS EXTINCTIONS: SELECTION BY A NEW SET OF RULES. We should distinguish selection by *new rules* from selection by *no rules*. Both of these can occur during catastrophic events. Selection by "no rules" refers to totally random extinction of organisms without regard to any of their traits. For example, a massive meteorite impact would obliterate all individuals within a large area for many square miles. It would not matter if individuals were the fastest, smartest, or most aggressive—each would die just as every other individual under these conditions. In his book, *Extinction: Bad Genes or Bad Luck?* (1991), David Raup calls this **field of bullets selection** because it is totally random in who it kills, just as a hail of bullets would be.

In contrast, selection by "new rules" means that selection is not random. However, the new conditions created by the catastrophe select individuals in ways that differ from normal times. Raup calls this **wanton selection** because the traits that determine survival evolved during normal times and "just happen" to be beneficial during the new demands of a particular catastrophe. For example, recall how mammals survived the end-Mesozoic mass extinction, perhaps because of their small size. Because natural selection cannot plan ahead, the small size of mammals evolved to meet the needs of their environment during normal times. Perhaps many were nighttime foragers. When the end-Mesozoic catastrophe occurred, this small size may have fortuitously been advantageous. Assuming a meteorite impact, the resulting food web collapse may have favored small animals with fewer food needs and rapid reproductive rates.

Interestingly, the new rules used by wanton selection may be roughly the same in at least some of the five mass extinctions. Table 4–4 shows traits that have been described by paleontologists as promoting survival during mass extinctions. These have much relevance for today because at least some of these traits appear to promote survival during the ongoing mass extinction of life caused by humans (see Chapter 5). Cold-adapted organisms apparently fared better because they could tolerate the cooler temperatures that caused so many mass extinctions. Cold-adapted organisms probably migrated south with cooling temperatures; tropical organisms, on the other hand, may have had nowhere to go if equatorial habitats became cold. Groups that are widespread have tended to do better because they have more representatives in different places, maximizing the chance of escaping the radical effects. Small-sized organisms require less food and space, usually have much higher population densities, and can reproduce much faster than larger organisms. Deep water groups tend to survive better than shallow water groups in marine organisms, perhaps because they are buffered better in their deeper habitats. Terrestrial organisms sometimes seem to survive more often than marine organisms.

These survival traits in Table 4–4 are only *limited generalities*. For one thing, the traits reflect only preferential survival in a statistical way. Having smaller body size or being widespread seems to improve a species' chances, but plenty of smaller species still become extinct. Another limitation is that this preferential survival is not evident in all mass extinctions for each trait. For instance, smaller body size is most often cited as an advantage at the end-Mesozoic mass extinction, especially in reference to smaller mammals surviving over dinosaurs. However, for some of these traits there is no confirmed evidence that they enhanced survivorship. Of the traits

TABLE 4–4. Traits That Seem to Promote Survival During the "Wanton Selection" of at Least Some Mass Extinctions, Including the Current Mass Extinction Being Caused by Humans

WINNERS	LOSERS
Cold-Temperate habitats	Tropical/Reef habitats
Widespread range	Narrow range
Small body size	Large body size
Deep water habitat	Shallow water habitat
Terrestrial habitat	Marine habitat

listed in Table 4–4, the most generally useful ones seem to have been cold/temperate temperature tolerance and being widespread. At least these are the two traits for which there is currently the most evidence in most of the mass extinctions.

Finally, the traits in Table 4–4 are sometimes *not independent* of one another. Species in cold and temperate climates, for example, tend to have wider ranges than tropical species. (This is called Rapoport's rule.) The reason seems to be that tropical species are more specialized for specific, stable habitats. In contrast, the yearly seasonal changes present in cold and temperate zone habitats mean that those organisms must stay more broadly adapted and are therefore able to range over a wider area. The importance of this to mass extinction survival is that organisms that have one of the newly favored traits may often have others. It may be that this "covariation of traits" enhances survival considerably beyond just having one of the traits.

In summary, mass extinctions act as bottlenecks through both field of bullets and wanton selection. Even though only about 5% of all extinct species died out during the five mass extinction events, the bottleneck effect occurred because over 50% of species *at that time* were killed. This introduced a profound influence on the direction of biosphere evolution. A classic example is the very existence of humans, and the dominance of mammals in general: Without the wanton selection introduced by the end-Mesozoic mass extinction, the elimination of dinosaurs and consequent rise of mammals would likely not have occurred.

BACKGROUND EXTINCTIONS: SELECTION BY THE OLD RULES. While mass extinctions exercise much influence over the direction of biosphere evolution, extinctions occurring during the much longer normal (background) times also exert much influence. Indeed, they provide the primary, underlying directionality into which the influence of mass extinctions is only temporarily interjected.

The specific causes of background extinctions are much more varied than mass extinctions, involving many kinds of disturbances of local and regional environments. An especially interesting cause that is not encountered in mass extinctions is environmental change induced by biological agents. For instance, exceptionally effective predators can eliminate prey species under some circumstances, as has been shown when humans introduce predators into new environments. Thus, the evolutionary appearance of predators has probably led to the demise of some prey species. Similarly, competition among species for food or other resources can cause the extinction of less effective competitors in some situations in modern ecosystems. Thus, the evolution of new, more effective adaptations might be expected to lead to the extinction of less effective competitors. We already mentioned how the three great phases of plant evolution apparently included competitive replacement of older groups.

Another example of biological causes of extinction may involve both predation and competition. The **Great American Interchange** occurred about 3 million years ago when North and South America became connected from geologic activity creating Central America. Many kinds of organisms were permitted to migrate from South America to North America and vice versa. This led to the extinction of formerly isolated groups, especially in South America. Figure 4–44 shows the example of hoofed grazing animals (ungulates): North American forms such as horses and camels replaced South American forms. This is thought by many to have occurred because North American grazers were competitively superior. It has also been suggested that immigrating North American predators, such as cats and dogs, were more effective hunters, further diminishing the native grazers. The suggested reason that South American organisms were poorer competitors and predators is that South America (unlike North America) had been isolated from the rest of the world for many millions of years. No one can prove that North American organisms were more effective competitors and predators, but the circumstantial evidence is persuasive to many. For instance, while many South American organisms became established in North America, they caused relatively few extinctions, apparently finding unoccupied ways of life.

Group Extinction Patterns

Our discussion of extinction has thus far focused on a very coarse view, that of the entire biosphere; in other words, all species considered together. However, when we observe the extinction of specific groups through time, we also find very interesting patterns. These provide a finer-scale view of extinction.

GROUP LONGEVITY. Figure 4–45 shows that each group tends to have a characteristic lifespan for each species. For instance, trilobite, insect, and mammal species tend to last an average of only 1 to 2 million years, as shown on the horizontal axis of Figure 4–45. They thus have much higher background extinction rates than species of other groups, such as corals or bivalves, which have species' averages of over 10 million years.

While there are probably many factors that govern such average lifespans, one important generalization often made is that higher extinction rates tend to be found in more complex organisms. Recall that trilobites, mammals, and insects are anatomically more complex than, say, many invertebrates, such as bivalves, corals, and the others shown.

CORRELATION OF GROUP ORIGINATION AND EXTINCTION RATES. If the species that comprise a group, such as mammals, last only a short time, then how does the group persist through geologic time? The answer must be that groups with high extinction rates must also have high origination rates in order to replace what was lost. In other words, origination rates are correlated with extinction rates. We said above that life persists only because there is a slight surplus of origination over extinction and the same must hold true for the groups that compose the biosphere. Indeed, this has been shown with fossil data. Groups with high group origination rates, as discussed earlier, also have high group extinction rates.

Given that this correlation must exist for a group to persist, we can next ask

Bears
Camels
Cats
Deer
Dogs
Elephants
Horses
Peccaries
Rabbits
Raccoons
Skunks
Tapirs
Weasels

Anteaters
Armadillos
Capybaras
Glyptodonts
Monkeys
Opossums
Porcupines
Phorusrhacid
Sloths
Teratorns
Toxodonts

UNGULATE GENERA IN SOUTH AMERICA

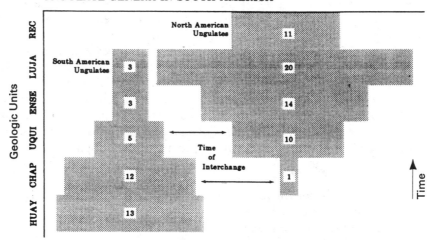

Figure 4–44. The Great American Interchange led to the replacement of South American groups by North American groups. Width of bar in graph shows number of ungulate (hoofed mammal) genera living in South America with most recent time at top. Top: From R. Cowan, *History of Life* (London: Blackwell Science, 1990), p. 370; bottom: modified from *Paleobiology* magazine, 1991, p. 272, Fig. 2.

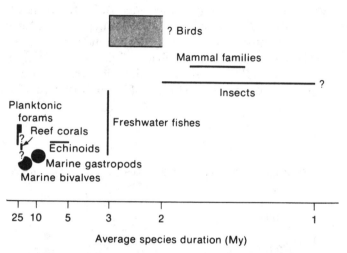

Figure 4–45. More "complex" groups tend to have species that become extinct sooner. Modified from S. Stanley, *Macroevolution*, p. 231.

why it exists: What is it about organisms with high extinction rates that also causes them to originate more often? One cause is related to pseudoextinction, wherein species often become extinct by evolving into another species. Groups with species that evolve (originate) this way often will also have species that become extinct often because the origination of one species by definition involves the extinction of the ancestor. In addition, there are a number of other reasons for the correlation that evolutionary biologists are now debating. In general, it seems that traits that lead to frequent origination from branching evolution also tend to lead to frequent extinction. For example, poor dispersal leads to new species because a population is easily isolated. At the same time, this leads to frequent extinction because the poorly dispersing group is often not very widespread. Complexity of the organisms may also play a role. We noted above that complex animals may tend to have higher origination rates for various reasons. The correlation of origination with extinction would therefore imply that complex organisms also tend to have higher extinction rates.

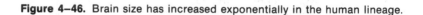

PATTERNS OF HUMAN EVOLUTION

HUMAN BIOLOGICAL EVOLUTION

Directional Patterns

The human lineage shows many directional trends over the last few million years, from *Australopithecus* to modern *Homo sapiens*. Virtually all of these are, directly or indirectly, associated with our increasing reliance on cultural adaptations. Examples of direct associations with culture include trends toward better manual dexterity for tool-working and numerous pelvic and limb modifications for upright walking (to free the hands). The most direct biological trend from cultural reliance is of course increasing brain size. Figure 4–46 shows that this increase, like nearly all

Figure 4–46. Brain size has increased exponentially in the human lineage.

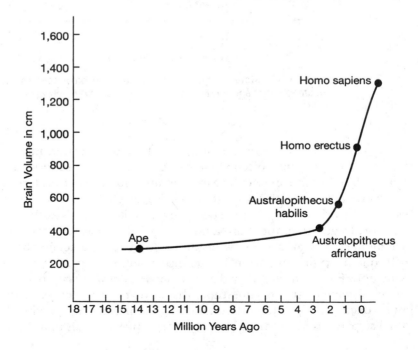

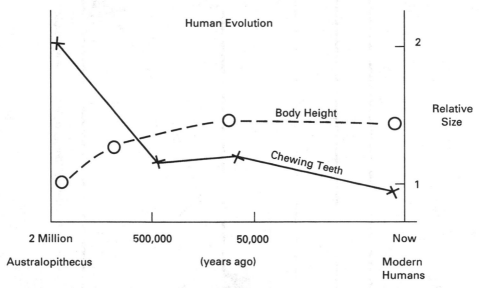

Figure 4–47. Two trends in human evolution: increasing body size and decreasing brain size. Data from C. Brace, *The Stages of Human Evolution* (Englewood Cliffs, NJ: Prentice-Hall, 1988).

evolutionary trends, has not been a simple straight-line change. Rather, it is strongly curved upward, indicating a progressively more rapid increase with time. Such an exponential pattern occurs when positive feedback processes create a "snowball effect": each increase in brain size led to selection favoring even greater brain size and more cultural adaptation.

Biological trends also occurred from more subtle effects of increasing cultural adaptation. Examples are trends in body and tooth size. As shown in Figure 4–47, humans have generally increased in body size since *Australopithecus*, from an average adult height of about 3 to 4 feet tall to around 5 feet. Many factors influenced our increasing size, but two of the most important are related to culture: (1) size advantages in hunting of larger game, and (2) our delayed juvenile (learning) period led to a longer growth phase. A second trend seen in Figure 4–47 is decreasing tooth size. This was caused by an increase in the amount of softer foods consumed, both from a change in diet (more meat) and the preparation of foods (such as cooking with fire).

Rate Patterns

Human body and tooth size trends differ from brain size in an important way: instead of a curve of ever-increasing rate of change, the body and tooth size curves show a tendency to level off. After a period of fairly rapid change, the rate becomes much slower. Such leveling off is common in many evolutionary trends. It occurs

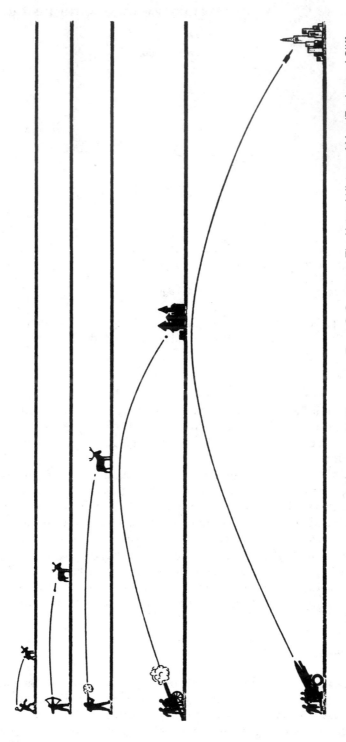

Figure 4–48. Evolution of weapons: increased range and killing power. From C. Swanson, *The Natural History of Man* (Englewood Cliffs, NJ: Prentice-Hall, 1973), p. 296.

because, after an early period of initial benefits, the processes that cause the trend eventually encounter limiting factors. For example, human body size can only become so large before gravity and other forces counteract the advantages. Similarly, if tooth size is reduced too much, even soft foods are not chewable. In the future, brain size increase will likely level off as well, if it has not already begun to do so. This is largely due to limits on the female pelvic size. It has been suggested that the growing use of computers will even reduce the need for biological storage of information in the brain (see Chapter 5).

CULTURAL EVOLUTION

Directional Patterns

The primary trend in cultural evolution has been toward increasing complexity. The most obvious manifestation of this, which is visible in the archaeological record, is that of more complex technology. From unmodified natural objects, humans have created an ever more complex repertory of tools from many materials. We are surrounded by technology of all kinds today, but for most of our past the technology was concerned with weapons, both for killing food animals and for defense against other humans. Thus, Figure 4–48 shows the evolution of increasing complexity in weapons, from simple hand-thrown objects to bows and arrows, guns, cannons, and missiles.

The trend toward technological complexity and increasing efficiency is, of course, bought at a price. Technology of all kinds, from cars to missiles, requires

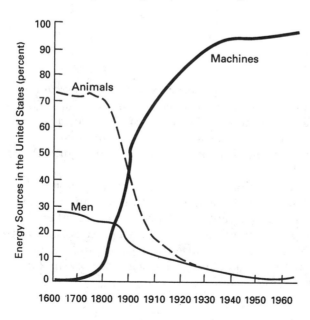

Figure 4–49. Exponential change: increasing use of machine energy, decreasing use of human and animal energy. From C. Swanson, *The Natural History of Man*, p. 305.

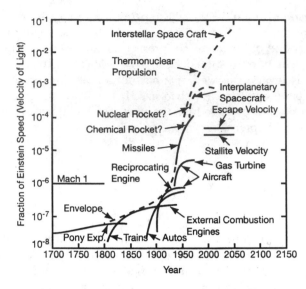

Figure 4–50. Exponential increase in transportation speeds. Dashed lines are extrapolations, which may not be justified. Modified from R. Prehoda, *Your Next Fifty Years* (Ace Books, 1978), p. 35.

energy, so there has been another trend in the amount of energy harnessed from the environment. The anthropologist Leslie White has argued that the best objective way to measure cultural evolution is by the amount of energy a society uses. An example of this is the decrease in animal and human energy in the last few hundred years in the United States. As shown in Figure 4–49, this decrease was accompanied by an increase in machines, which require much more energy from the environment than humans or animals.

Rate Patterns

Cultural trends show the same nonlinear patterns as human biological trends. As shown in Figure 4–50, the rate of change in transportation speed has so far shown the same exponential increase as brain size (Figure 4–46). However, like all trends, we would expect this, and all technological curves, to eventually encounter limiting factors. In this case, the ultimate limit is probably the speed of light, although we may be limited long before that.

There are many other accelerating cultural trends that have not yet begun to level off. For example, global human population continues to increase exponentially, as does our consumption of minerals, petroleum, food, and other resources. We will discuss these in Chapter 5 because such curves graphically summarize the current human predicament: How can we cause these trends to level off, on our own terms, without having nature impose its own, much more severe, limitations, such as famine and ecological destruction?

SUMMARY

Evolution is change through time. Evolutionary patterns can be subdivided into two categories: rate and direction of this change. Patterns of biological evolution can be divided into **originations** (the creation of new species or other groups), or **extinctions** (the termination of species or other groups). There are two ways by which evolutionary rate can be measured: **anatomical rates** are determined by direct measurement of a single anatomical trait of a group, while **group origination rates** are determined by observation of a group's appearance in the fossil record. Group origination rates measure change in many traits and can be used to characterize large-scale patterns and allow rate comparisons among different kinds of organisms. Both anatomical rates and group origination rates vary over time. Periods of **stasis** exist during which no evolutionary change occurs in a species. **Absolute rates** measure change over a defined period of time, while **relative rates** measure change over undetermined time intervals.

A major debate in evolutionary theory is whether most evolutionary change occurs gradually over long periods of time (**gradualistic evolution**) or rapidly over relatively short periods of time (**punctuated equilibrium**).

Correlated traits are groups of traits that tend to evolve together, such as increased brain size and body size in humans. **Mosaic evolution** describes traits that evolve at different rates. Similarly, **correlated trends** are trends in a trait that accompany trends of another trait. **Allometry** is the study of correlated traits that utilizes **allometric plots** to directly compare the amount of correlation between the traits.

The development of molecular biology has provided new methods for determining origination rates. **DNA hybridization,** for example, allows DNA strands from different species to be mixed and measurement of their similarity to be calculated. **Amino acid sequencing** directly compares molecules that make up proteins from different species. Even **molecular rates** of evolution vary with time. However, some molecules are biochemically regulated and have a constant evolutionary rate, as described by the **neutral theory.**

Intrinsic factors are characteristic of the organism, and include mobility and complexity; **extrinsic factors** include environmental processes. Both factors affect evolutionary rates.

Key innovations can lead to **adaptive radiations** and rapid evolution as a group adapts to a previously unoccupied environment.

Two types of evolutionary trends are **lineage trends,** seen in an evolving lineage, and **biosphere trends,** seen in the evolution of life as a whole. There are two types of lineage trends: with **divergent trends** a group evolves into two distinct directions, and with **convergent trends** unrelated groups follow the same trends and develop anatomical similarities. Divergent trends lead to **homologous organs** while convergent trends lead to **analogous organs.**

Two biosphere trends are an increase in diversity and increase in ecological complexity. Whether this apparent increase in diversity is real or an artifact of a biased fossil record (as proposed by the **equilibrium diversity hypothesis**) is debated, but most think it is real. Increased diversity has been influenced by **niche packing, competitive replacement,** and **ecological release.** Two trends that accompany increased diversity are increases in size and complexity.

Ecology is the study of interactions between organisms and their environments. Examples of ecological trends involve competition and predation. A predominant trend in predation is the development of a complex predator-prey feeding system defined by **food chains, food webs,** and **food pyramids. Coevolutionary trends** occur in which interacting organisms push each other in certain evolutionary directions, such as predator-prey organisms.

The **Gaia hypothesis** describes organisms and their physical environment as being highly interconnected to the point of forming a living "superorganism" maintained by **homeostasis** mechanisms that produce optimal conditions for life.

Extinction occurs when environmental change is too great for the amount of variation that exists. There are three types of extinction: **pseudoextinction, mass extinction,** and **background extinction. Extinction rate** is measured by the number of groups that go extinct per million years based on observation of the fossil record. **Reverse rarefaction** allows estimation of the number of species that became extinct during particular extinction events.

Background extinctions are caused by relatively minor changes in the environment with more species becoming extinct as more habitats are affected, as described by the **species area effect.** Background extinctions can also be caused by biological agents. An example is the **Great American Interchange.**

Mass extinctions are caused by environmental changes on a global scale such as sea level or climate change. During these events random extinctions of organisms may occur without regard to their traits **(field of bullets selection),** or organisms with traits beneficial in the new environment may preferentially survive **(wanton selection).** Whether mass extinctions are qualitatively different from background extinctions or just larger scale events is a debated question in evolutionary theory. In general, extinction rates are correlated with origination rates.

Trends in human biological evolution are a result of cultural adaptations. Most trends level off when some **limiting factor** is reached.

The primary trend seen in cultural evolution is an increase in technological complexity. Others are the increase in human population size and the utilization of natural resources.

KEY TERMS

rate	group origination rate
direction	stasis
origination	punctuated equilibrium
extinction	gradualistic evolution
anatomical rate	correlated trait

mosaic evolution
living fossils
DNA hybridization
amino acid sequencing
molecular rates
neutral theory
intrinsic factor
extrinsic factor
adaptive radiation
key innovation
increasing diversity hypothesis
niche packing
niche
Paleozoic fauna
modern fauna
competitive replacement
ecology
food chain
producer
consumer
food web
food pyramid
coevolutionary trend
arms race
Gaia hypothesis
homeostasis
extinction
pseudoextinction
mass extinction

background extinction
ecological release
lineage trends
biosphere trends
Bergmann's rule
orthoselection
random walk
divergent trend
convergent trend
homologous organ
analogous organ
allometry
allometric plot
Cope's rule
equilibrium diversity hypothesis
extinction rate
end-Paleozoic mass extinction
end-Mesozoic mass extinction
early-Paleozoic mass extinction
mid-Paleozoic mass extinction
early-Mesozoic mass extinction
reverse rarefaction curve
species area effect
refugia
return time
kill curve
field of bullets selection
wanton selection
Great American Interchange

REVIEW QUESTIONS

Objective Questions

1. Define origination and extinction.
2. Differentiate between anatomical rate and group origination rate.
3. Give two examples of anatomical rates.
4. Distinguish between absolute and relative rates.
5. What is punctuated equilibrium? How does this differ from gradualism? Which pattern is observed in the fossil record?
6. What are three advantages of group origination rates?
7. How are DNA hybridization and amino acid sequencing techniques used to measure similarity of species?
8. How do key innovations make adaptive radiations possible?

9. Define convergent and divergent trends. With which type of trend is each associated?
10. What correlated traits are observed in horses?
11. What is the equilibrium diversity hypothesis and how is it related to bias in the fossil record?
12. Give an example of each: primary producer and secondary producer.
13. How is "superorganism" defined by the Gaia hypothesis?
14. Differentiate between the following: pseudoextinction, background extinction, and mass extinction.
15. How many mass extinctions have occurred during the Paleozoic? When did they occur? What were possible causes of each event?
16. What evidence suggests that extinction occurs in cycles?
17. What evidence suggests that extinction is random?

Discussion Questions

1. Define and give an example of each: lineage trend, biosphere trend.
2. How do random walk experiments influence our interpretation of evolutionary trends?
3. What is the increasing diversity hypothesis and how is it related to the concept of niche packing?
4. How does the concept of survival of the fittest relate to background extinctions? How does it relate to mass extinctions?

SUGGESTED READINGS

On earth history patterns

VAN ANDEL, T. 1985. *New Views on an Old Planet*. Cambridge Univ., Cambridge.

On evolutionary patterns

SIMPSON, G. 1983. *Fossils and the History of Life*. Scientific American Library, NY.
ALLEN, K. and BRIGGS, D. 1989. *Evolution and the Fossil Record*. Belhaven, London.

On origination and trends

BONNER, J. 1988. *The Evolution of Complexity*. Princeton Univ. Press, Princeton.

On extinction

RAUP D. 1991. *Extinction: Bad Genes or Bad Luck?* Norton, NY.
ELDREDGE, N. 1991. *The Miner's Canary*. Prentice-Hall, Englewood Cliffs, NJ.

5

Evolution, Present and Future: Extinctions, Biotechnology, and ?

OVERVIEW: EXTRAPOLATING THE PAST

Having discussed the processes, products, and patterns of evolution, we will now look towards the future. However, studying the past is much easier than trying to predict the future. For instance, as shown in Figure 5–1 (top), many people are tempted to use simple **extrapolation** to predict future patterns. Extrapolation is the direct extension of past patterns, without allowing for any major change in the trajectory of events. Such extensions can be based on past events that have linear or curvilinear patterns (Figure 5–1).

Extrapolation has great appeal, but, in reality, it only works for systems that are very simple, such as the growth of bacteria in a test tube. In the real world, such as in ecological or social systems, there are always a huge number of complex factors that affect the future trajectory. For example, we saw in Chapter 4 how patterns of evolution often show an "S-shaped" curve as conditions changed, causing a formerly exponential process to level off. Perhaps the best example of all is the stock market, since more time and energy is expended on predicting that pattern than any other in the world (for obvious investment reasons). As shown in Figure 5–1 (bottom), complex systems have many fluctuations and radical rises or falls are unpredictable from past changes. For instance, no one could have foreseen the Cuban missile crisis before 1962 and its effect on the market. A biological example would be the impact of a giant meteorite on the biosphere, like the one that may have killed the

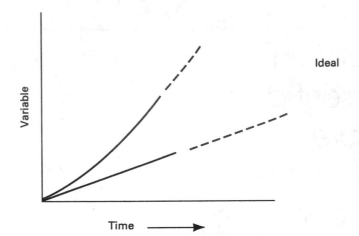

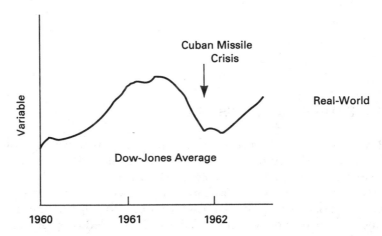

Figure 5–1. Change through time in an ideal simple system versus what we usually see in the real world.

dinosaurs. A graph of species diversity up to that point would be meaningless in predicting the massive diversity loss that could occur from such a catastrophe.

PREDICTION AND SPATIAL SCALE. The farther ahead we look, the less relevant present conditions become, making prediction more difficult. The faster change occurs, the less farther ahead we can predict with confidence. But what do we mean by "fast"? Discussions of change are meaningless unless we specify our scale of reference. This is because, as shown in Figure 5–2, change occurs according to many spatial scales. For instance, global scale changes "normally" occur in the range of decades to

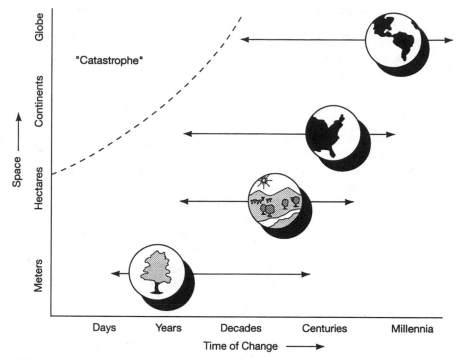

Figure 5–2. Relation of change in space to time. Modified from *Bioscience* magazine, Sept. 1991, p. 576.

millenia, while regional and smaller scale changes occur more rapidly. However, as Figure 5–2 shows, **catastrophes** do occur, defined as changes that are exceptionally rapid for that spatial scale. Thus, a change taking decades is normal for local areas of the earth, but is a catastrophe if it is global in scale.

We have seen that catastrophes have occurred naturally at the global scale in the geologic past (see Chapters 3 and 4). In this chapter we describe the many ways that humans are causing catastrophic changes on global scales, in both the physical and biological spheres of earth. These changes have largely caught humanity by surprise because of the exponential growth of the causes (for example population changes and technological growth). Unfortunately, even though humanity is becoming more aware of these problems, the ability to cope with them is inhibited by the very large spatial scales involved. In their book *New World, New Mind* (1990), Robert Ornstein and Paul Ehrlich discuss how the human mind has difficulty fully understanding changes that occur at scales outside our immediate environment. This has led to a "putting out fires" approach to problem-solving that is not conducive to strategic planning for global changes in our environment.

The extent that humans can successfully cope with global rapid change remains, of course, to be seen. However, one of the major reasons for our study of earth's and life's past is to find information (such as processes and patterns) that can help predict, and sometimes even control, future events. Perhaps learning about past changes across large time and spatial scales can help each of us focus more on problems beyond our small, personal scales of time and space.

PROSPECTS FOR PHYSICAL EVOLUTION

We start by assessing the future of the universe. We then narrow our view to the future solar system and earth. Recall from our discussion of patterns of physical evolution that physical processes eventually tend to "run down" due to the second law of thermodynamics (the law of entropy). Therefore, it should be no surprise that the ultimate fates of the universe, solar system, and the earth will produce conditions quite different, and probably much less hospitable for life, than those today. In all physical systems, when usable energy is completely lost, the resultant environment is clearly not conducive to life: stars burn out, planets freeze and disintegrate. Yet the entire universe is not necessarily doomed to eternal oblivion. One possible future is that the universe will be renewed in another Big Bang.

Future Evolution of the Universe

Recall that since the Big Bang, about 15 to 18 billion years ago, the universe has been expanding (Chapter 3). Thus, common sense would indicate only two possible futures for the universe: either it will continue its expansion forever (the so-called **open universe**), or it will eventually stop expanding and, due to gravitational attraction of its parts, reverse direction and collapse back upon itself (**closed universe**). These are graphically shown in Figure 5–3. The current evidence tentatively supports the open universe, although this evidence is far too weak to allow much confidence in it. However, in either case, there is no need for anyone to worry about his or her immediate fate being affected. The universe will continue very much as it is for many millions of years to come.

But this same massive scale that makes the universe irrelevant to your individual fate also gives it deep philosophical and religious implications. For instance, can it be true that no matter what we and our descendants do, human existence is destined to be snuffed out? Many such questions that have occupied thinkers of all cultures for centuries are directly affected by what **cosmologists** (scientists studying the origin and fate of the universe) will find. For the first time in human history, we can finally accumulate direct information to answer these questions, instead of relying on pure logic, revelation, or some other means.

THREE MAJOR LINES OF EVIDENCE. Three major lines of evidence are being closely examined to determine whether the universe is closed or open. One is the current size of the universe. As seen in Figure 5–3, an open universe might be larger at this time than a closed universe. A main problem with this evidence is that even the best

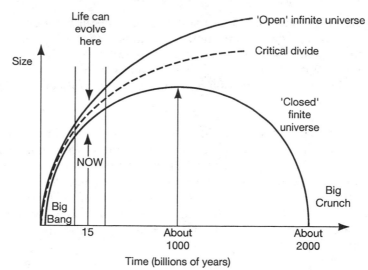

Figure 5–3. Size of the universe through time in an open versus a closed universe. An open universe expands forever; a closed universe will come back together in a Big Crunch. From J. D. Barrow, *Theories of Everything* (New York: Oxford University Press, 1991), p. 167.

existing devices cannot detect the edge of the universe because it is so far away. However, very refined instruments, such as the recently launched Hubble Telescope, are being developed. These instruments will be used in space, where the absence of earth's atmosphere will make them capable of "seeing" much farther.

A second line of evidence is the speed and direction of motion of the galaxies. By making precise measurements of light and other kinds of radiation, we can determine the speed of the galaxies and direction in which they were moving in the past and now. We can then estimate how fast galaxies are continuing to move apart from one another today and if such expansion is diminishing. Such comparisons presently indicate that there is no slowing down and that the universe will continue to expand forever. However, the extreme difficulty of taking such precise measurements makes this inference very tentative. Again, the recent deployment of the Hubble Telescope, which orbits the earth, will hopefully make more precise measurements possible.

The third major line of evidence on the universe's fate concerns the amount of matter in the universe. The main factor determining whether or not the expansion of galaxies will stop is how much gravitational attraction exists between them. Strong attraction will eventually slow their separation down and cause it to reverse. Since gravity increases with mass, the more matter there is in the universe, the greater the amount of gravitational attraction. A survey of the number of visible objects (galaxies, number of stars, and so on) has shown that there is not nearly enough matter to stop expansion. However, closer inspection has indicated that much of the universe is not visible to us. Much of the universe is composed of so-called **dark matter** that does not emit light but still has a great deal of mass. If we cannot see it, how do we know dark matter exists? There are many clues. For example, galaxies of a certain mass should rotate at a certain speed. Yet many galaxies rotate much too fast for the amount of matter that we can see. Further, from the velocity of rotation we can estimate how much of the matter is unseen. Such estimates indicate that as much as 90% of the universe is composed of such dark matter.

What could the mysterious dark matter be? Scientists are uncertain, though there are many possibilities (all of which could be correct): Black holes, very dim stars, large planets, and even tiny subatomic particles, are the most often mentioned. Yet even if the universe is 90% dark matter, there is still not enough to create the gravity to stop expansion. Thus, this third line of evidence, like the second, also indicates that the universe is open. However, this too is uncertain because the 90% estimate is very approximate. For the moment, whether the universe is open or closed is undecided.

WHY SHOULD WE CARE? What do these two alternative futures mean to our descendants in a practical sense? Is either one better than the other? An open universe gives our species a much longer potential existence. Assuming we can survive the demise of our planet and sun, described shortly, the *open universe will last about 100 trillion years*, before all the stars in it are dead. Thus, even though our own sun will burn out in about 5 billion years, our descendants could find energy sources for a huge amount of time into the future if they could devise ways to travel between the stars. Ultimately, however, even the dead stars would disintegrate and the universe would become a completely dead sea of unusable heat radiation. With no energy sources available, no life could possibly exist. The fate of an open universe would be a total victory for entropy, the physicist's description of true oblivion.

In contrast, a closed universe would give our descendants much less time to evolve and develop technology. *Expansion of the universe would continue until about 1 trillion years from now*, and then it would reverse itself. In contracting, the universe would produce a mirror image of the events that created it in the Big Bang. It would take about 1 trillion years to contract, just as long as it took to expand to its outer limit. As all the matter became crowded into an ever smaller amount of space, greater gravity would accelerate the compression. The greater density of matter would cause great increases in temperature and pressure, just as seen in the early Big Bang. These would eventually break up the particles (such as atoms, then protons, and smaller units) that had formed when the universe expanded and cooled after the Big Bang. It would eventually become so dense that the known laws of physics would fail as all the billions of galaxies in the universe became compressed into a tiny pinpoint of space. This scenario has been called the **big crunch.** It may be that such a process would create another Big Bang if the compressed matter reexploded. If this is the case, our universe may be just one of many that has formed and will form in the future. For this reason, many people find the closed universe a more philosophically pleasing scenario. If offers the prospect of renewal. Of course, it is only the universe itself that is renewed and not human beings. Any life in this present universe will be completely obliterated in the cosmic crunch; all information content will be erased as particles break down to pure energy. Nevertheless, a renewed universe would seem to offer more room for hope than a certain unending oblivion.

Future Evolution of the Sun

In our discussion of how stars form and die (Chapter 3), we noted that when stars use up their hydrogen, they begin to collapse. This creates great temperatures and pressures at the core, which causes helium atoms to fuse to form carbon. Since this nuclear reaction produces so much heat, the star expands to form a red giant. When the sun begins these death throes, it will become so large that it will encompass the orbit of Venus and maybe even the earth itself. Life on earth will surely end from the massive radiation at this time. After this, the available helium fuel will soon be used up and the sun will slowly shrink. It will enter the quiescent white dwarf and slowly dissipate away its remaining heat until it is but a black cinder. A medium-sized star like the sun uses up its hydrogen in about 10 billion years. Since the sun is about 5 billion years old, it should be about *5 billion years before our sun begins to die*. Hopefully, by then, humans will have developed the technology to travel to other stars.

Future of the Earth

The long time frames and the sheer size of the universe and even the sun make their future evolution seem far removed from our daily human activities. In contrast, the future evolution of the earth is not only very immediate, but humans have a great capacity to determine the course of its evolution themselves. We have seen that earth has had a long period of natural evolution which would continue without the presence of humans. There have been changes in the positions of continents (lithosphere), changes in the composition and location of seas (hydrosphere), and changes in the composition and temperature of the global atmosphere. No doubt these would continue their evolution without human interference. However, our influence will greatly alter these past patterns in all three spheres of the earth. With our technology, based on the use of massive reservoirs of energy and matter, our species will undoubtedly cause more changes on earth in just a few hundred years than would occur naturally over many millions of years. The focal question is whether these changes will lead to the continued survival and improvement of the earth, its life and humans, or whether we will render the earth uninhabitable for ourselves and other life forms. For ease of discussion, let us look at the future of each of the earth's three spheres separately: land (lithosphere), air (atmosphere), and water (hydrosphere).

FUTURE OF THE LITHOSPHERE. The major force determining the future of earth's continents is the internal heat engine causing plate tectonics. Since the early earth formed, this engine has been cooling down so that plate movement is slowing. Nevertheless, *continents will continue to move for many millions of years* at the average rate of one to a few inches per year. Given the projected rate of motion and the current direction of movement of the earth's plates, it is easy to predict where the continents will be in a few million years. As shown in Figure 5–4, North America and Europe will continue their motion away from each other, increasing the size of the Atlantic Ocean. Australia will continue its movement northward until it eventually collides with Asia, creating huge mountain ranges much like the Himalayas, which were caused when India collided with Asia.

FUTURE OF THE ATMOSPHERE. Humans have used the earth's atmosphere as a "dumping ground" since they built the first fires. It is ironic that such an ancient practice is perhaps the greatest threat to the global environment today. Fire (combustion) combines oxygen from the air with carbon in wood to form carbon dioxide:

$$O_2 + CH_2O \rightarrow H_2O + CO_2$$

An excess of carbon dioxide appears to be a major environmental threat.

Figure 5–4. Prediction of where continents will be about 50 million years from now, based on current speed and direction of plate tectonics. Dashed lines are present position. From E. Chaisson, *Universe* (Englewood Cliffs, NJ: Prentice-Hall, 1988), p. 441.

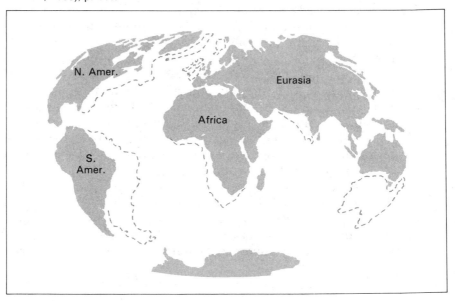

The release of carbon dioxide is not new: It occurred naturally for millions of years. The very act of breathing by animals releases it (metabolism is a kind of combustion). However, much of the carbon dioxide is "soaked up" in the ocean (to produce limestone) and by plants which use it in photosynthesis (reversing the chemical reaction above). Thus, carbon dioxide has accumulated only in very small amounts and has been a trace gas, much less than 1% of the atmosphere, for a very long time. However, as humans began to burn more and more fuel (e.g., wood, coal, and oil), ever greater amounts of carbon dioxide became emitted into the atmosphere. With the growth of industrialization in the West and in developing countries, the amount of carbon dioxide released because of factories, cars, and so on has increased dramatically (Figure 5–5). This has been aggravated by the cutting down of vast tracts of tropical forests, reducing the number of plants to "soak up" the carbon dioxide.

What is the harm in increasing a trace gas in the atmosphere? The problem is that carbon dioxide has properties that make it transparent to sunlight but not transparent to heat. It thus acts like glass in a greenhouse by permitting sunlight to penetrate the atmosphere, trapping the resulting heat that rises when the light warms up the earth. Appropriately enough, this is called the **greenhouse effect.** As

Figure 5–5. Gigatons (billions of tons) of carbon released into the atmosphere since 1860. After leveling off between the two World Wars, the rate since World War II has skyrocketed. Modified from *American Scientist* magazine, July 1990, p. 313.

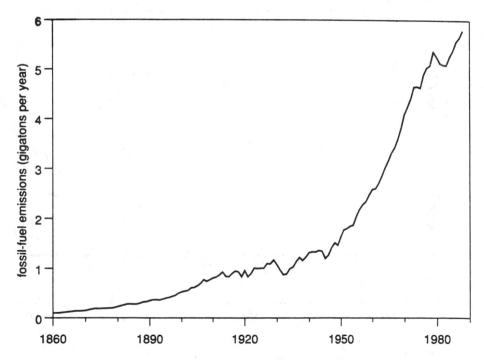

long as carbon dioxide is relatively rare in the atmosphere, the earth is warmed to generally hospitable temperatures. However, as more and more carbon dioxide is added, more heat is trapped. This much is well established from theory. What is not known is how much carbon dioxide must be added to significantly heat the earth's climate. The average global surface air temperature has risen, along with carbon dioxide content (Figure 5–6). However, nature is full of so-called "spurious correlations" wherein trends between phenomena jointly occur by sheer coincidence. Therefore, while we can be sure that the average global temperature has risen about 0.5 degrees centigrade over the last 100 years, we cannot be sure how much was caused by the emission of carbon dioxide.

In addition to spurious correlations, measuring the effect of carbon dioxide is complicated by many other factors that also influence global climate. One such complication is that other "greenhouse gases," besides carbon dioxide, are being released, and they, too, cause the atmosphere to heat up. For instance, methane is produced in great quantities by cows. As humans have greatly increased cattle populations, methane release has increased dramatically. Other greenhouse gases, including water vapor, are released by volcanoes. Another complication is the effect of solar output, which varies through time, changing the amount of radiation reach-

Figure 5–6. Projected global temperature change based on scenarios of continued growth of emission rate of greenhouse gases (A); fixed emission rate (B); drastic reduction of emissions (C). This is just one of many modeled projections. Modified from *Trends in Ecology and Evolution* magazine, March 1989, p. 66.

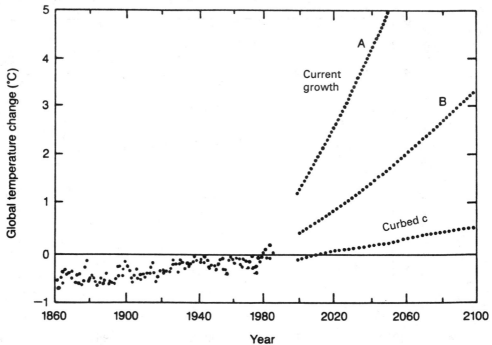

ing the earth. Yet a third complication is *clouds*. As the earth heats up, the number of clouds formed will be affected. More clouds would reflect more light into space, which could partly counteract the greenhouse gas effects of heating. A fourth factor is that the oceanic absorption of carbon dioxide seems to be increasing as more carbon dioxide is added to atmosphere. Thus, the earth may have a way of "adjusting" to changes.

These factors (and still others) make it very difficult to predict exactly what will happen with continued output of carbon dioxide and other greenhouse gases. Many complex computer models have been made to account for the many variables, and most show an inevitable rise in temperature as these gases are added. One likely scenario is shown in Figure 5–6. Global warming will no doubt be one of the major environmental topics for decades to come. Indeed, it may well be *the major environmental problem* affecting the earth for the foreseeable future and you will likely be regularly hearing and reading about it in the news media for the rest of your life.

Global warming involves much more than a simple warming of local temperatures all over the earth. Some areas will experience a much greater increase in temperature than others. This will affect wind and rainfall patterns, which are largely caused by differential heating of the earth's surface. While it is difficult to be certain, most computer models show that there will be increased rainfall in equatorial and polar areas and decreased rainfall in many temperate areas. This would be very bad for agriculture, since most of the world's grains and many other foods are produced in these areas. For instance, the American Midwest, the "bread basket" of the world, could become a virtual desert when diminished rainfall is combined with increasing heat. Overall, mass migration of human populations will probably occur, as they adjust to such major changes in local weather.

In discussing global warming, we have focused mainly on human influences over the next few thousand years. Yet, throughout this book, we have emphasized the vastness of geological time and the relatively tiny time span that humans have existed. Let us consider the very long-term future. What might happen long after humans stop heating the earth? As shown in Figure 5–7, a human-caused climatic warming is really just an *exaggerated natural warming period* (super-interglacial) that the earth has been in since the last glaciation ended about 10,000 years ago. Thus, it may be that the earth will enter another cooling period (Ice Age) a few thousand years after we stop heating the earth. This cessation of heating could occur if humans become extinct or if technology advances to where we rely on energy sources other than fossil fuels, such as nuclear or solar energy.

Another major potential human-induced atmospheric hazard is the destruction of **ozone** in the atmosphere. Another minor gas in the atmosphere, ozone is produced when sunlight strikes oxygen atoms and causes them to temporarily combine. This ozone has a key role in filtering out much of the high-energy radiation that bombards the earth from the sun. A reduction of it will increase the incidence of sunburn, skin cancer, and other illnesses. More importantly, harmful changes in the global ecosystem may occur, such as interference with photosynthesis. It has only been within the last few years that scientists have discovered a

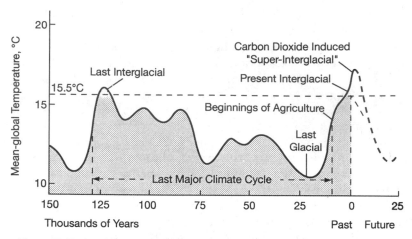

Figure 5–7. Average global temperature for the last 150,000 years. We are in one of many interglacial warming periods, made warmer by the Greenhouse Effect.

great "hole" in the ozone layer over Antarctica. There is much current research on whether this hole will spread to other areas, or if it is only a temporary phenomenon. (By 1992, it had spread well beyond Antarctica into South America.)

How are humans destroying the ozone? Again it is by the release of gases. In this case, these gases destroy the ozone molecules. The most well-known of these gases are the chlorofluorocarbons (CFCs), used in spray cans, plastics, refrigerators, and air conditioners. Fear caused by the rapid disappearance of the ozone layer has led to international agreements, such as the Montreal Accord in 1987, to phase out the use of CFCs. Unfortunately, it could well be too late to avoid major damage to the biosphere. CFCs take about 15 years to drift up to the upper atmosphere and, once there, cause damage for a long time. *The average CFC molecule destroys about 100,000 ozone molecules.* Furthermore, leakage from old refrigerators, spray cans, and many other discarded sources will continue to add new CFCs to the atmosphere.

Nuclear weapons pose the most potentially catastrophic threat to the atmosphere. Recent political events have greatly reduced the chances of a major nuclear exchange. However, computer studies indicate that even a few small tactical weapons could have major effects on the atmosphere. The "mushroom cloud" given off by nuclear blasts releases a great amount of dust, soot, and other particles far into the atmosphere. Being very small, these particles will stay suspended in the atmosphere for many months, where they will block sunlight, halting photosynthesis and cooling climate. Hence, this phenomenon is called the **nuclear winter.** There may certainly be some similarities between the potential damage of a nuclear winter and the meteorite impact that may have killed off the dinosaurs. In both cases, a large quantity of particles would be thrown into the atmosphere by great energies, blocking sunlight. The entire global ecosystem would be upset, not only because so many organisms would directly die of temperature changes, but many plants (including plankton in the ocean) would die because lack of light would not allow for photosynthesis. Hence, we once again see that study of the past can be relevant for current problems by providing "case studies" of past "experiments" that clearly cannot be replicated in the lab or a computer model.

It is also possible that a meteorite strike could cause a nuclear winter, just as it may have at the end of the Mesozoic. These are usually portrayed as science fiction, but insurance agencies estimate that your chance of dying from a meteorite strike are about 1 in 6,000. This is greater than your chance of dying from tornadoes, earthquakes, or even an airplane crash. While large meteorite strikes are much rarer than these events, they would kill many more people. Nor are these strikes so rare. Several large asteroids pass within a few million miles of earth every year. In 1991 a 30-foot wide asteroid missed the earth by about 100,000 miles, less than half the distance to the moon. A space program to prevent incoming asteroids from hitting earth has been proposed and is technologically very feasible.

FUTURE OF THE HYDROSPHERE. The three spheres of the earth are not isolated from one another. The temperature changes in the atmosphere discussed above can have strong effects on the hydrosphere by affecting the water cycle. When the global climate cools, more water precipitates as snowfall and accumulates as ice, especially at the poles. This ice serves as a reservoir for water, taking the water out of the water cycle, causing the sea level to drop.

We have already discussed how this occurred in earth's past, with the widespread glaciation causing decreases in sea level and exposing the continental shelves. If humans were to set off another glaciation, perhaps by a nuclear winter, such a sea level drop would occur again. As shown in Figure 5–8, for the United States sea level could lower from 300 to 400 feet, adding over 2.5 million square miles to the world's coastlines.

Alternatively, it is more likely that continued human activity will increase the greenhouse effect, causing warming of the atmosphere. This will have the reverse effect of a glaciation. The amount of water stored as ice will decrease (the polar ice caps will melt), releasing millions of gallons of water (about 2% of the world's water is now in the ice caps, and mainly Antarctica's ice caps). This released water will raise sea level about 200 feet, and many low-lying areas will be flooded (see Figure 5–8). For transportation reasons, many major cities are located on coastlines—the flooding would cause over a billion people to lose their homes. In addition, over one-third of the worlds' crops would be lost. It has been suggested that the slow pace of melting (tens to hundreds of years) would allow time to build dikes and other engineering mechanisms, as has been done in the low-lying areas of Holland. However, the cost of doing this on a global scale would be truly enormous, if it were possible at all.

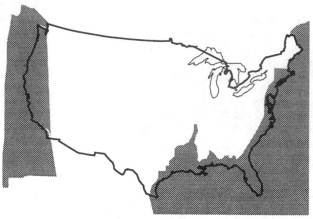

(a)

IF ALL THE ICE MELTED

Scientists estimate the sea level of the earth's oceans would rise 200 to 400 feet if all ice in the world melted.
The map above shows how a rise of 250 feet would change the coastlines of the continental United States.

(b)

OR THE EARTH COOLED

During the Ice Age, a heat loss of 3 to 4 degrees in the earth's temperature created gigantic glaciers, which lowered the sea level 300 to 400 feet. As a results lands now under the oceans were exposed. Over 2.5 million square miles of land were added to the coastlines.

Figure 5–8. Effects of melting or freezing of all the earth's ice. From D. Cargo and B. Mallory, *Man and His Geologic Environment* (Reading, MA: Addison-Wesley, 1977), p. 313.

PROSPECTS FOR BIOLOGICAL EVOLUTION

OVERVIEW

We saw in Chapter 4 that the evolution of life has involved a twofold process of origination and extinction. As summarized in Figure 5–9, both origination and extinction are driven by the same process: environmental change. If the change is too great, the existing species dies out. If the change is slow enough, and there is

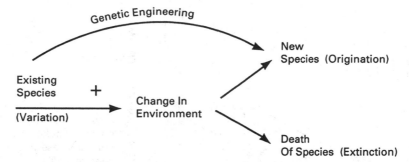

Figure 5–9. Environmental change acting on an existing species will eventually cause either (1) extinction or (2) origination of a new species. Humans can "short-circuit" the process and create new kinds of organisms by (3) directly altering the genes with genetic engineering.

enough variation in the gene pool, the existing species can adapt and become a new species. In this section we will discuss how humans, because of our technology and vast numbers (over 5 billion people on earth), are accelerating environmental change, leading to an acceleration of both origination and extinction. Furthermore, humans are accelerating origination in ways other than rapid environmental change. The ability to directly manipulate, or engineer, genes in the laboratory and exchange them between species has led to a veritable explosion of new life forms that could never be achieved in nature.

We now examine the effects of origination and extinction of humans on the future of biological evolution. First, we discuss the rapid increase in extinctions. Second, we review how humans are causing a rapid increase in originations through environmental change and genetic engineering. Third, we briefly consider an interesting "thought experiment" that sheds insight on the evolutionary process: What would happen to biological evolution if humans did not exist? What kind of organisms would evolve millions of years from now?

Mass Extinction: Human Decimation of the Biosphere

Most of us are familiar with extinctions. We go to museums to see dinosaurs, sabertoothed cats, and other long-dead animals. However, until very recently, the notion that humans were causing many extinctions was not widely held. Most people were dimly aware that the dodo bird and passenger pigeon had been driven to extinction, but these were considered to be rare, exceptionally fragile species. Yet, as shown in Figure 5–10, even these early extinctions, before industrialization and human populations recently accelerated, were more numerous than one might think. Furthermore, Figure 5–10 only shows extinctions for birds and mammals. No one is certain how many species of invertebrates (especially insects) and plants died out in earlier times, though it was probably a large number, especially as humans colonized islands. As the increasing exponential curve of Figure 5–10 implies, extinctions are

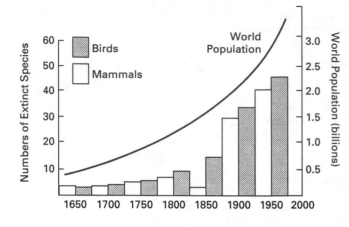

Figure 5–10. Rapid increase in human population is paralleled by rapid increase in extinction of birds and mammals. From D. Botkin and E. Keller, *Environmental Studies*, 2nd ed. (Columbus, OH: Merrill, 1987). p. 26.

increasing rapidly at a rate that will create, within a few decades, a mass extinction rivaling in size the five past mass extinctions we discussed in Chapter 4.

TROPICAL EXTINCTIONS: WHY THEY ARE THE WORST. The major reason for the rapid rise in extinction rate is the increasing destruction of the tropical rain forest. Tropical rain forests contain the most species by far of any ecosystem on earth. As long as industrialization and its accompanying environmental disturbance occurred mainly in nontropical areas (e.g., North America and Europe), the number of species lost was relatively low because these areas are relatively poor in species diversity, called **biodiversity.** Now that tropical countries such as South America, Asia, and Africa are firmly on the road to "development," destruction of tropical ecosystems has begun in earnest, and many more species are lost per acre of land disturbed simply because many more species inhabit it. For example, in the Amazon Forests of South America as many as 300 species of trees can occur in a typical hectare (about 2.5 acres). In Southeast Asia about 200 tree species occur per hectare, and in Central America, the number is about 120 species. In striking contrast, the average hectare of forest in North America (and Europe) ranges from only one species to a maximum of about 12 tree species.

Other organisms show the same pattern. One insect specialist recently counted 43 species of ants on one tree in Peru, equal to about the total number of ant species in the entire British Isles. Figure 5–11 shows just how dramatically diverse insects are in the tropics: They make up an estimated 96% of all arthropods (mainly insects). This percentage translates into a huge number of species because insects comprise the large majority of animals (Figure 5–11). The proportion of other organisms contained in the tropics is substantially less than the insects, but it is still highly significant. The point is that while rain forests cover only about 7% of the earth's land surface, they contain well *over one-half of the earth's land species*. Because about 90% of all species live on land, this is a large majority of the earth's total species.

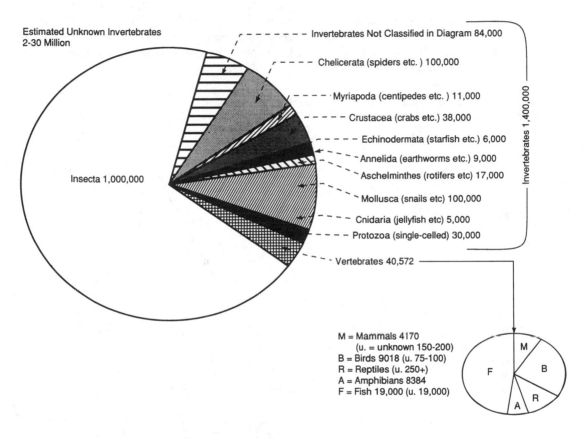

Estimated Unknown Invertebrates 2-30 Million

Insecta 1,000,000

Invertebrates Not Classified in Diagram 84,000

Chelicerata (spiders etc.) 100,000

Myriapoda (centipedes etc.) 11,000

Crustacea (crabs etc.) 38,000

Echinodermata (starfish etc.) 6,000

Annelida (earthworms etc.) 9,000

Aschelminthes (rotifers etc) 17,000

Mollusca (snails etc) 100,000

Cnidaria (jellyfish etc) 5,000

Protozoa (single-celled) 30,000

Vertebrates 40,572

Invertebrates 1,400,000

M = Mammals 4170
(u. = unknown 150-200)
B = Birds 9018 (u. 75-100)
R = Reptiles (u. 250+)
A = Amphibians 8384
F = Fish 19,000 (u. 19,000)

Percentage* of World Species in Tropical Forests

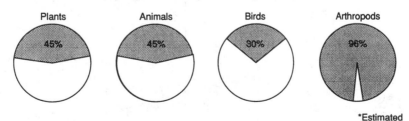

Plants — 45%

Animals — 45%

Birds — 30%

Arthropods — 96%

*Estimated

Figure 5–11. Most animals are insects. Most insects by far occur in tropical forests. Modified from L. Durrell, *State of the Ark* (New York: Doubleday, 1986), p. 99, and G. Lean et al., *Atlas of the Environment* (Englewood Cliffs, NJ: Prentice-Hall, 1990), p. 135.

The general increase in number of species as one moves toward the equator is called the **latitudinal diversity gradient.** The gradient holds true for many kinds of animals, including ants, clams, lizards, and birds (Figure 5–12). It also applies to most plants, as well as organisms in the oceans. For example, tropical coral reefs tend to be the most diverse marine ecosystems. These are similarly threatened by humans, from sewage, construction, and other activities along the nearshore shallow waters where the large majority of marine species live. (This tropical richness is also why mass extinctions in the geological past were often caused by climatic cooling: The diverse, warmth-loving land and sea organisms were especially vulnerable to the cold.)

Figure 5–12. Number of species increases toward the equator for most groups. Modified from M. Begon et al., *Ecology*, 2nd ed. (London: Blackwell, 1990), p. 832.

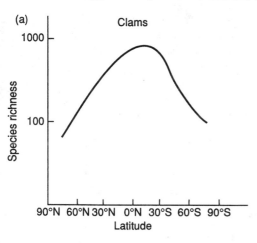

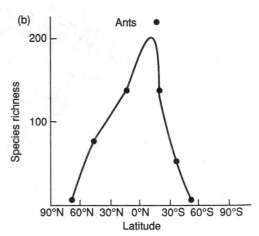

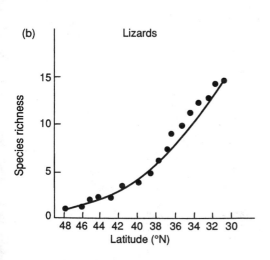

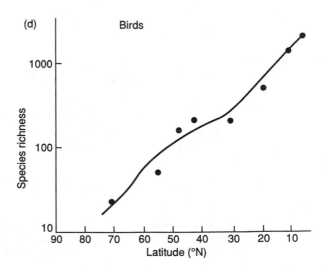

What is the cause of the latitudinal diversity gradient? Ecologists are not sure, but a combination of at least three factors seem to be involved. One, equatorial areas are more climatically stable than more northern areas. A lack of pronounced seasons may allow species to become more specialized to a specific set of environmental conditions. (Recall Rapoport's rule from Chapter 4.) This leads to niche packing (more species per area). Two, the tropics have been relatively free of the more drastic effects of the recent Ice Age. Advancing glaciers may have periodically eliminated species to the north so that they must now recolonize and readapt, keeping diversity lower. Three, the equatorial areas receive more sunlight (per unit area). This allows plants, at the base of the food chain, to generate more food for consumers in tropical food chains.

CAUSES OF TROPICAL DEFORESTATION. There are two basic reasons why tropical rain forests have recently begun to be destroyed at rapid rates. The first reason is sociopolitical. In terms of political boundaries, just 13 tropical countries contain over 60% of the world's species. These countries are all "developing" economically so that incomes are low and the population is skyrocketing. The result is massive exploitation of resources in a desperate attempt to raise living standards. The second reason for rapid deforestation relates to the particularly destructive nature of this exploitation. Because rainfall is so great in the rain forests, nutrients are quickly washed out of the soil. Therefore, virtually all nutrients must be retained in the living tissues of the ecosystem itself. This has led to **slash and burn agriculture** whereby the forest is cut down and burned. The high-nutrient ashes are then used to enrich the soil for crops. Unfortunately, the nutrients are soon washed away, and the farmers must move on, leaving a relatively barren wasteland compared to what existed before. Worst of all, biologists have found that the land does not revert back to its previous state for a very long time, if ever. We will see below that habitat destruction is the leading cause of extinction, so this kind of agriculture is the worst kind of environmental disturbance. The fact that the farmers must continually move to new rain forest land virtually guarantees that huge tracts will be destroyed.

RATES OF MODERN EXTINCTION. As of 1992, it is estimated that 45% of the earth's rain forests have been destroyed, and just in the last few decades. Figure 5–13 shows deforestation in Latin America. Brazil has the most rain forest in the world remaining today, although it also has one the highest rates of destruction. Using satellite photos and other methods, it has been estimated that about 15 hectares per minute are being destroyed in Brazil alone. Globally, an area of rain forest totaling about the size of the state of West Virginia is being destroyed each year. At these rates, all remaining rain forest is expected to be eliminated by sometime in the next 100 years, perhaps as early as the year 2020, if the rate continues to increase as human populations soar.

These high rates of deforestation translate into an estimated extinction rate of at least 10 species per day, although most estimates are higher, and some researchers have calculated the loss of several thousand species per day. For compari-

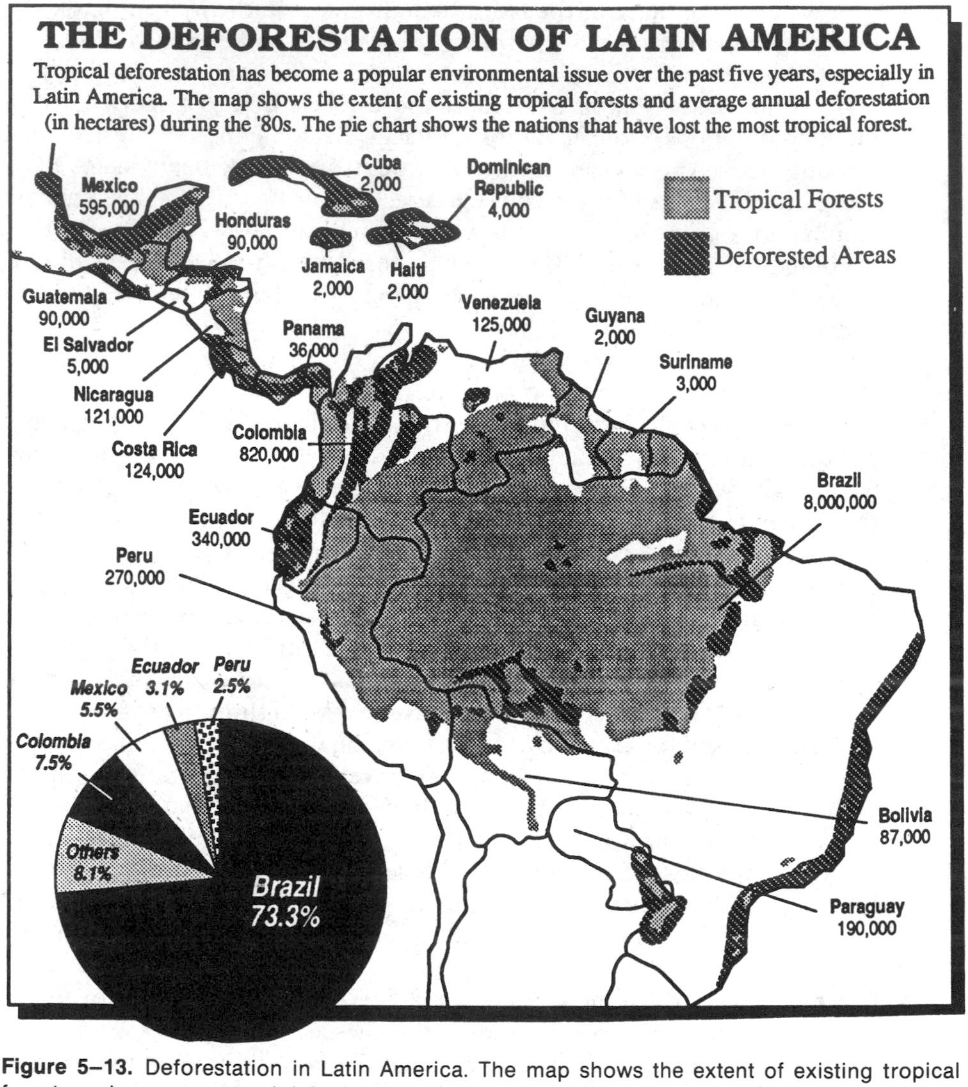

THE DEFORESTATION OF LATIN AMERICA

Tropical deforestation has become a popular environmental issue over the past five years, especially in Latin America. The map shows the extent of existing tropical forests and average annual deforestation (in hectares) during the '80s. The pie chart shows the nations that have lost the most tropical forest.

Cuba
2,000

Dominican
Republic
4,000

Mexico
595,000

Honduras
90,000

Tropical Forests

Deforested Areas

Jamaica
2,000

Haiti
2,000

Guatemala
90,000

El Salvador
5,000

Nicaragua
121,000

Panama
36,000

Venezuela
125,000

Guyana
2,000

Suriname
3,000

Costa Rica
124,000

Colombia
820,000

Brazil
8,000,000

Ecuador
340,000

Peru
270,000

Mexico
5.5%

Ecuador
3.1%

Peru
2.5%

Colombia
7.5%

Others
8.1%

Brazil
73.3%

Bolivia
87,000

Paraguay
190,000

Figure 5–13. Deforestation in Latin America. The map shows the extent of existing tropical forests and average annual deforestation (in hectares) during the '80s. The pie chart shows the nations that have lost the most tropical forests. From *Knoxville News-Sentinel*, Oct. 13, 1991, p. F-1. Data from World Resources Institute.

son, estimates from the fossil record indicate an average extinction rate of about one species per year. Thus, even if the minimum estimates of modern extinction rate are used, they are still thousands of times higher than the "normal" extinctions of geologic time (3,650 species per year vs. 1 per year). At these rates, which are increasing, it is estimated that up to one-third of all species on earth will be extinct by the early 21st century. As mass extinctions are defined as short-term episodes where over 50% of species die out (Chapter 4), it would seem that a mass extinction will be achieved by the end of the 21st century. This would be the first one in over 60 million years.

Most of these extinct species will be tropical insects simply because they constitute such a vast proportion of life on earth (see Figure 5–11). However, Table 5–1 shows that virtually all groups of organisms are showing high extinction rates. Nearly half of all turtles, for example, are threatened. Furthermore, Table 5–1 shows that extinctions are not limited to the tropics. For instance, almost half of Australia's mammals are threatened. Over 90% of the native forests in the United States have been destroyed. Long-term surveys by bird watchers show that many migrating songbirds in the eastern United States, such as warblers, flycatchers, and thrushes, have been declining in population by about 3% per year since 1980.

CAUSES OF MODERN EXTINCTIONS. Recall that extinction is caused by a change in environment that is too great for the existing variation to adapt to. Today, we can isolate

TABLE 5–1. Observed Declines in Selected Animal Species, Early Nineties

SPECIES TYPE	OBSERVATION
Amphibians[1]	Worldwide decline observed in recent years. Wetland drainage and invading species have extinguished nearly half of New Zealand's unique frog fauna. Biologists cite European demand for frogs' legs as a cause of the rapid nationwide decline of India's two most common bullfrogs.
Birds	Three fourths of the world's bird species are declining in population or threatened with extinction.
Fish	One third of North America's freshwater fish stocks are rare, threatened, or endangered; one third of U.S. coastal fish have declined in population since 1975. Introduction of the Nile perch has helped drive half the 400 species of Lake Victoria, Africa's largest lake, to or near extinction.
Invertebrates	On the order of 100 species lost to deforestation each day. Western Germany reports one fourth of its 40,000 known invertebrates to be threatened. Roughly half the freshwater snails of the southeastern United States are extinct or nearly so.
Mammals	Almost half of Australia's surviving mammals are threatened with extinction. France, western Germany, the Netherlands, and Portugal all report more than 40 percent of their mammals as threatened.
Carnivores	Virtually all species of wild cats and most bears are declining seriously in numbers.
Primates[2]	More than two thirds of the world's 150 species are threatened with extinction.
Reptiles	Of the world's 270 turtle species, 42 percent are rare or threatened with extinction.

[1] Class that includes frogs, toads, and salamanders.
[2] Order that includes monkeys, lemurs, and humans.
Source: Worldwatch Institute.

three basic categories of environmental change that cause nearly all extinctions:

1. habitat disturbance
2. introduction of new species
3. hunting

It is not necessary for these causes to eliminate every individual of a species. Even if many individuals survive one of these three disruptions, the breeding population may still not be able to recover. There are two main reasons for this. One is increased inbreeding levels. This can lead to too many genetic defects in offspring. This kind of "genetic bottleneck" has recently been discovered in the cheetah, which has much lower genetic variation than most cats. It is also a major problem in the Florida panther, which has only 30 to 50 individuals left. If severe enough, the species cannot recover and all remaining individuals die. The second problem facing small populations is that their odds of being wiped out by a random environmental fluctuation, such as a hurricane, storm, freeze, and so on, is increased. As Figure 5–14 shows, the probability of extinction increases exponentially with smaller populations.

The **minimum viable population (MVP)** is the minimum population size necessary for the species to survive these two problems of inbreeding and environmental fluctuations. If population falls below this size, the species will probably become extinct. What is this minimum level? There is much debate among ecologists. Some argue that about 50 individuals (25 breeding pairs) will permit a species to rebound. Others hold that 500 individuals (250 pairs) are necessary. The main problem is that there is no real magic number that is valid for all species. The MVP will vary with the species and the conditions. For example, in highly territorial species, where males stay in and defend one area, breeding becomes much more difficult with low populations. In contrast, species that breed freely will have fewer problems at low population densities.

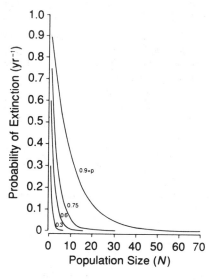

Figure 5–14. Probability of extinction increases exponentially with decreasing population size. Modified from R. Ross and W. Allmon, *Causes of Evolution* (Chicago: University of Chicago Press, 1990), p. 119.

Habitat disturbance is the alteration of a species' physical and biological environment (habitat). This is the *most common cause of extinction* worldwide because of widespread environmental changes brought about by humans. It may also have been the main cause of extinction in the past, although it is difficult to prove. Natural examples of extinction by habitat alterations include cooling climate, meteorite impacts, volcanoes, and many other natural ways of physically changing surrounding conditions. Sometimes habitat disturbance involves outright destruction of the environment caused naturally (such as meteorite impacts) or by human interference (e.g., bulldozers, fires, and other means of clearing an area). Whether natural or humanmade, it is easy to see why such habitat destruction causes extinction: The species simply has no place to live.

Extinction also occurs due to minor alteration of the habitat. Often these alterations are surprisingly small, such as the emission of very small amounts of pollutants into the air, water, or soil. Some species will be much more sensitive to such pollutants than others, and may consequently become endangered or extinct in that area. For example, sulfur air pollution (from the burning of coal) has led to **acid rain,** which increases the acidity of lakes. Some fish, such as trout, are much less tolerant to increasing acidity in lakes than other fish. Another common example of disturbance is the selective removal of some part of a habitat. For example, the ivory-billed woodpecker was once a common bird in the forests of southeastern United States. It thrived on insects that live in rotting trees. Then the timber industry began a practice of cutting and removing trees before they became old and decaying. Thus, the southeastern forest continues to thrive, but the ivory-billed woodpecker, having had a key component in its habitat destroyed, is extinct in the United States. Similarly, the spotted owl of the northwestern United States also inhabits only "old growth" forests, and controversy surrounds the possible extinction of this bird.

Where minor habitat disturbance initially eliminates only one or two species, the loss can cascade through the food web and cause other species to go extinct. This is called **secondary extinction.** For instance, the loss of a major prey species can cause a predator to go extinct. Other examples are more subtle: With the extinction of the dodo bird, the Calveria tree became unable to reproduce (Figure 5–15). The dodo ate the seeds and through digestion removed the seeds' outer coverings, which was necessary for new trees to grow. This also transported the seeds. It is striking how long species removals can affect ecosystems. Scientists have only recently discovered that the extinction of large plant eaters, such as the woolly mammoth, at the end of the last glaciation (which occurred 10,000 years ago) is still affecting the plants on which those animals fed. The most extreme secondary extinctions occur where ecosystems contain **keystone species.** These are species so interdependent with other species that when the keystone species is removed, these other species become extinct too. For instance, in some ecosystems just one or a few plants serve as food for many animals during certain times of the year. If these plants are removed, many of these animals also may become extinct.

A second major cause of modern extinctions is introduction of new species

Figure 5–15. The extinct dodo bird, by digesting its seeds, was crucial in the reproductive cycle of the Calvaria tree. From R. Augros and G. Stanciu, *The New Biology*, New Science Library, p. 111, Fig. 4.5.

into an ecosystem. As humans have migrated widely throughout the world, they have often brought other organisms with them, both on purpose and inadvertently. Cats and dogs, brought as pets, seem harmless and cuddly to us, but they have devastated native life forms on many islands. For example, on the island of New Zealand, many species of flightless birds were killed off because they proved to be easy prey. (A single German shepherd recently escaped from a farm in New Zealand

and killed dozens of native animals in just a few days.) The problem is that all organisms have evolved to meet the adaptive needs of a specific set of environmental conditions. By introducing a new species, the environment is altered and old traits may no longer be adequate for survival of some native species. Flightlessness was useful when no major ground carnivores were around, as in prehuman New Zealand, but led to extinction when the predators were introduced.

Other introductions have occurred when organisms have been imported for sport or food. The introduction of rabbits into Australia has led to a massive population explosion. In this case they are detrimental to native species because they compete for food with the native kangaroos and other marsupials. Similar stories are found in virtually every part of the world. Tropical environments are especially sensitive to such introductions because species tend to be more specialized. Many lake and forest ecosystems in Florida, for example, are rapidly changing from the introduction of plants, some secreting toxic chemicals, that invade native habitats, such as the Everglades. The "walking catfish," an example of an introduced animal from Asia, has become a widespread nuisance in Florida.

So far we have focused on examples of ecosystem disruption from new predators or competitors. Introduced diseases and parasites can also upset ecosystems. A well-known example is the parasitic fungus from China that devastated the American chestnut tree in the early 1900s. Island ecosystems, such as Australia and Hawaii, are most susceptible to such introductions because they are often isolated for long periods of time with little contact with other ecosystems.

Humans are not immune to the direct effects of new species. The American Indian suffered mortality rates of 90% or more of some tribes because they lacked immunity from European diseases such as smallpox. Nor are human impacts limited to disease; imported species are a major source of economic loss. The zebra mussel is a small European clam that was inadvertently released into Lake Erie in 1986 by a tanker flushing its ballast. By 1989 they had become so common that a single cubic yard of water contained, on average, over a half-million larvae of this species. Because they eat the plankton that form the base of the food chain, the sport fishing industry has been seriously damaged. In addition, the clams attach to water intake pipes in the Lake and clog them. Unless some method is found to stop them, engineers estimate that over 26 million people around the Great Lakes could lose their water supply.

In summary, even though habitat disturbance is probably the main source of extinctions now and in the past, the effects of species' introductions in causing extinctions have been very great. Indeed, the historian Alfred Crosby, in his book *Ecological Imperialism* (1990), argues that the introduction of European plants, animals, and diseases into Africa, Asia, Australia, and the Americas was perhaps the main reason for success in European colonization. The overall impact of introduced species is even more impressive when we realize that only about 10% of introduced species actually survive in their new environment and only about 10% of these cause direct harm to other species. Thus, only about 1% of introduced species ever become "pests," so there must be many more introductions that do not succeed.

However, as with habitat disturbance, species' introductions have not always been instigated by humans. Many times in the geologic past, large land masses have come together allowing the life forms of continents to interact after many millions of years of separation. The long isolation of Australia has caused its marsupial species to be especially sensitive to introduction. While humans have ended this isolation, nature would have eventually done so. The northward drift of Australia will ultimately cause it to unite with Asia. The fossil record shows that species' introductions between continents begin long before the continents collide. Many species swim, float on debris, fly, and find other ways to disperse. A classic example is the linkage of North America with South America, which occurred about 5 millions years ago when the Isthmus of Panama developed. A massive exchange of organisms occurred. This led to the extinction of many South American species, probably because South America had been isolated so long from the mainstream of mammalian evolution (Chapter 4).

The third major cause of modern extinctions is human hunting. This cause of extinction is well known to many of us because it involves the larger, better known kinds of animals. Very few people can name any of the thousands of insect and plant species going extinct every year in the tropics (especially from habitat destruction or disturbance), but many know about how the huge herds of American bison were nearly hunted to extinction. A modern example is the many large African and Asian animals, such as the rhinoceroses, which have been hunted to near extinction (Figure 5–16). It seems unlikely that the Javan or Sumatran rhino will ever rebound. Similarly, elephants once numbered in the millions and are now down to about 700,000, and are projected to drop to an astoundingly low 25,000 within 40 years. Wolves, large cats of all kinds, and many other large animals all over the world have been hunted to extinction or near extinction.

While killing for sport was common in the past, hunting for economic reasons has become more common recently. Most of the large species live in the developing tropical countries where people are poor. Large predators are often killed because they eat domesticated food animals. Many others are killed for what seem like less justifiable reasons. Elephants are killed by poachers who strip them of their tusks. These ivory tusks are then sold to make billiard balls, carvings, and many other "fine" articles. Rhinos are usually poached for the horn, which is prized in some countries as a knife handle and in other countries as an aphrodisiac. The hunting of whales for food, oil, and other products has finally been reduced to a small minority of countries, such as Japan.

As with other causes of extinction, hunting can remove keystone and other species that will cause cascading effects to other species in the ecosystem. Ranchers, for instance, have been killing wolves, large cats, foxes, and other predators for centuries to protect their stock, only to find that the natural prey animals of the predators undergo a population explosion, competing with their animals for food. Deer in forested areas multiply out of control and will denude all forage when the wolves and large cats are killed off. Similarly, the removal of large herbivores can

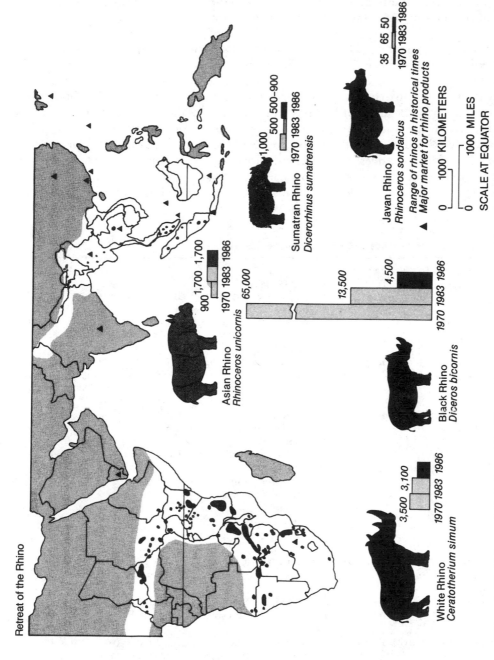

Retreat of the Rhino

Asian Rhino
Rhinoceros unicornis
900 1,700 1,700
1970 1983 1986
65,000

Sumatran Rhino
Dicerorhinus sumatrensis
1,000
500 500–900
1970 1983 1986

Javan Rhino
Rhinoceros sondaicus
35 65 50
1970 1983 1986

Range of rhinos in historical times
▲ Major market for rhino products

0   1000 KILOMETERS
0   1000 MILES
SCALE AT EQUATOR

Black Rhino
Diceros bicornis
13,500
4,500
1970 1983 1986

White Rhino
Ceratotherium simum
3,500 3,100
1970 1983 1986

Figure 5–16. Rhinos originated over 50 million years ago and more than 30 fossil species are known. Of the five left today, all have greatly reduced ranges and are threatened or endangered species. From J. Turk, *Introduction to Environmental Studies* (Philadelphia, PA: Saunders, 1989), p. 148.

drastically affect the plant distribution of the ecosystem because the plants rely on certain species to distribute them and govern their growth rate by cropping them.

SUSCEPTIBILITY OF SPECIES TO EXTINCTION. We discussed in Chapter 4 how species are not equally likely to become extinct. For many reasons, some species are more able than others to survive environmental change. A well-known living example of a very resilient group are cockroaches, which have existed for over 300 million years and will probably be alive for millions more. Table 5–2 shows *seven major characteristics* that make species susceptible to extinction today. Some of these characteristics are strictly related to human activity. For example, if a species is highly prized for sport or food, its chances of becoming extinct are greatly increased. However, we will see that some of these traits making organisms susceptible to extinction today are related to traits that allowed groups to survive extinctions in the past.

TABLE 5–2. Characteristics of Extinction Susceptibility

CHARACTERISTICS THAT CAUSE SOME SPECIES TO BE SUSCEPTIBLE TO EXTINCTION	REASON CHARACTERISTICS TEND TO CAUSE EXTINCTION	EXAMPLES
Island species	Unable to compete with invasion from continental species.	More than half of the 2,000 plant species in Hawaii
Species with limited habitats	Some species are found in only a few ecosystems.	Woodland caribou, Everglades crocodile, millions of species in the tropical rainforest
Species that require large territories to survive	Large-scale habitat destruction in the modern world	California condor, blue whale, Bengal tiger
Species with low reproductive rates	Many species evolved low reproductive rates because predation was low, but in modern times, people have become effective predators against some of these species	Blue whale, California condor, polar bear, rhinoceros
Species that are economically valuable or hunted for sport	Hunting pressures by humans.	Snow leopard, blue whale, elephant, rhinoceros
Predators	Often killed to reduce predation of domestic stock.	Grizzly bear, timber wolf, Bengal tiger
Species that are susceptible to pollution	Some species are more susceptible than others to industrial pollution.	Bald eagle (susceptible to certain pesticides)

Modified from J. Turk, *Introduction to Environmental Studies* (Saunders, 1989).

Most of the susceptibility characteristics in Table 5–2 are obvious upon reflection. We have mentioned why island species become extinct so often: They are sensitive to the introduction of new species because they have been isolated for a long time. Species with limited habitats become extinct easily simply because they have so little habitat to destroy or disturb that human activity can quickly eliminate it. For instance, one species of tropical insect will often be specifically adapted to only one part on one kind of local plant. When the plant is eradicated, so too is that insect species, along with others that are also adapted to other parts of the plant. Species with large territories die off quickly because they need lots of area to support them. Thus, even if only some of that land is developed, what is left may not offer enough for the species to survive. This has been repeatedly shown when apparently large game reserves are set up and yet wide-ranging species (such as large predators) still die out. When only one individual needs many square miles to forage, it takes a very large reserve to preserve enough individuals to maintain the species. Low reproductive rates make it difficult for a species to rebound from habitat disturbance, hunting, or other causes of population declines. Economic and sport value cause species to be sought after by hunters. Predators are generally high on the food pyramid so that they are relatively less abundant than many other organisms. Pollution-sensitivity is another trait leading to extinction.

These seven traits are not exclusive of one another: many species have more than one of the traits and are therefore that much more likely to become extinct. For example, the food pyramid tends to concentrate pollutants so that predators, in addition to being relatively rare, often die from pollution. (An example is DDT poisoning of eagles.) Furthermore, predators are often highly prized for sport so that is yet another trait working against them.

Note also that some of these seven characteristics can be related to the traits that paleontologists believe made groups susceptible to mass extinctions (Chapter 4): tropical habitats, local distribution, large body size, and marine (especially shallow water) habitats. The first two are related to limited habitats. Local distribution is often caused by having a limited habitat. Similarly, tropical species tend to have limited habitats because they tend to be specialized, with much niche packing (Chapter 4). Large body size is related to the last five characteristics of Table 5–2. Big animals need more territory, reproduce slowly, are usually hunted as predators or for sport, and are often most sensitive to pollution. The one mass extinction trait that may differ from modern susceptibility is that of marine habitats. Because we are land dwellers, we may be more effective at causing land instead of marine extinctions. Nonetheless, marine extinctions are clearly occurring, and where they are, shallow marine waters are usually the hardest hit, in agreement with past extinctions.

REASONS TO STOP EXTINCTIONS. Why worry about extinctions anyway? Most city dwellers have little need of elephants, exotic tropical insects, plants, and so on. Evolution has produced humans and the earth is our domain, with "lesser" animals existing for our use. To some, this is an absurd attitude. They take the opposite view and want to preserve nature for its own sake. They see humans as the intruders and any extinction is wrong. Yet while there may (or may not) be some philosophical validity to this, it is also true that human beings have rights too and any solution must consider their survival as well. In addition, it is not completely correct to view humans as "unnatural" intruders. We are products of biological evolution, just as all other species. Nevertheless, since humans are the active agents of destruction in so many cases, it is clearly up to us to do something about it, if species are worth saving.

The first step then is to determine why we should bother to save nature's biodiversity. Four classes of reasons are often put forth to save species, called the FOUR Es:

1. esthetic
2. ethical
3. economic
4. ecological

The *esthetic* reasons to save species state that biological diversity makes life more enjoyable and enriching. Indeed, an avid naturalist might say that it makes life worth living. Given the mental and physical rejuvenation many of us experience after a long hike or other form of outdoor recreation, this argument alone would seem to have much validity. Surely such recreation would be less rejuvenating if we had only the same few species to see, smell, and experience all the time. The *ethical* reasons to save species state that humans have no right to destroy other species to the point of extinction. Conversely, animals have rights of their own, including to live a life that is relatively free from pain.

The *economic* reasons to save species are the most immediately persuasive to most of the world's people because the human condition in developing countries is so poor that survival must, to them, take precedence to esthetics and ethics. When faced with life or death, few people give much thought to philosophical considerations. For convenience, we can divide these practical, economic reasons into two groups: food and nonfood uses of biodiversity.

First, we discuss food uses. Historically, humans have used about 7,000 plant species as food. However, as Figure 5–17 shows, only about *30 plant species provide 95% of the world's nutrition.* Just four of these (wheat, corn, rice, and potatoes) provide most of the world's food and these four have been subjected to decades of inbreeding. The effects of this reliance on a few species can have many bad consequences. One is a matter of taste. Do we want our diet to be so bland and monotonous? Botanists estimate that at least 75,000 edible plant species exist, many superior in flavor and nutrition to the ones we use. Other consequences are more serious. Low diversity makes organisms more susceptible to change. For example, the high inbreeding of most crop species makes them notoriously susceptible to diseases and insects. The infamous potato famine, which killed many thousands of people, was caused by a blight that wiped out a very inbred strain of potato in wide use. Wild,

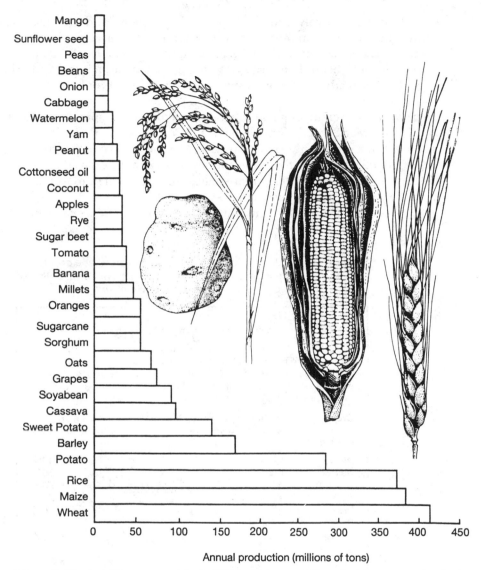

Figure 5–17. Just 30 crops contribute 95% to human nutrition. Just four crops dominate the world's diet. From A. Cockburn, *An Introduction to Evolutionary Ecology* (London: Blackwell, 1991), p. 306.

related species can often be interbred with closely related cultivated forms to improve resistance to disease, general hardiness, and so on. For example, this has been used to improve corn, using a rare wild strain just found in Mexico. An estimated $300 million per year was saved as a result. Improvement using wild strains is not limited to species that can interbreed. The new science of genetic engineering, discussed below, involves surgically removing of genes and directly inserting them from one species to another, even where they are unrelated. This increases the value of wild strains even more because we can combine useful traits among many species without being limited to close relatives that can naturally interbreed.

It is thus advantageous to maximize diversity not only within crops that we grow, but to have as many different crop species at our disposal as possible. Yet if current rates of extinction continue, some 25,000 plant species are expected to die out by the year 2000, before we have a chance to study them. This has led to the proposal that **gene banks** be set up, as a reservoir to save seeds, spores, sperm, and other genetic materials for those species that cannot be saved in the wild. Figure 5–18 shows that seed banks contain only a small fraction of most plant varieties and that many wild species have yet to be collected.

We have focused on plants, but very similar arguments can be made for more

Figure 5–18. Seed banks contain only a small percentage of wild species; a large percentage of wild species have not even been collected. Modified from G. Lean et al., *Atlas of the Environment*, p. 131.

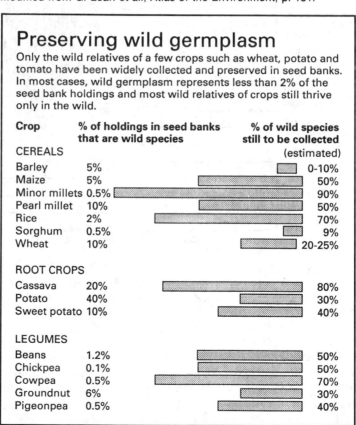

Preserving wild germplasm

Only the wild relatives of a few crops such as wheat, potato and tomato have been widely collected and preserved in seed banks. In most cases, wild germplasm represents less than 2% of the seed bank holdings and most wild relatives of crops still thrive only in the wild.

Crop	% of holdings in seed banks that are wild species	% of wild species still to be collected (estimated)
CEREALS		
Barley	5%	0-10%
Maize	5%	50%
Minor millets	0.5%	90%
Pearl millet	10%	50%
Rice	2%	70%
Sorghum	0.5%	9%
Wheat	10%	20-25%
ROOT CROPS		
Cassava	20%	80%
Potato	40%	30%
Sweet potato	10%	40%
LEGUMES		
Beans	1.2%	50%
Chickpea	0.1%	50%
Cowpea	0.5%	70%
Groundnut	6%	30%
Pigeonpea	0.5%	40%

diversity in domesticated animal species. Breeding of cattle with buffalo ("beefalo") to strengthen the stamina of cattle breeds and their resistance to predation is one of many examples where wild genes have been useful. Similarly, there are many wild animal species that are themselves potentially tasty and healthy food sources, even without being interbred with existing food species.

Nonfood uses also provide cogent economic reasons to save species. Wild species directly provide many nonfood products, ranging from medicines to rubber. A large number of modern medicines began as plants used by premodern societies (aspirin is the best known). With so many undescribed plants, it is certain that many potentially useful medicines will be lost. For instance, the rare Pacific Yew was recently found to be useful in fighting cancer. Many products imported by industrial societies would be lost if native ecosystems were destroyed because they do not grow well in domesticated conditions. For instance, rubber trees are a highly valuable resource in the tropics, but rubber can only be economically extracted from living trees year after year. Resource economists call such products *forest-sustainable* items because they contrast with products that call for the forest to be cut down. Other forest-sustainable resources are exotic fruits, oils, and fiber. A recent report has shown that, in many cases, such products bring in more money than the wood, so that countries may be better off leaving the forests intact (discussed below).

The fourth reason for saving species is *ecological* importance. In many ways, this is the ultimate reason because it states that species compose the ecosystems that provide us with many of life's essentials, beyond food, furniture, and other daily economic items. In other words, ecosystems are environmental support systems that provide us with things that we now take for granted: oxygen to breathe (from plants), drinkable water (purified by microbial activity), and many other natural chemical cycles often called *ecosystem functions*. The key point is that if we remove too many species from an ecosystem, it will collapse and ecosystem function will be impaired. If too many ecosystems collapse, we are obviously in danger.

How to Stop Extinctions. Assuming diversity is worth preserving for the reasons just discussed, how do we go about stopping extinctions? While it is easy to feel frustrated, there are actually many steps that can be taken. A tentative outline would include the following steps.

1. *Research and Description of Species.* With only 1.7 million species described out of a possible 50 million or more, it is essential that we begin to gain a better understanding of what species are out there. The job will be enormous. Most of these are tropical species, mainly plants and small animals, and especially insects. One expert, E. O. Wilson, has estimated that it would take the entire lifetimes of 25,000 specialists to describe them all. Currently there are about 1,500 specialists with the required knowledge in tropical biology.

Given the short time available, biologists are considering salvage operations to obtain brief surveys of the many unstudied species before they become extinct. If possible, such surveys should include description of life habits, anatomy, and so on.

To expedite this, biologists have classified species into five (nonexclusive) categories:

1. indicator species
2. keystone species
3. umbrella species
4. flagship species
5. vulnerable species

Indicator species are those with known tolerances to specific pollutants and other kinds of environmental damage. They thus provide important information on how much damage has occurred and what might happen. **Keystone species,** as we discussed earlier in the section, play a pivotal role in ecosystems. Thus, exceptional attempts ought to be made to save them. **Umbrella species** have large area needs. Usually these are large animals. They are called umbrella species because if enough area is preserved to save them, then many other species will also be placed under the umbrella of protection. **Flagship species** are the charismatic organisms that attract so much media attention and appeal to many people. The Florida panther is an example. It is important to remember that for each flagship species, there are thousands of ugly, small, slimy, or otherwise unattractive species that often play major roles in ecosystems. **Vulnerable species** are those most likely to become extinct. They possess those qualities already discussed: limited habitat, large size, and so on.

2. *Establishment of Preserves.* Even before completing the assessment of species diversity, it is essential to preserve some area of habitat for those that are endangered. It is not enough to try to reestablish the species after the habitats have been destroyed because virtually all attempts to do this have failed. It is much more desirable therefore to simply preserve some area before it is disturbed. Preserves are well established in many parts of the world, such as in the developed countries and in Africa and Asia. However, they are badly lacking in Central and South America. Even where established, there is a major problem of enforcement, since most of the tropical countries are poor and often cannot afford adequate policing of the preserve from poachers. Even if they could be policed, many more preserves are needed, especially in the rain forests. Currently *only 1% of the earth's land surface is legally preserved* in this way.

Preserves involve more than just setting aside any section of land. It must often include prime nesting areas, migratory pathways, and so on. Furthermore, the **preserve design** is critical. Figure 5–19 shows that species preservation is maximized by each of the following: larger areas, clustering, connection "corridors" for migration, and minimum circumference to minimize "edge effects." Large edge effects mean that organisms from the surrounding area can diffuse into the preserve too easily. Preserving a sufficiently large area of the natural ecosystem is usually the biggest problem. Bigger animals usually have more territorial needs: They need more food per individual and more space to forage. Therefore, it is often necessary to set aside very large tracts of land to save them, such as Kruger

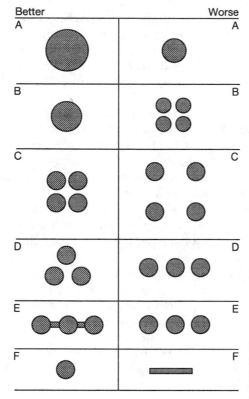

Figure 5-19. Each reserve design on the left will preserve a larger breeding population than that on the right.

Park of South Africa. Failure to set aside large enough tracts has often led to the slow decline of large species within those preserves.

3. *Laws Protecting Endangered Species.* It is very difficult to catch poachers, especially in large preserves of poor countries. Hence, a more effective tactic has been to make it illegal to trade, transport, and sell products made from endangered species. In 1975, 81 countries signed the Convention on International Trade in Endangered Species (CITES), which outlaws such trade in endangered species' products. This approach has been much easier to enforce in developed countries and has been very effective in making poaching less enriching. Perhaps the greatest success story with this approach has been a 1989 world ban on selling or trading ivory from elephant tusks. Before the ban, elephant ivory in Africa sold for about $100 a pound and elephant populations dropped about 50%. Ivory now sells for only $5 a pound on the black market. Until more effective bans are implemented for other organisms, however, poaching will continue. The sale of outlawed products such as leopard-skin coats and crocodile shoes globally generates an estimated 2 to 5 billion dollars per year. A major factor is public desire to avoid such products; this was crucial in the success of the ivory ban.

Protection of species within the United States is also a growing problem. Some biologists predict that up to 4,000 species within the United States are in danger of extinction by the year 2000. The main legal apparatus to protect species in the United States is the **Endangered Species Act** of 1973. This directs the U.S. Fish and Wildlife Service to maintain a list of species that are endangered or threatened. **Endangered species** are in immediate danger of extinction. **Threatened species** are likely to be endangered soon. This act was scheduled for renewal in 1992 and a great controversy surrounds the cost of saving increasing numbers of species. The act has worked fine when corporations only had to make minor changes in construction plans to avoid building on the habitats of endangered species. But what about the northern spotted owl habitat, where logging on over 8 million acres in the Northwest has been halted, locking up billions of dollars in timber and thousands of jobs? This requires very difficult social decisions. These decisions are complicated by the extreme rigor of the act, which seeks to protect not only species but subspecies and populations. Genetic analysis often shows that a subspecies that has cost millions to protect is extremely similar (with over 99% of the same genetic variation) to another subspecies that is common and not endangered. For example, the red squirrel is very similar to the gray squirrel. Efforts to save the red squirrel in Arizona have been costly and have obstructed the building of the world's most powerful telescope. Such situations support the argument that instead of making very costly efforts to save every population and subspecies, the act should try to be more realistic and take social and economic considerations into account. Furthermore, they argue that these costly efforts are often futile: *Most officially listed species are closer to extinction now than when originally listed.* Thus, emphasis is moving away from species to saving entire ecosystems.

4. *Breeding in Captivity.* This is less desirable in many ways than the previously described steps, but there is often no other choice if the genetic diversity is to be preserved. It is less desirable for the species itself because captive conditions are often unnatural and the species often does not do well, is not happy, and does not reproduce. It is possible to artificially impregnate females, but up to 1989 this had been successful on less than two dozen wild species. If conditions become really desperate it may be possible to freeze the sperm and egg of endangered species just to have their genetic material available for genetic engineering, discussed later in this chapter. Even if captive species do reproduce well, the offspring often cannot survive in the wild, if suitable habitats are found, because they were not raised under natural conditions. Finally, captive breeding is less desirable from the human point of view because it is very expensive. Consider that *all the zoos in the world could fit within the boundaries of the District of Columbia.* It is doubtful if any modern zoo could maintain healthy breeding populations of more than 900 species, and probably a lot less.

5. *Reduce Socioeconomic Causes of Extinction.* Ultimately, to stop extinctions, we must treat the basic social and economic causes. The most basic of these is rapid human population growth. The population in developing, mostly tropical, countries is growing at very high rates due to better medical care and many other reasons. As population increases, so do pressures to exploit more resources. Consider the

example of a person living in tropical Africa, where the average income is around $100 to $300 per year. In 1986 an average rhino horn could be sold wholesale for nearly $2,000, or up to 20 years worth of income! If a person has a family—and often a large one—to support, what would the rational person do? Similar arguments could be made for poor families trying to make a living from the soil by slashing and burning in Brazil or elsewhere. Unfortunately, slowing human population growth is proving to be very difficult. It reached five billion humans in the late 1980s and shows no sign of slowing.

Second, we must treat the economic incentives that make species extinction profitable. We have already mentioned acts that outlaw certain practices, but this can also be done with the profit motive, perhaps more effectively. For example, public reluctance to buy tuna caught with driftnets that killed many dolphins resulted in a major reduction of dolphin deaths, shown in Figure 5–20. On a much larger scale, a rapidly growing branch of economics, called **ecological economics,** has shown ways to make ecosystems more profitable if left intact than if destroyed. An example mentioned earlier involves forest-sustainable products derived from tropical rainforests. Table 5–3 shows that drugs, entertainment (such as movie-making), and tourism can be extremely profitable ($2 trillion for tourism alone). This does not even include such items as exotic fruits and other food items. All told, these forest-sustainable uses seem to be much more valuable than destruction by the usual forestry practices of harvesting trees. Certainly it would seem more profitable than slash and burn agriculture, which only produces a subsistence diet for a few years.

Dolphin kills down

Since activists first raised the issue two decades ago, the number of dolphins killed accidentally in tuna driftnets has declined. Since 1990, most U.S. firms haven't used tuna caught in dolphin-snaring driftnets. Yearly number of dolphins accidentally killed by fishing boats selling to U.S. firms:

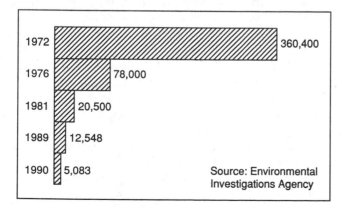

1972	360,400
1976	78,000
1981	20,500
1989	12,548
1990	5,083

Source: Environmental Investigations Agency

Figure 5–20. Rapid decline in dolphin deaths in the last 20 years. From *USA Today* newspaper, May 31, 1991.

TABLE 5–3. The Financial Size of Different Industries in Rain Forests

INDUSTRY	ESTIMATED WORLD MARKET (US$) BILLIONS	GROWTH RATE
Forestry		
Total	85	1–2%
Tropical	7	
Pharmaceuticals	?	?
Tourism[a]	2000 (1987)	4–5%
Entertainment[b]	150 (1988)	10–15%

[a] Tourism is estimated to employ 6.3% of the global work force.

[b] In 1983, American film studios earned $800 million from video cassettes, against $2.6 billion at the box office; the comparable figures for 1988 were $4.5 and $2.9 billion.

Source: "How to Pay for Tropical Rain Forests," *Trends in Ecology and Evolution*, 1991, p. 348.

Unfortunately, rapid population growth makes the need for a subsistence diet so pressing that long-term economic incentives may not have time to work. Therefore, an increasingly common practice is the **debt-for-nature swap.** This is when a country agrees to set aside a certain area of pristine land as a preserve in exchange for reduction in the amount of money (debt) owed to another country. Thus, richer countries lose money on loans made to the poorer countries, but natural preserves are gained for the world's benefit. Similarly, the World Bank has recently begun to stipulate that environmental considerations be incorporated into any projects financed by loans made to developing countries.

New Diversity: Humans Creating Species

Awareness of extinction has-increased dramatically over the last decade. Articles appear almost daily in magazines and newspapers documenting the loss of species. Yet, humans are also causing new forms to originate. Recall from Figure 5–9 that this origination occurs in two ways. One, when we modify an environment (through deforestation, pollution, and so on), some organisms do not become extinct, but are able to *naturally adapt* to the change. In the process of adapting, these organisms often undergo evolutionary changes. Two, humans *directly manipulate* the gene pools of species, creating new combinations of genes and therefore new kinds of organisms. The simplest way of doing this, selective breeding, has been done for over ten thousand years, and is still important today. However, a much more powerful method of combining genes has become available over the last few years: Genetic engineering allows humans to combine genes from radically different organisms (such as bacteria and frogs), which could never be accomplished in nature or even through selective breeding.

Whether human-created species diversity is "better" or "worse" than natural evolution is largely a value judgment. The main point is that, like it or not, our biological environment—from pets to farm animals to house plants to urban landscapes to diseases—is becoming increasingly dominated by organisms created by the indirect (environmental change) and direct (gene manipulation) actions of humans. As this continues, the number of difficult ethical, social, and ecological questions will increase as well.

First, we discuss the indirect human alteration of species through environmental change. Following this, we examine the direct creation of new species by breeding and modern genetic technology.

NATURAL ADAPTATION TO HUMAN-MODIFIED ENVIRONMENTS. Some organisms not only cope with, but are able to thrive in, the environments created by humans. For example, the spread of cities across the countryside has greatly reduced or doomed many native species in North America, such as the bald eagle, yet other species have flourished within the urban environment. Consider the many songbirds, squirrels, trees, and plants that occupy any undeveloped city lot.

Why can some organisms thrive while others become extinct when humans modify environments? The answer lies in each organism's *unique evolutionary past*. Just as evolution has endowed some species with traits that predispose them toward extinction, so has evolution endowed other species with traits that allow them to take advantage of human-modified environments. Two basic categories of organisms have been successful in human environments: (1) opportunistic species and (2) species specialized for human-created environments.

Opportunistic species evolved to take advantage of natural disturbances in ecosystems. These tend to be small, fast-growing organisms that have a wide range of environmental tolerances, including a broad diet. They also tend to be very mobile and adept at dispersal. Examples would be mice, rats, oppossums, many kinds of insects, and the many "weedy" plants that typically invade our suburban lawns and gardens. In nature, these organisms are usually among the survivors of natural disturbances, such as forest fires, hurricanes, and so on. They are also among the first to recolonize areas that have been completely devastated by such disturbances. As the ecosystem matures and more specialized competitors and predators move in, these opportunists become more rare. Thus, the reason opportunistic species flourish in human-affected environments is that we are disturbing natural ecosystems everywhere, in many ways. Suburban homes represent a highly disturbed ecosystem wherein the natural landscapes have been bulldozed, with most native species being run off or killed. The average homeowner fights a constant battle against roaches, mice, rats, and other pests in the home as well as fast-growing, hardy "weedy" plants that constantly take root in our manicured lawns and gardens. These natural opportunists are simply doing what evolution has endowed them to do, pioneering disturbed environments, paving the way for other organisms.

In contrast to the hardy, generalized, opportunistic organisms are the many highly specialized organisms found in nature. Their relatively slower growth, specialized food, and other resource needs make them more likely to become extinct in human-modified environments. However, a few specialized organisms were fortunate enough to have evolved needs that humans provide, usually inadvertently. We might say that these organisms were "preadapted" to humans and our favorite environments. The most obvious examples are the various microorganisms and parasites that infect humans directly. The **AIDS** (Acquired Immune Deficiency Syndrome) virus, which is now threatening to become a global epidemic, originally infected only monkeys in Africa—that is, until a mutated form infected human populations 30 to 40 years ago. From Africa it spread to Haiti, the United States, and elsewhere. Other beneficiaries of our success are those species that are specialized to live in environments that we create and maintain for ourselves. For example, squirrels and many songbirds that populate the urban and suburban environments are specialized to rely on certain trees and other plants that we cultivate mainly for aesthetic reasons. These plants, such as oaks, provide both food (nuts, berries, and seeds) and shelter.

Many of these "human-preadapted" organisms, such as the songbirds, are considered desirable by us for aesthetic or other reasons. Even though they are but a small remnant of the original, native ecosystem, such creatures are pleasing symbols of the natural world. Of course, some are not so pleasing, such as the various diseases specialized to live on humans and our pets. Most of the opportunistic organisms are considered pests (e.g., weeds, rats, and most insects). In spite of our best efforts to eradicate them, it is likely that such pests are going to be with us for a long time, accompanying us wherever we go and evolving, as we carry them into new environments. For example, the Norway rat has been carried all over the world by trading ships, and it will undoubtedly form new races and even species in different environments, given enough time.

By carrying species into new environments, we create new opportunities for natural selection. Alternatively, when we try to eradicate species with pesticides, we accelerate their evolution even more because we cause strong directional selection, as shown in Figure 5–21. The species will eventually evolve a resistance to the chemicals. The importance of this is illustrated in Figure 5–22, which shows how the number of insect species resistant to one or more insecticides has dramatically increased over the last few decades. The chemical industry is constantly producing new pesticides to counteract this, but many experts believe that they are fighting a losing battle.

DIRECT CREATION OF LIFE FORMS BY HUMANS. There are two ways that humans directly create new life forms. The oldest and simplest method is **selective breeding.** This occurs when humans determine which individuals in a plant or animal species are allowed to breed. Selective breeding began at least 10,000 years ago. Sometimes our criteria for breeding were practical, as with more docile and productive meat animals. Other times our criteria were simply aesthetic or even whimsical, as with the creation of trendy dog or cat breeds. The profound power of human favoritism in evolution is seen in the overwhelming populations of domesticated species compared to wild ones. Consider that over 50 million domestic cats live in the United States alone, compared to about 30 to 50 Florida panthers in the Everglades. Yet the ancestors of domestic cats were at one time no more common, and perhaps less

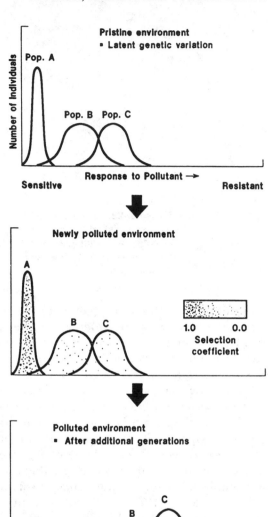

Figure 5–21. Selection favors individuals that can tolerate pollution. The higher the selection coefficient, the more individuals are selected against. Eventually, only pollutant-resistant individuals comprise the population. From *Trends in Ecology and Evolution* magazine, Sept. 1988, p. 234, Fig. 1.

so, than most of the larger forms now on the endangered species list. They were minor components of a few ecosystems.

Selective breeding is still important today. New strains of crops and animals are being continuously developed to improve agricultural yields and new pet breeds still appear. However, selective breeding is being rapidly replaced by a much more effective method. **Genetic engineering** is the use of highly advanced biochemical

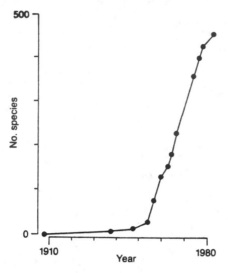

Figure 5–22. Cumulative number of species resistant to one or more insecticides. From *Trends in Ecology and Evolution* magazine, Nov. 1989, p. 336, Fig. 1.

and microsurgical techniques to directly transfer genes (DNA) between different kinds of organisms. This process is also called **gene splicing.** The technology is called **recombinant DNA technology,** or more briefly, **biotechnology.** The potential of this technology for drastically altering your world cannot be overemphasized. Biotechnology has been called the *greatest revolution in human history*, even greater than the industrial and agricultural revolutions before it. Like all such revolutions, it will likely create major upheavals in society, raising complex economic and ethical issues. Its great potential is that, unlike selective breeding, genetic engineering is not limited to working within existing gene pools (organisms that can interbreed). Instead it permits the transfer of genes between different kinds of organisms, called **transgenesis.** This greatly increases the number of combinations, and thus new life forms, that can be created. For example, DNA for making human insulin have been inserted directly into the genes of bacteria. In contrast, with selective breeding inserting the genes of one species into another would never happen because the two species could not interbreed.

METHODS OF GENETIC ENGINEERING. The following steps outline the basic techniques of genetic engineering:

1. **Restriction enzymes** locate and cut DNA strands at specific sites in a donor organism. These enzymes were first discovered in 1970, leading to the idea that certain genes could be "snipped out." Over 400 restriction enzymes are now known, which can recognize and remove over 100 different DNA sites.
2. **Vectors** are free-floating DNA molecules that attach to the cut DNA strands in the donor and transfer the DNA strands to the genes of another cell. A number of vectors are used, especially **plasmids,** which are DNA molecules that occur naturally within bacterial cells, and **viruses,** which are DNA molecules that force cells to reproduce the viral DNA. Splicing of the cut DNA into the vectors and new cells is facilitated by **DNA-ligase enzymes.** These enzymes bind DNA strands together.

Figure 5–23 illustrates this two-step process in the transfer of human DNA into the DNA of a bacterial cell. In this case, the DNA represents the human gene for producing insulin, a crucial protein used to treat diabetes. By inserting the gene into the bacteria, the microbe will produce human insulin. The unique code of the human gene serves as a blueprint for assembling the insulin molecule. The *inserted*

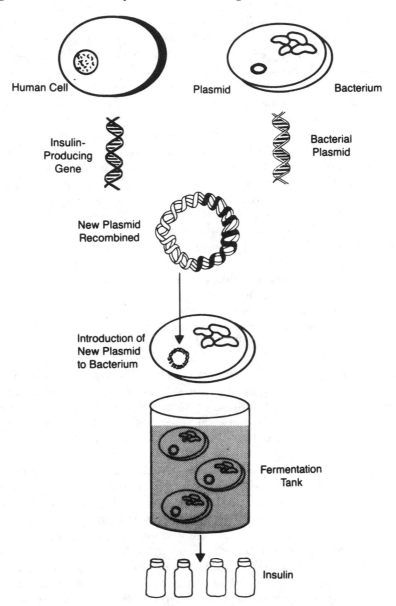

Figure 5–23. Splicing a human gene for insulin production into a bacteria's genes yields bacteria that can produce human insulin. From R. Brennan, *Levitating Trains & Kamikaze Genes* (New York: Wiley, 1990), p. 51.

gene is passed on to the offspring of the original bacterium, being copied along with the other genes during reproduction. This permits large amounts of human insulin to be produced by the manufacturer.

DNA transfer to a multicellular animal or plant differs from Figure 5–23 in that the foreign genes are inserted into a newly fertilized embryo. Usually this is carried out through **microinjection** of the vectors into the embryo. (It is called microinjection because fine-scale microscopic instruments are used.) More efficient delivery methods, such as particle guns, are now being developed to replace the "shotgun approach" of microinjection. Because the newly fertilized embryo is a single cell, like the bacterium, the inserted DNA can be copied to the other cells of the animal or plant as it grows by cell division. Some of these cells can become reproductive cells (sperm or egg) so the new DNA is passed on, becoming a permanent part of the species.

There are many technical problems and limitations to genetic engineering, which this simplified discussion omits. Among the most important is that some genes simply do not function well in other kinds of organisms. This often occurs when traits are polygenic, where the trait is governed by many genes. For instance, a gene for growth hormone production in one species may not function well in another kind of organism if the growth hormone needs to precisely interact with other genes that are lacking or somehow vary in the new organism. This is seen in experiments with pigs that were engineered with growth hormone genes from cattle. The pigs grew faster and were leaner, as desired, but they also suffered from stomach ulcers, arthritis, enlarged hearts, and kidney disease because they differ from cattle in other genes that interact with growth hormones.

USES OF GENETIC ENGINEERING. The potential uses of biotechnology are vast—they include medical uses, agriculture and industrial applications, and environmental clean-up projects. This can be seen in the rapid growth of biotechnology companies. Genentech, the first U.S. biotechnology company, was formed in 1976. Its work proved so promising that when the second biotechnology company, the Cetus Corporation, was formed and went public, it raised the largest sum ever recorded on Wall Street. As of 1990, there were over 400 biotech companies in the United States with sales of over $1 billion per year. Let us look briefly at how genetic engineering is applied in each of the major areas.

About 25% of the biotech companies specialize in the medical field, employing biotechnology to make over 80 drugs and chemicals. These are usually created with bacteria that are genetically engineered to produce human gene products, such as insulin. One notorious obstacle in the drug industry is the long period of testing necessary before the drugs are allowed on the market. Thus, in spite of the many human gene products being researched, only four were on the market by 1987: human insulin, human growth hormone (for treating some dwarfisms), interferon (to treat some cancers), and a hepatitis vaccine. Among other drugs that may soon be available are: clotting factors for hemophiliacs, a drug for dissolving clots, drugs to treat high blood pressure and heart failure, and cancer therapy products. Only a few more drugs have been produced since 1987, shown in Table 5–4.

TABLE 5–4. Some Human Genetic "Drugs" Cloned into and Produced Primarily by Bacterial Cell Cultures. These Compounds are Currently in Use for Disease Control and Clinical Studies.

HUMAN "DRUG"	DISEASE
Alpha Interferon	Hairy Cell Leukemia
	Non-Hodgkin's Lymphoma
	Kaposi's Sarcoma
	Multiple Myeloma
Human Insulin	Diabetes
Interleukin II	Various Immune Deficiencies
Human Growth Hormone	Various Growth Deficiencies
Blood Clotting Factor VIII	Hemophilia
Tumor Necrosis Factor	Malignant Myeloma
	Various Cancers

Agricultural uses of biotechnology are another major application. Usually the engineering is used to increase the productivity of domesticated animals and plants. This is most often accomplished in two ways: (1) increasing the growth rate of the plant or animal, or (2) increasing the environmental tolerance of the plant or animal to disease or other stresses. For example, genes for human growth hormone have been inserted into cattle, chickens, and pigs to produce faster-growing, leaner animals. A more common approach is the transfer of genes between related animals. For instance, a gene that controls growth in rainbow trout was introduced into a carp. Although the carp has its own growth gene, this added gene produced a 20% increase in growth rate in the carp without affecting quality of the fish as food. There are also many examples where animal resistance to disease has been improved. Indeed, the first license issued by the U.S. Department of Agriculture for a genetically engineering organism (in 1986) was for a vaccine to prevent herpes in pigs.

Because of the difficulty of isolating specific genes, such as that for growth hormones, geneticists are planning to produce genome maps for cattle, pigs, chickens, and other food animals. A **genome map** is the description of all the genes of a species, specifying the location of each gene on a chromosome and what traits each gene determines. These will be costly projects, but will ultimately make genetic engineering much faster and efficient. A genome map for humans is also being made (discussed below).

Gene transfer in plants has advanced somewhat more slowly than in animals, in part because of greater difficulties in finding appropriate vectors. However, rapid progress is now being made and enormous agricultural benefits are being realized. As with animals, increased productivity results from increasing growth rates and promoting resistance to environmental stresses. Resistance is especially important in plants because they are prone to weeds and herbivores of all kinds (especially insects) that farm animals are not subject to. Thus, in addition to disease resistance, genetic engineering in plants has been used to aid in the fight against weeds and plant-eating pests. For example, weeds are combatted by inserting genes

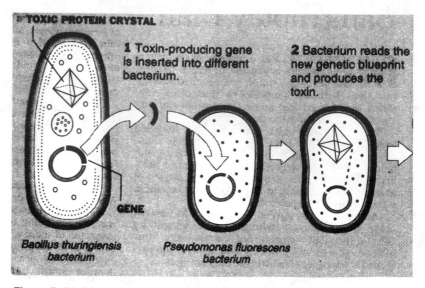

Figure 5–24. Making safer pesticides: Splicing a toxin-producing gene into another bacterium that dies and is then sprayed onto plants. From *New York Times*, July 17, 1991, p. C7.

into crop plants that promote tolerance to poisons that kill weeds. Insect resistance has been created by transferring genes from bacteria that kill harmful insects. The bacterial genes that produce the insect toxins have been successfully transferred to tomato, tobacco, and cotton plants. The experimental plants now produce the toxins themselves and suffer much less predation from insects. The resultant crop yields for such "supertomatoes" have increased 20 to 30% and overall costs went down as well. Genetic engineering can not only increase crop output but has been used to improve crop quality. Flavor and nutrient content have been altered in corn and many other foods by altering the oil, starch, and other chemical properties. Nor is plant genetic engineering limited to food plants. Current gene research is aimed at improving plant fibers used in clothing and other materials. There is also much promise for the many plants used in producing drugs, especially since mammalian proteins have been produced by plants through genetic engineering. Because of all these uses, genome maps for plants, similar to those of animals, are being developed.

Finally, applications of genetic engineering are useful in solving industrial and environmental problems. Most of these applications will involve modification of bacteria and other microbes to produce useful biochemicals. Examples of industrial uses seem almost limitless, including applications in the chemical industry (everything from paints to fertilizers), in the food industry (such as with fermentation), and even current research into using bacteria to extract metals from ores. It is ironic that there are many potential environmental benefits from genetic engineering, considering how much focus has been on its potential for environmental damage.

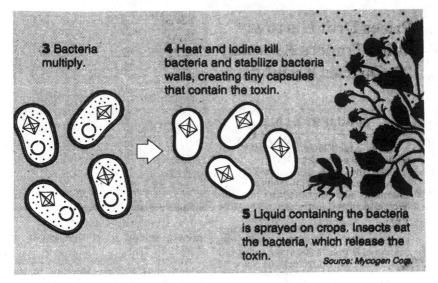

3 Bacteria multiply.

4 Heat and iodine kill bacteria and stabilize bacteria walls, creating tiny capsules that contain the toxin.

5 Liquid containing the bacteria is sprayed on crops. Insects eat the bacteria, which release the toxin.

Source: Mycogen Corp.

Figure 5–24. (*Continued*)

Thus, "oil-eating" bacteria were among the first organisms to be genetically engineered. They have been successfully deployed on oil spills to help digest the oil, which normally takes a long time to break down, causing much harm to other organisms. Similarly, there is even experimentation with bacteria to break down toxic wastes, such as cancer-causing PCBs. Many microbes can be engineered to kill plant pests, thus replacing environmentally harmful chemical pesticides, as shown in Figure 5–24. They can also be modified to replace some fertilizers, which now cause much of the pollution of our waterways.

ETHICAL, LEGAL, AND ENVIRONMENTAL PROBLEMS. While the potential uses of biotechnology are vast, the ability to modify life in such novel ways also creates an enormous set of new problems for society. We can divide these problems into three types, for convenience: ethical, legal, and environmental.

Ethically, many people question whether humans have the right to "tamper with life" in this way. Such ethical questions become even more difficult when you consider issues that involve gene manipulations in humans. Legally, many questions have arisen concerning who owns the genetically engineered organisms. Are they "inventions" that can be patented? For economic reasons, patents must be granted to the groups (such as Genentech) that create such organisms to pay for the high cost of equipment and experimentation, if such research is to continue. On the other hand, many argue that life cannot be patented because no one can possess the rights to the genes of a species. However, at least for now, the U.S. Supreme Court has ruled that microorganisms can indeed be patented and in 1987 the U.S.

patent office began issuing patents for genetically engineered species. In a more recent and controversial development, patents have been granted to those researchers that simply discover the location of the genes.

In spite of such ethical and legal problems, it is the potential for environmental disturbance by genetically engineered organisms that many people find the most distressing. While some of these fears are unjustified, there is a real basis for others. For example, we have seen how human introduction of species into ecosystems (such as Australia) can be very destructive. While fears of "super" disease organisms or other major disasters are highly unlikely, unforeseen disturbances of native ecosystems must be an important consideration where genetically engineered organisms may escape or be purposely released into the environment. Often such potential disruptions can be very subtle. For example, botanists in Germany spliced a corn gene into a petunia plant to produce a petunia colored like an ear of Indian-red corn. While this was praised by florists for its beauty, environmentalists argued that if the plant were allowed to grow wild, the strange color might confuse bees. It could conceivably upset the fine balance that has evolved between certain species of bees and natural petunias. Only ecological studies can tell. Thus far, such impact studies, by restricted field testing, have been very effective in estimating ecological effects. By 1990, eight outdoor releases of genetically engineered organisms had occurred in the United States, with no major ecological ill effects. The first release consisted of altered bacteria that reduced frost damage when sprayed on strawberries in California. The bacteria reduced frost damage by 80% and have had no side effects or spread beyond the area of application.

An interesting technical solution to problems of environmental disturbance by genetically engineered organisms may be **kamikaze genes.** These are "suicide genes" that, when switched on, destroy the organism. They occur naturally in many, and perhaps all, organisms and help regulate their life cycles. In genetically engineering an organism, it may be possible to "program" it to die when its job is over by inserting kamikaze genes into the organism. For example, bacteria that secrete helpful enzymes for plants might be programmed to die when the enzymes reach a certain concentration.

It is unlikely that ethical, legal, and environmental problems will prevent widespread use of biotechnology. The practical advantages of designing organisms for our purposes are too great. History has shown that whenever major technological advances have occurred, they inevitably become widespread in spite of resistance. Even if some groups or countries can agree to restrain their use, some other one will not feel so restrained. In so doing, they gain such a practical advantage that the other groups must then use the new technology to stay economically competitive.

Complicating the difficult ethical, legal, and environmental problems is the very rapid increase in the uses of this technology. This forces societies to make decisions even before the last set of problems has been resolved. As shown in Table 5–5, less than 20 years after Watson and Crick discovered that the DNA helix existed, scientists had already developed a way of manipulating that DNA. Now, less than 20 years after discovering that genes could be cut, a huge multimillion

TABLE 5–5. A Brief History of Biotechnology

1953	Watson and Crick discover the DNA helix.
1970	Nathans and Smith discover the first enzyme that can cut DNA strands at precise locations.
1973	Boyer and Cohen perform the first genetic engineering experiment, splicing toad genes into a bacterium.
1976	The first guidelines for recombinant DNA are set by the National Institutes of Health.
	The first U.S. biotechnology company, Genentech, is formed.
1980	The U.S. Supreme Court rules that microorganisms can be patented.
1982	The U.S. Food and Drug Administration approves the first sale of a drug (human insulin) made by genetically engineered microbes.
1988	The first authorized outdoor release of genetically engineered bacteria occurs. These microbes protect strawberry plants from frost.
1989	The U.S. government approves the introduction of foreign genes into humans, as a form of cancer treatment.

dollar industry has developed, with over 400 companies (in the United States alone) manipulating the genes of many types of organisms. Over 200 patents for engineered organisms, mostly bacteria, had been issued by 1990. Many products are already on the market, many outdoor releases have already occurred, and medical gene therapy on human beings is a major industry, as we shall see later in this chapter.

Future Evolution Without Humans

A major point in our discussion of patterns of evolution (Chapter 4) was that chance has played a large role in the history of life. Thus, if humans seriously damage or destroy all complex life via nuclear war or some other major ecological disaster, it is impossible that the exact species, including humans, would reevolve; nor is it even certain that intelligent life would reevolve. The exact conditions that led to the history of life as we know it would never be exactly duplicated.

While the exact outcome of future evolution without humans is a matter of speculation, let us consider it briefly, as an intellectual exercise to gain insight into the evolutionary process. Also, it is important to remind ourselves that evolution will occur in the future, just as it has in the past, and that it will do so whether or not humans exist. Probably the most complete examination of future evolution without humans is in the book, *After Man: A Zoology of the Future* (1981) by Dougal Dixon. Most of the hypothetical species depicted are mammals that inhabit a wide variety of environments, from tropical forests to polar seas. Figure 5–25 shows some examples of these.

In reading Dixon's book, two important points stand out that may help you imagine your own interpretation of future life. One, there is an extremely rich variety of life forms. While many seem exotic to us today, this is only because they

SWIMMING MONKEY
Natopithecus ranapes

HELMETED HORNHEAD
Cornudens horridus

Figure 5–25. Speculative mammal species of the future. From D. Dixon, *After Man: A Zoology of the Future* (New York: St. Martin's Press, 1981), composite of various figures.

are unfamiliar to us. If we remove the human influence on evolution—selective breeding, genetic engineering, and, especially, global destruction of habitats—it is clear that such a rich variety of species has a much greater chance to evolve. Two, in spite of its seemingly exotic nature, all future species must ultimately have ancestors in today's organisms. Even though the organisms will be altered by adapting to new environments, as they arise, many basic traits would be recognizable to someone alive today. As discussed earlier in the book, evolution can only modify the raw material that exists at any given time. Only those mammals that exist today can form the pool of raw material for future evolution.

PROSPECTS FOR HUMAN EVOLUTION

OVERVIEW: WE WILL EVOLVE; THE QUESTION IS "WHERE"?

Students often ask if evolution has stopped. The answer is emphatic: No! We have just discussed how evolution will continue to shape the future of the biosphere, whether or not humans are involved. Similarly, if humans survive into the future, it is very likely that our evolution will continue. Through time, various traits will be selected over others, and the human species will therefore change as new traits are favored and grow in abundance. But this is not to say that the evolutionary process will occur as it had before humans. As with the future of life's evolution, the primary selective force acting on humans will no longer be solely natural selection. Instead, artificial selection will often determine which traits are favored. This selection will act on humans in the same two forms that we discussed for evolution: (1) unplanned selection, as different traits favor survival in our rapidly changing manmade environment, and (2) planned, human manipulation of genes, in an effort to consciously steer the future of the human species.

Future Human Biological Evolution

Science fiction often depicts humans of the future as having large heads, relatively small bodies, and fewer and smaller teeth (Figure 5–26). This is also the most common perception by the average person of future humans. The origin of this perception lies in the fallacy of extrapolation, discussed at the beginning of this chapter; it assumes that previous and current trends in human biological evolution will continue. Thus, we have seen how brain size has increased from selection for intelligence, and tooth size has decreased, as food has become softer (see Chapters 3 and 4). It is assumed that an advancing technological environment will continue these trends. The projection of smaller body size for future humans is based on the assumption that modern technology will continue to make large physical size and strength less advantageous than it once was.

Figure 5–26. A speculative, if unflattering, view of future humans. This commonly proposed puny, big-headed specimen is based on the unwarranted notion that past trends will continue. From C. Brace, *The Stages of Human Evolution* (Englewood Cliffs, NJ: Prentice-Hall, 1988), p. 141.

While these assumptions seem reasonable, we have already pointed out that the simple extrapolation of past trends rarely holds true for very long. Unpredictable events will eventually occur. For example, some speculators now believe that the development of computers will eliminate selection for larger brains because information storage is being transferred from biological brains to microchips, which can hold much more information. Indeed, some predict that human intelligence will diminish as we become more dependent on machines. Furthermore, even if we discount unpredictability in general, we have already seen that nearly all evolutionary trends eventually encounter forces that oppose them, causing the "S-shaped" curves of Chapter 4. For instance, Figure 5–27 shows that when past human brain evolution is extrapolated, it is predicted to evolve to *three times its present size* over the next one million years. Yet, most human biologists agree that human brain size at birth is already about as large as it can possibly be, without routinely killing the mother. Here we have another case of a possible limitation being placed on a past trend. It is very difficult for most biologists to imagine how evolution could overcome this severe limitation, at least to the extent that our brains would increase by a massive 300%. However, the point is not which prediction is correct, but that virtually anything is possible, and therefore all kinds of wild speculation can be made. The small, toothless, big-headed human of Figure 5–26 may become a reality, but it could just as easily be that humans will not resemble that at all. Indeed, if humans manage to colonize other worlds, a diversity of human forms is possible, as we adapt to greatly different environments.

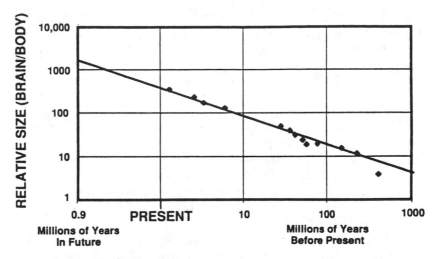

Figure 5–27. An unjustifiably simple extrapolation of past evolutionary brain size increase yields humans with huge brains of three times present size. From G. Seielstad, *At the Heart of the Web* (Orlando, FL: HBJ, 1989), p. 185.

Whatever the future holds for the exact path of human biological evolution, there are some broad generalizations that we can still make. For example, we can be sure that the selective pressures determining human evolution can be categorized into the two basic groups noted above: (1) unplanned selection, which is a byproduct of rapid, human-induced environmental change, and (2) planned selection, resulting from conscious efforts by humans to alter the human species. We turn first to unplanned selection.

UNPLANNED SELECTION IN FUTURE HUMAN EVOLUTION. We already have a powerful array of methods that allow us to control our own evolution. Genetic engineering of human genes is only the latest technique for determining which traits are passed on to new human generations. Such "high tech" methods are really just an extension of methods used in the past to control our own evolution. For instance, the many class and caste systems found in all human societies, since long ago, are basically ways that humans control the interbreeding and, thus, flow of genes to the next generation.

Considering such methods for controlling human evolution, why have this discussion on unplanned selection in future human evolution? The answer is that humans have never had, and probably never will have complete control of evolution. There are two basic reasons for this. One, there are so many factors that influence which traits are passed on that it is impossible, as a practical matter, to control all of them. Imagine the extreme regimentation (and bureaucracy) that would have to be implemented to dictate which individuals would be allowed to breed. Not the least of these would be the many problems entailed in keeping vast records for the genetic "bookkeeping." This does not even allow for the strong social resistance that would occur if human beings were assigned breeding partners. Two, even aside from practical problems of implementation, there would be the many problems of deciding which schemes to implement. Who would decide which genes should be favored?

For these and other reasons, the future evolution of humans will almost surely have many influences outside the range of conscious human planning. What might such influences be? One influence that is important today is **relaxed selection** toward diseases and other physical disabilities. Relaxed selection refers to the transmission of genes that cause diseases and various disabilities because modern medicine allows individuals with those genes to survive. Modern medicine participates in "relaxed" natural selection by allowing such individuals to live and pass on those traits, when they would not have survived in earlier times. A prominent example of relaxed selection is childhood diabetes. Treatment with insulin and other drugs allows people with genes for this disease to live long enough to pass on the genes to offspring. A common example of relaxed selection toward genes for physical disabilities is poor eyesight. Eyeglasses permit people with poor eyesight to function normally in our society, whereas they would have had great difficulty in a primitive society of hunters and gatherers. This is actually still apparent when we observe human races that have only recently become "civilized." Australian Aborigine males have only a 2% rate of color blindness compared to 5 to 10% for European populations. More importantly, the Aborigines have a much lower rate of nearsightedness, farsightedness, and other forms of visual acuity problems than Europeans and other populations who have not relied on hunting for hundreds of years. Once these populations settled down to a life of farming, the stringent selection for visual skills needed in hunting was relaxed. If life becomes less physically strenuous and medical technology continues to improve, genes for diseases, handicaps, and other physical disabilities will continue to accumulate in the human species. Such relaxed selection has led many people to worry that the human species is physically deteriorating, as genes for diseases and handicaps accumulate. However, as long as technology and treatment keeps pace, people afflicted with such genes are often able to function as effectively as anyone else. Furthermore, as we discuss below, new genetic technologies may allow us to counteract or even eliminate such genes directly.

Of even greater concern to some is the relaxed selection that occurs in modern society toward mental diseases and disabilities. For many years, there has been a small but persistent outcry that the overall intellectual quality of the human race is declining because people of lesser abilities are not only permitted to survive in modern society, but tend to produce more offspring, thereby loading down the gene pool with genes for "lower intelligence." The evidence cited for this is the well-known economic principle that people tend to have smaller families as their socioeconomic condition improves. This argument has most recently been raised in a number of articles written by Harvard psychologist Richard Herrnstein. He argues that I.Q. scores in the United States are dropping by about 1 point per generation from this effect.

However, while there is superficial appeal to this logic, there are at least two serious flaws visible upon closer inspection. One, there is no sound evidence that lower classes are less intelligent. While it is true that genes play an important role in intelligence, it is not proven (or even likely) that lower classes have a greater proportion of genes for stupidity or genius than upper classes. In general, people of lower socioeconomic status owe their position to reduced opportunities for education and other social disadvantages rather than reduced mental abilities due to inherited genes. Two, even if there was a difference in I.Q. scores among classes, there is no quantitative evidence that lower classes have more offspring. As shown in Table 5–6, what evidence exists on I.Q. and reproduction indicates that very low I.Q. individuals tend to have the lowest reproductive rates, while very high I.Q. parents tend to have relatively more offspring than most. While more studies of this kind are needed, the evidence at the very least indicates no strong tendency for lower classes to have significantly more offspring than other classes.

A second major kind of unplanned selection that will operate on future human evolution is **urban stress selection.** This refers to selection favoring genes that help individuals cope with the mental and physical stresses of modern urban life. The human species spent over 99% of its evolutionary period living in small, mobile bands of hunters and gatherers. As civilization in the modern world advances, there is a clear shift away from a mobile lifestyle, with small social groups, to a sedentary lifestyle, with very large groups of people in high population densities. This shift creates a new environment that favors traits different from those which our heritage has prepared us for. For instance, new kinds of psychological stresses arise in densely crowded social conditions. Thus, individuals with genetic tendencies for antisocial behavior are at an obvious disadvantage. Similarly, individuals who can better tolerate job-related stresses, such as boredom or repetitive activities, may, in the long run, leave more offspring behind. The fact that one-fourth of all hospital admissions are for mental illnesses and that one-third of the U.S. population suffers from serious addiction of some kind at some point in their lives attests to the profound power of psychological urban stresses. Just how much such stresses will affect human evolution depends on how well medicine can treat mental illnesses and how long human beings are willing to put up with such an existence.

In addition to psychological stresses, urban stress also includes many physical

TABLE 5–6. IQ and Number of Offspring per Individual, from a 1963 Study in Michigan

IQ RANGE	SAMPLE SIZE	AVERAGE NUMBER OF OFFSPRING
120 and above	82	2.6
105–119	282	2.2
95–104	318	2.0
80–94	267	2.5
69–79	30	1.5

Eugenics Quarterly Vol. 10, pp. 175–87.

stresses. This includes the ability to tolerate polluted air, which has a strong genetic component. For instance, certain people are clearly more predisposed than others toward getting emphysema from smoking or living in a city with poor air quality. This arises because tissues in their lungs are deficient or less rich in various proteins that promote lung elasticity, which is largely controlled by genes. Similarly, some people are more predisposed toward getting cancer from various environmental stimuli (such as foul air, smoke, or contaminants in water). Physical stress is also affected by nutrition. In the wealthier countries, such stress involves overeating or eating rich foods, which leads to early death through arteriosclerosis (hardening of the arteries), heart disease, and other problems. Genes play a large role in many of these diseases, so that some people can tolerate a bad diet much more than others. In the poorer, overpopulated countries, nutrient stress selection is different: It involves the ability to survive on a poor diet, consisting of insufficient amounts or lacking in various nutrients. Genes play a large role in tolerating this kind of stress, so that some people can survive on lower amounts of key nutrients. For example, it is advantageous to have a naturally low metabolism in areas where food is scarce.

PLANNED SELECTION: IMPROVING THE HUMAN BODY. The dream of creating perfect humans goes back long before Darwin proposed a mechanism for evolution. Plato, for instance, spent much of his life trying to develop the blueprint for an ideal society. However, Darwin's theory of natural selection provided a powerful set of tools for carrying out the idea of improving humans. If selection of traits, such as artificial selection, can change a species, then the same method might work just as well with humans. Thus, if selective breeding can be used to improve a breed of pigeon or horse, could not the same method be used to improve the human species? **Eugenics** is the term used to describe the selective breeding of humans. This idea was first popularized by Sir Francis Galton, a cousin of Charles Darwin, in the late 1800s. Galton's ideas for eugenically improving humans were very influential. The Eugenics Education Society, founded in 1907, included many famous and powerful people, including the novelist H. G. Wells, the playwright George Bernard Shaw, and many members of the British nobility. However, it was in the United States that the ideas of eugenics were most actively pursued. In the early 1900s, the United States and state congresses passed laws that forced tens of thousands of "feeble-minded" people to be sterilized so that their genes would not be passed on. Furthermore, restrictive immigration laws were enacted to keep out "inferior" immigrants. However, it was in Nazi Germany of the 1930s and 1940s that eugenics was carried out to the fullest extent yet seen. Hitler's belief that Germans (and other Aryans) were superior led to the encouragement of breeding between the most "pure-blooded" and "superior" German citizens, whose babies were raised in state-run programs. In addition, people marked as "inferior," most notably the Jews, were actively executed.

The criminal excesses and propaganda of Hitler's attempts at eugenics led to such moral outrage that the ideas of eugenics have fallen into disfavor throughout the world since the 1940s. However, the general notion of improving the human species is one that continues to be raised by many groups, and will almost certainly always be with us. Eugenics is not an outmoded concept by any means, and in fact new methods of genetic technology are being developed at a rapid pace. In an earlier section we discussed the ability to genetically engineer plants and animals using recombinant DNA technology. The ability to apply the same technology to human beings raises even more difficult issues than selective breeding of humans because it is an even more precise way of controlling human evolution. Let us turn first to a brief discussion of planned evolution of human physical traits, then to the more controversial topic of directed evolution of human mental traits.

In its most "watered-down" form, most people actually approve of eugenics. **Soft eugenics** is the elimination of genetic human disabilities and diseases. Over 4,000 such disabilities and diseases are known to be genetically caused. About 4% of all babies are born with a significant genetic defect. Some of the most common of these genetic disorders are shown in Table 5–7. Few people would argue in favor of keeping such genes in the human gene pool, so in this minor, "soft eugenics" sense, most people are in favor of directing human evolution. Indeed, much medical and genetic research is aimed at just this treatment and elimination of such disorders. A good example of soft eugenics is one of the largest scientific projects in the United States today, the **Human Genome Project.** This is a federally funded project estimated at spending $3 billion to completely map all 100,000 genes in the human genome. (Recall from above that the genome is the sum of all genes in a species.) As of 1990, about 2,000 genes have been mapped. The process of mapping is rapidly accelerating as technology improves, and it is estimated that the entire 100,000 gene complement should be mapped by about the year 2000. In mapping the genome, scientists will learn what specific genes control various traits, from body height to eye color. However, this is much more difficult than it sounds because many traits (such as body height and weight) are influenced by many genes (which interact in complex ways), as well as environmental factors such as diet.

Fortunately, many physical genetic disorders are simple to identify, being controlled by only one or two genes. Some of these have been isolated to specific sites on the human chromosome, including some of the disorders shown in Table 5–7. Of the last 400 genes identified by 1990, 53 were linked to specific health

TABLE 5–7. Genetic Physical Disorders

DISORDER	PREVALENCE	SYMPTOMS
Polycystic kidney disease	1 in 250	Kidney damage
Hereditary high cholesterol	1 in 500	Heart disease
Sickle cell anemia	1 in 1,471	Poor circulation
Cystic fibrosis	1 in 2,500	Lung/stomach impaired
Muscular dystrophy	1 in 7,143	Muscle degeneration
Hemophilia	1 in 9,091	Uncontrolled bleeding
Cancer (lung, colon)	common	Uncontrolled cell growth

Basic data: March of Dimes Birth Defects Foundation, 1990.

defects. A main benefit of identifying these genes lies in treating the afflicted person more effectively, by eliminating the defective gene instead of merely trying to compensate for its deficiencies. Methods are now being devised that replace defective genes in cells with a nondefective gene. For instance, many genetically caused diseases are caused by the lack of a key enzyme due to a defect in the gene that normally produces that enzyme or other biochemical. Muscular dystrophy results from lack of dystrophin production. Hemophilia is caused by the lack of a clotting factor. Even cancer may be caused by such defects, wherein a biochemical that normally suppresses tumor growth is not produced.

By replacing the defective gene with a functional one, the biochemical deficiency can be corrected. This treatment of genetic disorders with new genes is called **gene therapy.** One such method, shown in Figure 5–28, uses a harmless virus to deliver the gene to the cells. (Viruses in nature can attack the DNA in cells and alter it.) Other methods of delivery are being tried because there is concern that the virus may mutate or somehow get out of control. Also, research so far indicates that it is easier to add the new, functional genes rather than actually replace the defective one. In 1990, the U.S. government approved limited experiments with putting foreign genes into the human body. So far, such experiments have been successful. However, research is in its infancy and the ability to routinely replace defective genes is not likely to be a reality until at least the early twenty-first century. Until then, knowledge of the genes involved in disorders may actually make life worse for those who are afflicted. This is because once the genes have been isolated, it is possible to analyze an individual's DNA to see if he or she has the defective gene. Because many genetic disorders do not become a problem until people are older (such as with high cholesterol or cancer genes), insurance companies and employers can use this information against individuals. For example, insurance companies have refused to insure people who they know will come down with a costly, debilitating health problem in a few years. This raises many legal questions, such as who should have access to this information. Another problem is that many of these genetic diseases (such as some cancers) are often fatal, so that knowledge of the disease is essentially a death sentence. Therefore, many people refuse to be tested for these genetic diseases.

In evolutionary terms, the identification of debilitating genes may ultimately lead to their removal from the human gene pool by advanced techniques of genetic engineering. Gene therapy will not accomplish this because it only replaces genes in the body cells of the afflicted individual. To prevent the gene from being passed on, techniques must somehow go one step further and remove it from the sperm or egg cells of the person. **Genetic counseling** by a trained medical specialist offers couples advice on whether or not to have a child. If there is a great risk of a child inheriting an untreatable genetic disorder, they can choose to adopt or remain childless. This would help remove the gene from the gene pool.

In contrast to soft eugenics, which seeks to eliminate debilitating traits from the gene pool, **hard eugenics** is the active encouragement of increasing desirable

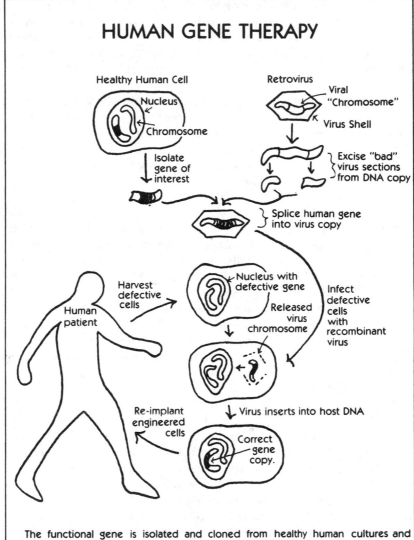

HUMAN GENE THERAPY

The functional gene is isolated and cloned from healthy human cultures and spliced into a retrovirus "chromosome" from which all "bad" sectors are removed. The virus is reconstructed and infects defective cells from the patient. The virus inserts itself (and the functional human gene) into the DNA of the patient's defective cells. The cells, with the corrected gene copy, are transplanted back into the patient.

Figure 5–28. How human gene therapy works. From *Analog Science Fiction & Fact* magazine, Feb. 1992, p. 80.

traits in the human gene pool. An example of this is Hitler's attempt to build a superrace including humans who were physically, as well as mentally, superior. There have been a few other attempts to physically improve the human race by selective breeding. Slaveholders throughout history have sometimes encouraged breeding on a very small scale. However, the ethical resistance against any kind of hard eugenics is much greater since it involves much more than removing genes that are obviously harmful. Instead, it requires us to define what traits we desire. While it may seem obvious that it is better to be strong than weak or fast than slow, the matter is rarely so clear-cut, since traits are often correlated with one another: A fast runner may have difficulties in other ways. More importantly, physical diversity is the key to success in any species. A physique that is better for some situations is much less useful for others.

Thus, while it is technically possible that genetic engineering could be used to create physically superior people, it seems unlikely that this will occur, at least in the near future. The problems of (1) agreeing on which traits to increase and (2) ethical resistance toward dictating human breeding are major obstacles to hard eugenics. This seems especially unlikely when one considers mental improvement of the human species, as we will do next.

PLANNED SELECTION: GENES AND THE HUMAN MIND. Beginning with Galton, most eugenics promoters have been more concerned with improving the human mind than the body. However, in addition to agreeing on which traits to select and the ethical difficulties just noted, improvement of the human mind poses two more problems: (1) mental traits are extremely difficult to measure, and (2) mental traits are often largely nongenetic.

Turning first to the problem of measurement, if we were to decide to genetically promote certain mental traits, we would have to identify and assess them in order to pass them on. Measuring mental traits is usually approached by various tests, such as the I.Q. test, which purports to assess intelligence. The I.Q. test has its roots in a test devised by the French psychologist Alfred Binet in 1905. Shortly thereafter it was adapted by U.S. psychologists for the assessment of mental attributes in immigrants and soldiers in World War I. In theory, it measures intelligence relative to other persons of the same age, with an I.Q. score of 100 being average. The I.Q. test and others like it, while still used, have come under severe criticism over the years. First, individuals have been unfairly declared as incompetent for reasons of clear prejudice or ignorance based on I.Q. test results. For example, in 1912 the test was used by U.S. immigration officials to declare that 80% of Hungarians and 79% of Italians were "feeble-minded" and should be deported. Second, the test itself is flawed. Most psychologists now agree that "intelligence" consists of a number of mental attributes, from memory to various data-processing abilities, and that trying to summarize such a complex composite of traits with a single number (the I.Q. score) is misleading, to say the least. In addition, there are cultural biases in the test that make it unreliable for testing groups comprised of individuals with different backgrounds.

TABLE 5–8. Genetic Mental Disorders

DISORDER	PREVALENCE	SYMPTOMS
Alzheimer's disease	1 in 100	Mental degeneration
Down's syndrome	1 in 1,000	Retardation
Phenylketonuria	1 in 16,667	Retardation
Huntington's disease	1 in 20,000	Mental degeneration
Alcoholism	very common	Addiction
Chronic depression	very common	Loss of mental acuity

Even if one could make a better test and accurately measure a person's mental traits, an even bigger problem still impedes the genetic improvement of mental traits: Many of our mental abilities are not completely genetic in origin. Rather, such abilities are often largely determined by how we are raised. The debate concerning which mental abilities are determined by genes has raged for many years and continues even now. It is often called the **nature-nurture debate,** with "nature" referring to the idea that genes control most mental abilities, while "nurture" refers to the notion that upbringing ("environment") is the major influence. In some behaviors, it is clear that genes do have a primary effect, such as shown in Table 5–8. In such cases, we say the trait has a high **heritability** rate and is largely determined by genes. However, aside from such disorders as listed in Table 5–8, most human behavior is derived from a very complex set of interactions between many genes and many environmental influences. The best way to determine whether genes or environmental factors have a greater influence in animals is to perform breeding experiments. An especially important experimental condition is to separate the offspring from the parents at an early age and observe a particular behavior. If the offspring still show the behavior, even under a variety of environmental conditions, then we have good evidence that the behavior is not learned but is genetic. Most instinctive behaviors are genetic: rage, fear, sleep, hunger. Figure 5–29 shows how sleep time can be increased or decreased in mice by selective breeding to favor genes that promote more or less sleep.

Because it is not ethical to perform selective breeding experiments on humans, this major method of measuring heritability of a trait is not available. Furthermore, human behavior is very complex, making it difficult to describe and analyze. (For example, imagine trying to study a behavior such as "inventiveness"!) Nevertheless, it is possible to gain some information on the genetic heritability of human behaviors and mental traits by using the natural experiments of *identical twins*. Identical twins originate when early cells of the growing embryo split into separate groups. Therefore, they come from the same fertilized egg and have all the same genes. If identical twins are subjected to significantly different environments, such as when each twin is adopted by different parents, then we can infer that any major mental or behavioral differences are due to upbringing and not genes.

Decades of twin adoption studies have demonstrated that human mental traits run the gamut from being highly genetic in origin (such as chronic depression) to

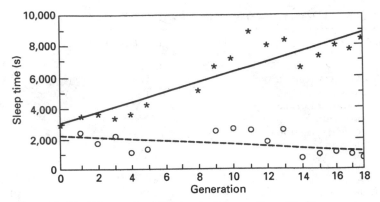

Figure 5–29. Effects of artificial selection on sleep time in mouse breeds. After 18 generations, there is a drastic difference in the amount of time each breed slept. * = average sleep time for one breed, ○ = average sleep time for second breed. Modified from *Science* magazine, vol. 248.

highly environmental (such as reading ability). Most controversial of these traits is "intelligence," which these studies repeatedly demonstrate is strongly determined by both genes and environment. In other words, *intelligence has an intermediate heritability*. This is shown in Figure 5–30, where I.Q. scores for identical twins are still similar (highly correlated). This indicates that genes play an important role in this

Figure 5–30. Correlation coefficients for IQ test scores from 52 studies. Identical twins reared together have a much higher correlation than unrelated persons reared apart.

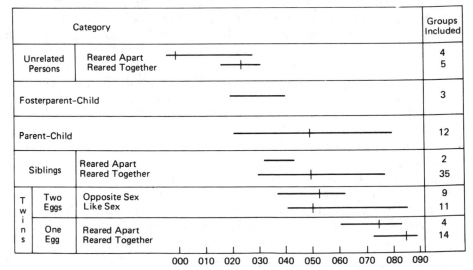

trait. However, the I.Q. scores for identical twins reared together are even higher than for those reared apart. This indicates that environment also plays a role. The most recent such study (by Christiane Capron and Michel Duyme in 1989 in the international science journal, *Nature*) was widely publicized for its scientific rigor. It confirmed once again that "intelligence," as measured by I.Q. tests, is both genetic and environmental. The average I.Q. score of adopted children was 12 points higher when raised in wealthy families, regardless of the background of the true parents, illustrating the importance of upbringing. In contrast, the I.Q. scores of adopted children raised in poor homes averaged 15 points higher when the true parents were from affluent families. This implies that heredity is a major factor as well.

Future Evolution of Society

The most fundamental future trend that would affect every aspect of human society is *continued population growth*. The dramatic increase of world human population in the last few decades, largely from improved medical care, is shown in Figure 5–31. We have gone from less than 1 billion in the mid-1800s to over 5 billion in 1987. At the present time, the population continues to grow exponentially, roughly at the rate of an extra 100 million people per year. Most predictions therefore predict that global population will reach 6 billion by the year 2000. Obviously, this growth

Figure 5–31. Past population growth and future projections.

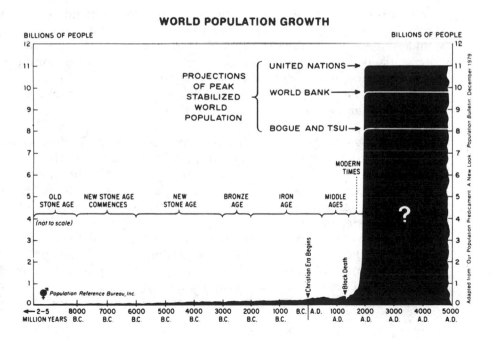

has to stop at some point because the earth has limited space and resources. The upper limit of population that the earth can support is called the **carrying capacity.** No one is certain what the carrying capacity of the earth is, but Figure 5–31 shows some of the predictions that estimate where the earth's population might stabilize. They range from 8 to 11 billion people or more. Most population growth is occurring in developing countries and it is thought that as economic conditions improve, family size will shrink as it has in wealthier nations. Clearly, unless changes are made soon there is great potential for ecological disaster and increased human suffering.

Notice how Figure 5–31 illustrates the basic evolutionary pattern of the "S-shaped" curve that we saw in many evolutionary patterns (refer to Chapter 4). We noted how the exponential growth of many trends (such as the evolution of life's diversity) eventually encountered some environmental limit that caused the trend to slow and then level off. Therefore, it is important to put our own predicament into its proper natural and historical context as just one more example of a very common pattern. Even with our large brains and technology, humans are not isolated from the same natural laws that ultimately govern all other life.

An excellent perspective of the dynamics of the "S-shaped" curve is provided by the **doomsday equation.** This was published in 1960 (in the journal *Science*) to predict population growth for the future:

$$\text{population} = \frac{(1.79)10^{11}}{(2026.87 - \text{time})^{0.99}}$$

It attracted wide attention in magazines and newspapers (including the comic strip *Pogo*)—people thought it was a joke because it predicted that the human population would go to infinity by the year 2026. Yet consider this crucial fact: At the present time, actual population growth has *exceeded that predicted* by the doomsday equation. Thus, even though we are only 36 years away from an "infinite" population, we are ahead of schedule! (Since we will never reach infinity, either because of catastrophe or restraint, this demonstrates the fallacy of extrapolation once again.)

SOCIAL EVOLUTION AND DIMINISHING RESOURCES. We have already discussed how humans are (1) making the earth less habitable (e.g., the greenhouse effect), and (2) causing the widespread extinction of other species. Yet a third unfortunate consequence of human actions is that our rapidly growing population is rapidly depleting the resources on which we directly depend. We can divide such resources into two basic categories, *energy* and *matter*.

Figure 5–32 shows the past, present, and estimated future production of crude oil for the world. It shows that U.S. production has already peaked and that world production will peak in a few years. Thereafter, diminishing reserves will lead to a loss of production where supply will not meet demand. Such estimates are obviously tentative, but the large majority of geologists predict that demand will

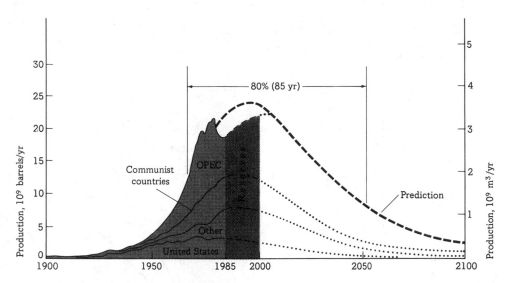

Figure 5–32. Crude U.S. oil production peaked in 1967. World production is estimated to peak in the early 21st century, if not before. From S. Judson et al., *Physical Geology*, 7th ed., p. 373.

exceed supply sometime in the first half of the twenty-first century. This does not mean that energy supplies will run low because other forms of energy are available, such as coal, nuclear power, and solar energy. There is enough coal for hundreds of years. However, both petroleum and coal are essentially nonrenewable resources for our purposes so we must realize that we are using up irreplacable resources, making them unavailable to our descendants.

More troubling in some ways is the rapid depletion of material nonrenewable resources (matter), such as ores from which metals are made. Unlike energy, for which various forms can be substituted, material resources often cannot be replaced at similar costs. For instance, the conductive properties of copper make it invaluable for electrical wiring. Not surprisingly, many of these earth materials are being depleted and their future supply curves are similar to the curve shown for petroleum production in Figure 5–32. Such materials can often be recycled, but this can be very expensive and will use much energy.

An important point about the depletion of both energy and matter is shown in Figure 5–33. It illustrates how the *per capita* consumption of earth materials has risen over the last few decades. This means that depletion is occurring not only because of the number of people using resources but because each person is using more matter and energy than people did in the past. The reason, of course, is that advancing technology has created more and more ways to use energy and materials.

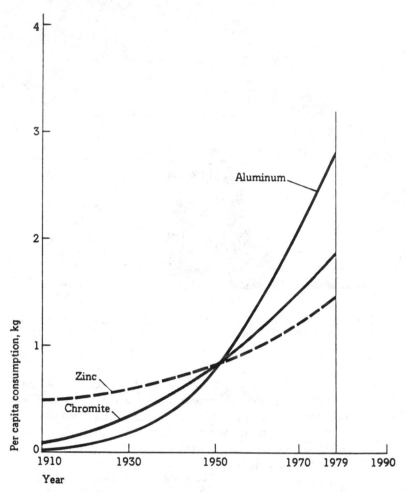

Figure 5–33. Rapid increase in per capita use is now common to virtually all minerals as world industrialization expands. From S. Judson et al., *Physical Geology*, 7th ed., p. 395.

For example, the rapid increases in weaponry and locomotion that we saw in Chapter 4 are made at the cost of increasing amounts of energy and materials per person using the technology.

SUMMARY

Attempts have been made to predict future evolution, but it is difficult to use **extrapolation** because complex systems have many unpredictable fluctuations, including unforeseen **catastrophes. Cosmologists** predict two possible outcomes for the universe: eternal expansion (**open universe**) or contraction and collapse in a "**big crunch**" (**closed universe**).

The future lithosphere will continue to be shaped by plate tectonics. As gases accumulate in the atmosphere, destruction of the **ozone** layer, global warming, and the **greenhouse effect** will continue. The future of the hydrosphere is largely dependent on global temperature as sea level fluctuations follow warming or cooling trends. A catastrophe such as **nuclear winter** would disturb the entire global ecosystem.

What is the future of human evolution? **Originations** and **extinctions** are driven by environmental change; therefore, biosphere destiny will be largely controlled by the amount that the environment is disrupted. One of the greatest disruptions of the environment is the destruction of tropical rain forests.

Biodiversity is greatest in tropical areas, following the **latitudinal diversity gradient.** Therefore, devastation of tropical rain forests causes loss of many more species per acre than in any other area. One of the main causes of tropical deforestation is **slash and burn agriculture.** Habitat destruction is the leading cause of extinction, with tens or hundreds of species becoming extinct daily. The extinction of **keystone species** is most severe because this can lead to **secondary extinction** of other species. Introduction of new species and human hunting are two other leading causes of extinction.

For a species to survive there must be a **minimum viable population** able to overcome environmental fluctuations and interbreeding. Species are not equally likely to become extinct. Characteristics such as pollution tolerance and reproductive rates determine the susceptibility of a species to extinction.

For esthetic, ethical, economic, and ecological reasons (the four "Es"), humans are becoming more convinced of the need to save species from extinction. Research, establishment of preserves, legislation such as the **Endangered Species Act,** and captive breeding are the first steps humans can take to stop species extinction. In the meantime, **gene banks** can serve as reservoirs for seeds, spores, and genetic material from species that cannot be saved in the wild. Researchers have classified species into five categories: **indicator species, keystone species, umbrella species, flagship species,** and **vulnerable species.**

Another means of slowing down human caused extinction of species is to make it economically unprofitable for destructive practices to continue. **Ecological economics** finds ways to make the ecosystem more profitable if it remains intact rather than if destroyed; one such method is the **debt-for-nature swaps.**

Humans are creating new life forms with **selective breeding,** which allows individuals with particular traits to interbreed, and **genetic engineering,** which is the direct transfer of genes between organisms (**gene splicing**). Using **recombinant DNA technology** or **biotechnology,** the transfer of genes between different types of organisms, or **transgenesis,** is possible. Recombinant DNA techniques involve the use of **restriction enzymes,** which produce DNA fragments. The fragments are attached to **vectors** such as **plasmids** or **viruses.** These vectors transfer the DNA molecules to another cell using **DNA ligase enzymes,** which bind DNA strands together. There are unlimited applications for biotechnology in areas such as medicine, agriculture, industry, and environmental cleanup. Although this technology has generated legal, environmental and ethical problems, its use will probably become even more widespread as the technology improves.

Human evolution will occur by (1) unplanned selections, as those possessing certain traits preferentially survive; and (2) planned selection with direct manipulation of genes. Processes of unplanned evolution include **relaxed selection** and **urban stress selection.** Relaxed selection allows humans with disease and disability to survive and pass on their genes due to the advances of modern medicine. This has led some to worry that the human species is physically and mentally deteriorating. Some want to implement a program of **eugenics,** or the selective breeding of humans, to generate a more physically and mentally fit species. **Soft eugenics** is selective breeding to eliminate genetic disabilities and disease, while **hard eugenics** will increase "favorable" traits. Other methods of implementing planned selection are **gene therapy** and **genetic counseling.**

The future of society is uncertain. The primary trend of social evolution is an alarming increase in global population. The **doomsday equation** has been used to estimate future population growth. Unless steps are taken, the population will continue to increase exponentially until some carrying capacity is reached. Accompanying this increase will be increased depletion of natural resources and environmental destruction.

KEY TERMS

extrapolation	vulnerable species
catastrophes	preserve design
open universe	Endangered Species Act
closed universe	endangered species
cosmologists	threatened species
dark matter	ecological economics
big crunch	debt-for-nature swap
greenhouse effect	opportunistic species
ozone	AIDS
nuclear winter	selective breeding
biodiversity	genetic engineering
latitudinal diversity gradient	gene splicing
slash and burn agriculture	recombinant DNA technology
minimum viable population (MVP)	biotechnology
acid rain	transgenesis
secondary extinction	restriction enzymes
keystone species	vector
gene banks	plasmid
indicator species	viruses
umbrella species	DNA-ligase enzymes
flagship species	microinjection

genome map   gene therapy
kamikaze genes   genetic counseling
relaxed selection   hard eugenics
urban stress selection   nature vs. nurture debate
eugenics   heritability
soft eugenics   carrying capacity
Human Genome Project   doomsday equation

REVIEW QUESTIONS

Objective Questions

1. What are two possible outcomes for the universe?
2. What is the ozone layer? How is it being destroyed by humans? What effect will this have on climate?
3. What is a likely outcome of global winter?
   a. glaciation
   b. rise in sea level
   c. global warming
4. What is the outcome of the greenhouse effect?
   a. drop in sea level
   b. global warming
   c. glaciation
5. What are three factors that contribute to the latitudinal diversity gradient?
6. What is the minimum viable population? What are two problems the MVP must overcome for a species to survive?
7. How might the extinction of a keystone species cause a secondary extinction?
8. What are three major causes of modern extinctions?
9. What are seven traits that make species susceptible to extinction?
10. Discuss four general reasons why species should be saved from extinction.
11. Describe five ways humans can stop extinctions.
12. Define the five categories used to classify species.
13. What is relaxed selection? Give an example.
14. How does urban stress selection affect human evolution?
15. Why are opportunistic species particularly successful in human-altered environments?
16. How are selective breeding and genetic engineering used to create new life forms?
17. Differentiate between soft eugenics and hard eugenics.
18. Has the doomsday equation accurately predicted the growth of human population? Explain.

Discussion Questions

1. Do you think that with increasing technology humans will be able to control evolution? Explain.
2. Do you think that the creation of a superior human will be possible through genetic engineering? Do you feel that it is ethical to "tamper with life" in this manner?
3. What are two problems associated with improvement of the human mind through eugenics?
4. Briefly describe the greenhouse effect. What problems cause difficulty in correlating an increased carbon dioxide level in the atmosphere with increased global warming?
5. Briefly describe recombinant DNA technology and give an example of its application to the following: medicine, agriculture, and industry.

SUGGESTED READINGS

On modern extinction

EHRLICH, P., AND EHRLICH, A. 1981. *Extinction*. Ballantine, NY.

On biotechnology

BRENNAN, R. 1990. *Levitating Trains and Kamikaze Genes*. Wiley, NY.

On future evolution

DIXON, D. 1981. *After Man: A Zoology of the Future*. St. Martin's Press, NY.

On the future of the environment

ORNSTEIN, R., AND EHRLICH, P. 1989. *New World, New Mind*. Simon & Schuster, NY.

On evolution and human nature

MORRIS, R. 1983. *Evolution and Human Nature*. Avon, NY.

6

Epilogue: Some Personal and Social Implications

Compassion and humor are the highest forms of understanding.
Richard P. Feynman

Few subjects have as many subtle but profound implications as evolution. The knowledge that we evolved from lower life forms has drastically changed how we perceive ourselves, the world around us, our purpose for being here, and many other basic ideas. All aspects of human thought, from religion and philosophy to literature, have been affected.

However, because these implications are often subtle, they are often purposely distorted, or simply misunderstood. Therefore, let us review some of the major facts of evolution and their implications, as explicitly as possible.

BASIC FACTS AND INTERPRETATIONS

Fact 1

The earth (with its three "spheres" of air, water, and land), its biosphere, and the human species, resulted from a very long sequence of events that took over 5 billion years to produce. It ultimately began with the "Big Bang" origin of the universe over 15 billion years ago.

INTERPRETATION #1: We cannot automatically assume, based on this information, that humans and our world were the goal of a preconceived evolutionary plan. Indeed, in examining the history of life we have seen no conclusive evidence of any inexorable "drive." Instead, evolution is full of reversals (where traits are evolved, lost, and then reevolve), chance events (such as meteorite impacts) that "wipe the slate," and many other contingencies that could have easily altered the outcome of evolution.

INTERPRETATION #2: This apparent lack of a specific direction does not mean that evolution is without *general* direction. Indeed, there are many general trends (statistical tendencies), such as increasing size and complexity. If we could go back in time and start over, just the slightest change in the sequence of events would change specific outcomes: Life, humans, and the world as we know it would not be exactly the same. However, some of the general trends would still occur: We could expect increasing size and complexity in some forms of life.

Fact 2

The processes that caused the evolutionary sequence leading to humans and our world were of three kinds—physical, biological, and cultural. Physical processes (the laws of physics and chemistry) created (1) a habitable world and (2) life very early in the earth's history (over 3.5 billion years ago). The biological processes of (1) reproduction with random mutation and (2) natural selection of some mutations then took over to create progressively larger and more complex forms of life, as some rare mutations proved favorable. Much later (about 5 million years ago), the biological processes created organisms with brains large enough to begin cultural processes of evolution. Processes passed on information more quickly because they were based on learning instead of genetic changes.

INTERPRETATION #1: The very early origin of life by physical (especially chemical) processes implies that life is readily produced by natural processes. Since the same laws of physics and chemistry apply throughout the universe, it may be that life is abundant throughout. However, the next step, that of producing intelligent life by biological laws (mutation and natural selection) took much longer (over 3 billion years). This implies that intelligent life may be much more difficult to produce and is therefore much rarer in the universe.

INTERPRETATION #2: Life was created by physical, biological, and cultural "laws" that characterize our universe. These laws have designed life as we know it; there is no agreed-upon evidence of a personified Creator. No one knows why these particular laws exist in this universe, or if other universes exist with different laws.

Fact 3

Evolution has not stopped. In fact, physical, biological, and cultural evolution are all being *accelerated* and *redirected* by human activities, especially as our population and technology increase. Biological evolution of humans and other species, while slower than cultural evolution, is still being accelerated and redirected by humans in two major ways: (1) pollution, pesticides, and other ways of altering the environment force other species and ourselves to adapt to these new conditions, and (2) genetic engineering (biotechnology) allows humans to directly alter the genes of humans and other species.

INTERPRETATION: Evolution is not a "dead" subject of no relevance. It not only gives us a crucial frame of reference and insight into our origins, but is an ongoing process that will determine the future of the earth, the biosphere, and ourselves.

SOCIAL AND PERSONAL IMPLICATIONS

These three facts have a number of major implications for society and for our individual perceptions of life. While the known facts of evolution certainly do not provide all the answers to many major philosophical and religious questions, they have created important limits on the speculation that existed before such facts were known.

Implications for Religion

In his popular book examining the religious and moral implications of evolution (*Created from Animals*, 1990), the philosopher James Rachels points out two basic religious and moral premises that are weakened by evolutionary theory:

1. Humans are made in the image of God.
2. Humans are the only uniquely rational beings on earth.

In showing that humans arose from a long, complex series of natural events, driven by natural physical and biological laws, evolutionary theory does not support the idea that we are created in the image of a personified Creator. Instead, creation occurs because of natural laws and also a good deal of "chance." Similarly, in showing that humans evolved from "lower" life forms, evolutionary theory detracts from the notion that humans are uniquely rational. When we look at monkeys, apes, and even other life forms, such as porpoises, we can see gradations of behavioral complexity as well as anatomical gradations in the brain that lead to the human brain. (Recall that our brains have no new organs or areas, only enlargements of those found in other primates.)

However, we have noted that this only *weakens* traditional values. Many natural scientists are not atheists, or even agnostics, but continue to believe in a "Supreme Being" or other metaphysical concepts. This is possible because in exposing the natural laws that created humans, we have not uncovered many other relevant unknowns, such as where those laws came from or why those particular life-generating laws exist and not others. Furthermore, just because there are gradients of behavior and intelligence in animals does not mean that our own rationality is not far above that possessed by other animals. Indeed, by most objective assessments, we are so far above other organisms in intellectual abilities that we truly are "qualitatively" different. This is common in nature: nonlinear relationships often lead to threshold effects. For example, heating a pot of water results in only hotter water up to the boiling point, when a qualitatively different event happens—the water evaporates.

Implications for Self-Reliance

The most basic evolutionary message for religion is that a personified God may exist, but there is no physical evidence of it thus far. Given this, the most immediate outlook is self-reliance. Humans cannot rely on some external source of wisdom or guidance. This obviously has profound implications in that we must keep our own house in order or expect to pay the price: Overpopulation, ecological destruction, wars, and other immediate problems must be addressed.

Such self-reliance is made all the more difficult as social and environmental conditions deteriorate, and the value of human life deteriorates with it. The biologist Garrett Hardin has used the phrase **lifeboat ethics** to refer to the extremely harsh set of rules that might have to be employed if conditions degrade to the point where some human lives must be sacrificed to save others.

Implications for Self-Understanding

Evolutionary theory cannot answer such deep questions as why we are here. However, it provides much information about how we got here. While this may not be as satisfying as knowing why we are here, it greatly helps us to better understand ourselves. For instance, evolutionary theory explains why some of us have blue eyes, why we see colors, why we have five fingers, and so on. Furthermore, it can explain much about our behavior. Why are we social animals? Why do we need to explore our environment?

Evolutionary theory explains so much because we can see these physical and mental traits as adaptations of a tool-using primate to a hunting and gathering lifestyle. Blue eyes are a fairly recent "neutral" mutation caused by eye depigmentation, color vision is important to fruit eaters, five fingers are part of a general vertebrate heritage, and social organization is crucial to a hunting and gathering way of life. Of course, there is much that we are still trying to explain. Most notable is the human mind, the most complex phenomenon in the known universe. This requires unraveling of the intricate interaction between learning and genes.

Implications for a Personal Philosophy

Aside from its implications for human society at large, evolutionary thought touches each of us as individuals, greatly affecting how we view ourselves and the world

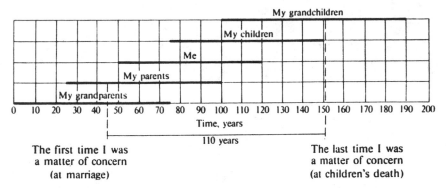

Figure 6–1. Our direct influence on society spans a few brief decades, compared to the over 3.5 billion year span that life itself has occupied. From A. Iberall, *Toward a General Science of Viable Systems* (New York: McGraw-Hill, 1972), p. 252.

around us. For instance, we know that our time on earth as individuals is a very tiny fraction of the total time that earth has existed. Think about how your view of yourself and your actions today, yesterday, or 50 years from now, is altered when you consider that vastness of time. How many of us know the names of our great-great-great-great-grandparents? Yet, on average, these people died less than a mere hundred years ago. What will be the importance of your actions and life thousands of years from now? Figure 6–1 illustrates this. Even the entire human species is humbled by this perspective. Recall that over 99% of all species are extinct and that the average mammal species lasts only 1 to 2 million years.

Knowledge of evolution can also help you understand yourself better. Many of your impulses (e.g., anger, hunger, and sexual drives) have their origin in the natural selection of genes that promotes such behavior. Knowing that such strong feelings are rooted in our evolutionary heritage gives us a better perspective for coping with them in inappropriate situations.

Glossary

ABSOLUTE RATE When accurate dating of rocks and fossils is possible and actual anatomical change can be measured correctly through time.

ABSOLUTE TIME Reconstruction of events where specific ages can be assigned.

ABUNDANCE Number of individuals per species.

ACHULEAN A tradition of stone making by *Homo erectus*. The basic tool was the hand ax.

ACID RAIN Rain that is acidic due to sulfur air pollution.

ADAPTATION A trait that helps an organism survive in its environment. Also, the process of adjusting to environmental change by evolution.

ADAPTATIONISM Tendency to attribute virtually every feature of an organism to some adaptive function.

ADAPTIVE RADIATION Rapid divergence of a group to exploit many unoccupied environmental **niches.**

AGNOSTIC Someone who believes the existence of God can't be proven.

AIDS Acronym for Acquired Immune Deficiency Syndrome, a potentially lethal virus that originally infected African monkeys. A mutated form of this virus, which suppresses the immune system's ability to combat disease, infected humans beginning in the mid- to late-1970s.

ALLELES Alternate forms of a gene found at a particular chromosome site.

ALLOMETRIC PLOT A plot that depicts the degree to which **correlated traits** are interconnected.

ALLOMETRY Study of evolutionary change among **correlated traits.**

AMINO ACIDS Molecules that serve as the building blocks for protein.

AMINO ACID SEQUENCING A process that compares the order of amino acids that compose protein molecules.

ANALOGOUS ORGAN Organs with differing origins that are similar anatomically because of adaptations to similar environmental conditions (for example, wings of birds and insects).

ANATOMICAL RATE Measurement of change in an anatomical trait through evolutionary time.

ANGIOSPERM Flowering plants that became prominent in the Cenozoic era.

ANKYLOSAUR A bird-hipped dinosaur with a heavy club on its tail used as a defense mechanism.

ANTHROPIC PRINCIPLE (Greek *anthropos*, human) The theory that life, planets, and stars would be unable to form if the basic physical and chemical forces of the universe were to change even slightly. Thus, there may be a "design" behind the universe.

APE Arose from monkeys, differing from monkeys by having larger body size; no tail; and larger, more complex brains.

ARCHAEOLOGY Study of human-made objects to reveal cultural evolution.

ARCHAIC REPTILE Early reptiles of the late Paleozoic.

ARCHAIC UNGULATES Early forms of hoofed mammals.

ARMS RACE Biological term for the continued competition for survival between predators and prey.

ARTIFICIAL SELECTION Occurs when humans are involved in the selection of individuals.

ASEXUAL Organisms that do not reproduce sexually but reproduce exact copies of themselves.

ATHEISM Belief that no God exists.

ATMOSPHERE Gaseous envelope surrounding a planet.

AUSTRALOIDS One of the four *Homo sapien* races. Characteristics include dark skin, ample body and facial hair, large teeth and jaws, and heavy brow ridges.

AUSTRALOPITHECUS (means "southern ape") The earliest definitive human ancestor, appearing about 5 million years ago.

BACKGROUND EXTINCTION Extinctions that occur through relatively minor environmental changes.

BAND The first and simplest of the four human social organizations, characterized by mobile hunters and gathering societies that have a sense of territoriality or private property (see also **tribe, chiefdom,** and **state**).

BARBARISM Proposed second stage of evolution in human cultures (see also **savagery** and **civilization**).

BERGMANN'S RULE Cooler climates tend to favor larger individuals.

BIG BANG Explosion that led to the creation of basic subatomic particles composing the universe.

BIG CRUNCH Theory that all matter in the universe was concentrated into a pinpoint area and that all current physical laws were nonexistent.

BIODIVERSITY Diversity (different types) of species.

BIOLOGICAL EVOLUTION The evolution of all life, caused through a process of natural selection of variation.

BIOSPHERE (life's sphere) All living organisms.

BIOSPHERE TREND Trend seen throughout the evolution of life as a whole (for example, size increase).

BIOSTRATIGRAPHY The use of fossils in correlation of sedimentary rock layers.

BIOTECHNOLOGY Technology that involves the combining of DNA from very different groups.

BIRD-HIPPED DINOSAUR Dinosaurs with hip bones similar in structure to those of modern-day birds; this group consisted of moderate-sized herbivores with specialized defense mechanisms. Included in this group were ornithopods, stegosaurs, horned dinosaurs, and ankylosaurs.

BLACK DWARF A small star that has completely cooled.

BLACK HOLE Star material in which the density has been so great that even neutrons have collapsed on themselves.

BLADE TOOL Method that produces long, thin flakes that can be readily retouched to make a number of specialized tools.

BODY FOSSIL Remains of the whole body or parts of the body.

BONY FISH Fish with internal bony skeletons; currently the most common fish on earth.

BOOK GILL Leaflike organs that can be modified to absorb oxygen in air when kept moist.

BRACHIOPOD Sessile invertebrates prominent in the Paleozoic era.

BRANCHING EVOLUTION A type of evolution in which a species branches out in different directions and creates a diversity of new species. This form of evolution is generally brought about through environmental changes occurring at a local level, not affecting the entire species.

BRONZE AGE The period when neolithic people learned to heat ores to high temperatures, leading to chemical reactions and allowing metals to be separated or combined.

BRYOZOAN Sessile invertebrates, which were prominent in the Paleozoic era.

CARBON FUSION Process in which carbon atoms are fused to create oxygen atoms.

CARRYING CAPACITY The limit at which the environment can support organisms.

CATASTROPHES Changes that are exceptionally rapid for a defined spatial scale.

CATASTROPHISM Early theory of evolution which proposed that new species were created as a result of a great catastrophe.

CAUCASOID One of the four **Homo sapien** races. Traits include light skin and hair, moderate body and facial hair, and blue eyes.

CEPHALOPOD Swimming invertebrate related to squids, first found in the Paleozoic era.

CEREBRAL CORTEX The outermost layer of the brain, containing all higher thought processes.

CHEMICAL REACTION Occurs when atoms share or exchange electrons and become electromagnetically attracted to one another and combine.

CHIEFDOM Usually found only in agricultural societies. The system is characterized by inherited—often theocratic—leadership positions, craft specialization, and an absence of egalitarianism (see also **band, tribe,** and **state**).

CHITIN Tough, brittle material that forms the exoskeleton of insects.

CHOPPER Rounded, water-worn pebbles about the size of a tennis ball that were given a cutting edge by knocking off a few flakes with another stone.

CHROMOSOMES Strands within the cell nucleus made up of genes.

CIVILIZATION Proposed third stage of culture evolution in human societies (see also **savagery** and **barbarism**).

CLADISTICS Method of classification which uses measurable traits as the criterion for grouping organisms.

CLAM Burrowing mobile invertebrate which first became common during the Mesozoic era.

CLINE A continuum of traits, usually associated with geographic variation.

CLOSED UNIVERSE The theory that the universe will eventually stop expanding and collapse into a **"Big Crunch."**

COAL Undecayed accumulated organic (fossil and plant) material.

COEVOLUTION The evolution of one group is dependent on the evolution of another (for example, flowers and bees).

COEVOLUTIONARY TREND The situation in which evolutionary development within a species can affect other closely interacting groups also to change and adjust.

COLONIAL HYPOTHESIS The theory that multicellular organisms evolved when reproduction in eukaryotic cells produced a number of identical offspring cells that stayed together and formed a colony.

COMPETITIVE RELEASE When a dominant competitor is eliminated and surviving species are allowed to radiate in diversity by exploiting new environments (for example, mammals replacing dinosaurs).

COMPETITIVE REPLACEMENT Replacement of one group by a competitively superior group.

CONIFER The most successful living gymnosperm, which can be identified by its needle-like leaves.

CONSUMERS Animals that are unable to synthesize their own food from sunlight and must eat their food directly.

CONTRACTION Occurs when spatial dust clouds contract around a center of gravity to form a single gaseous object.

CONVECTION Movement of liquid from heat (for example, earth's mantle).

CONVERGENT EVOLUTION Occurs when two unrelated groups converge on a similar appearance to exploit a similar ecological niche.

CONVERGENT TREND Trends where unrelated groups converge on similar anatomical appearances.

COPE'S RULE Lineages tend to increase in body size through time.

CORAL Sessile invertebrates which were prominent during the Paleozoic and later eras.

CORE The earth's center, composed of heavy elements, such as nickel and iron.

CORRELATED TRAITS Traits that are connected in some manner and tend to evolve together.

CORRELATION Matching of strata across long distances.

COSMIC BACKGROUND RADIATION An "echo" caused by radiation produced by the Big Bang.

COSMOLOGISTS Scientists who study the origin and fate of the universe.

COSMOLOGY Study of the universe and its origination and fate.

CREATIONISM Fundamentalist belief that generally rejects evolutionary theory. Proposes that the universe was created by conscious design of a creator.

CRUST Top layer of the earth, composed of light elements (especially oxygen, silicon, and aluminum).

CRUSTACEANS Organisms such as crabs, lobsters, shrimp, and relatives of these organisms.

CULTURAL EVOLUTION The evolution of culture, caused by the ability to learn.

CULTURE Elaborate learned behavior.

CYCAD Arose in the Mesozoic era. They are similar in appearance to palm trees, but are not related.

DARK MATTER Matter in the universe which does not emit light and remains unobserved by astronomers (for example, **black holes**).

DEBT FOR NATURE SWAP A country sets aside a preserve in exchange for reduction in the amount of money owed to another country.

DEISM A religious belief that God created the universe and left it to run "independently."

DERIVED TRAIT Traits that are recently evolved and usually limited to fewer organisms, such as the grasping hand of primate mammals.

DIFFERENTIATION Geological process in which elements become divided into layers. Occurred on earth when the heavier elements sank to the core and the lighter elements rose and floated on top as the **crust.**

DINOSAURS ("terrible lizard") Extinct, often gigantic reptilian creatures which were dominant in the Mesozoic era.

DIPLOID Nongamete cells that contain two chromosome pairs.

DIRECTION The tendency of a phenomenon to consistently increase or decrease over time.

DIRECTIONAL SELECTION Environmental selection on one extreme of the trait distribution leading to a shift in the distribution average.

DIVERGENT BOUNDARY A boundary where magma pushes plates apart from one another.

DIVERGENT TREND Splitting of lineage into descendant species as their gene pools become progressively more separate in time.

DIVERSITY Number of species.

DNA (Deoxyribonucleic acid) Basic chromosomal material that transmits hereditary patterns and has a helix shape.

DNA HYBRIDIZATION Method that allows the measuring of the similarity of DNA material between species. The process involves mixing single-strand fragments of DNA of the two species and studying the recombination patterns.

DNA-LIGASE ENZYMES Used to bind DNA strands together.

DOMINANT ALLELE A gene that has the capability to mask the appearance of a recessive gene.

DOOMSDAY EQUATION A mathematical equation that predicts future human population growth.

ECHOLOCATION The emission of high-pitched sound waves that bounce off objects and are detected by an "echo" or return signal.

ECOLOGICAL ECONOMICS Branch of economics that attempts to find ways to make ecosystems more profitable when left intact rather than destroyed.

ECOLOGICAL RELEASE The removal of a dominant group that releases another group to diversify and radiate (for example, a **competitive release**).

ECOLOGY Study of the interactions of organisms and their environments.

ECOSYSTEMS Organisms and their interactions with the physical environment.

EDIACARA FORMATION Special fossil formation, located in Australia, that contains rarely preserved soft-bodied creatures over 600 million years old. This was when multicellular life was rapidly evolving.

ELECTROMAGNETIC FORCE Causes positively charged nuclei to unite with negatively charged electrons to form atoms. It also causes atoms to unite and form molecules.

ELECTRON A negatively charged subatomic

particle that orbits around larger **protons** and **neutrons** in the nucleus.

EMERGENT PROPERTIES Properties that develop in an individual when a certain level of complexity has been reached.

ENDANGERED SPECIES Species in immediate danger of extinction.

ENDANGERED SPECIES ACT Legislation that directed the U.S. Fish and Wildlife Service to maintain a list of species that are endangered or threatened.

ENDOSKELETON Bones that support the inside of the body.

ENTROPY The second law of thermodynamics states that in any isolated system, there will be an inevitable tendency toward disorder and loss of usable energy.

EPICONTINENTAL SEAS Warm shallow seas that covered most of the continents, especially in the early and middle Paleozoic era.

EQUILIBRIUM DIVERSITY HYPOTHESIS The biosphere supports a relatively constant number of groups.

ERA Four major subdivisions of the **geologic time scale**—Cenozoic, Mesozoic, Paleozoic, Precambrian.

EUGENICS The selective breeding of humans.

EUKARYOTE Complex cells with an organized nucleus and specialized organs.

EVE HYPOTHESIS All humans are related to the same African woman who existed 200,000 years ago. This hypothesis supports the **out of Africa model**.

EVEN-TOED UNGULATE A mammal with two hooves on each foot (for example, pig and deer).

EVOLUTION (Latin, to unroll) Any change through time, usually viewed in a biological sense.

EVOLUTIONARY HUMANISM Philosophical belief promoted by Huxley which described evolution as a progressive increase in the ability to transcend limits imposed by the environment.

EVOLUTIONARY MYSTICISM Philosophical belief promoted by De Chardin which described evolution as driven by a spiritual force toward progressively "higher" goals.

EXOSKELETON Hard chitinous outer skeleton.

EXPANDING UNIVERSE Today's universe in which galaxies are moving away from each other.

EXTINCTION The termination of a species, population, or other group.

EXTINCTION RATE The number of extinctions per unit time (for example, number of families per one million years).

EXTINCTION SELECTION Process whereby groups in an extinction event are affected differently because of their different traits.

EXTRAPOLATION An attempt to predict the future by using past trends.

EXTRINSIC FACTORS Environmental factors that affect an organism's evolutionary pattern (compare **intrinsic factors**).

FERN Seedless plant with true roots and stems, first found in the Paleozoic era.

FIELD OF BULLETS SELECTION Extinction event that is totally random in who it kills.

FLAGSHIP SPECIES Organisms that attract media attention and appeal to people (for example, panda).

FOOD CHAIN Sum of all predaton in an ecosystem.

FOOD PYRAMID A way of showing energy flow in an ecosystem.

FOOD WEB Chart showing complexity of eating relationships among organisms.

FOSSIL (Latin, to dig up) Remains of prehistoric life.

FOUNDER EFFECT A nonrepresentative sampling of a parent species because of the small size of the founding population.

FROG A tailless amphibian.

FRONTAL LOBES The area of the brain where long-range planning and analytical abilities occur.

FUEL EXHAUSTION Death stage of a star, occurring when hydrogen has been depleted.

FUNDAMENTALISM Religious groups that adhere to traditional beliefs such as a literal interpretation of the Bible.

GAIA HYPOTHESIS Postulates that the biosphere is highly interconnected not only among its life forms, but also with the earth's atmosphere, oceans, and land.

GAMETOPHYTE Specialized plant that produces sperm and egg cells.

GENE Stores the information for an organism's development and maintenance.

GENE BANKS Reservoirs to save seeds, spores, sperm, and genetic material from species that cannot be saved in the wild.

GENE FLOW Movement of genes from one

gene pool to another, usually by migration of individuals.

GENE POOL The total of all genes contained in a species.

GENE SPLICING Method of using biochemical and microsurgical techniques to directly transfer genes between different kinds of organisms.

GENE THERAPY The treatment of genetic disorders with new genes (by **gene splicing**).

GENETIC COUNSELOR A medical specialist who can advise prospective parents on their likelihood of transferring genetically transmitted diseases to their offspring.

GENETIC DRIFT Random sorting of genes in a gene pool.

GENETIC ENGINEERING Human manipulation of evolution through use of advanced scientific techniques.

GENOME MAP Shows the location of each gene on a chromosome and what trait or traits each gene "codes" for.

GENOTYPE Allele combinations for a particular gene.

GENUS A taxonomic classification that groups similar species together.

GEOLOGICAL TIME SCALE A timetable that provides absolute dates for many geological and biological events.

GINKGO A gymnosperm with fan-shaped leaves.

GOLDILOCKS' PARADOX Planets cannot be too hot or too cold but "just right" for life to be created.

GRACILE (means "small bodied") One of two species that arose approximately 2.8 million years ago from the *Australopithecus* genus, was gracile and was a direct human ancestor.

GRADUALIST Believe that evolutionary change is a slow, continuous process.

GRADUALISTIC EVOLUTION Theory that evolution occurred through gradual, transitory stages.

GRAVITATIONAL FORCE The force that caused rocks and minerals to combine and form planets.

GREAT AMERICAN INTERCHANGE Migration of species between continents when geologic activity a few million years ago joined North and South America, creating Central America.

GREAT CHAIN OF BEING A belief that na-
ture is arrayed from nonliving simple organisms up to humankind. This chain was not evolutionary but was a static ranking toward godliness.

GREEN ALGAE A eukaryotic single-celled plant believed to be the ancestor of the first plant colonizer of land.

GREENHOUSE EFFECT Brought about when carbon dioxide and other gases are released in the atmosphere. These gases trap heat from the sun's rays.

GROUP ORIGINATION RATE The rate at which new species appear in the fossil record (for example, the number of families per million years).

GYMNOSPERM Nonflowering seed plant abundant in the Mesozoic era (for example, the **conifer**).

HALF-LIFE The time it takes for one-half of the radioactive parent material to decay into daughter material.

HAND AX The basic tool of the Acheulean toolmaking tradition.

HAPLOID Gametes (sperm and egg cells) that have only one chromosome pair (compare **diploid**).

HARD EUGENICS The active promotion of desirable traits in the human gene pool (compare **soft eugenics**).

HARD-SHELLED EGG The major factor enabling reptiles to live completely on land—water was no longer needed for development.

HARDY-WEINBERG EQUATION Equation used to predict the frequencies of phenotypes in populations.

HELIUM FUSION The fusion of helium atoms to create carbon atoms.

HERITABILITY A trait that is consistently passed on to offspring under a variety of environmental conditions has high heritability.

HETEROCHRONY Process in which the development of an organism is altered.

HETEROZYGOUS Combination of different alleles at one chromosome site.

HOMEOSTASIS The presence of mechanisms that produce and maintain optimal conditions for life.

HOMO Genus name for humans.

HOMO ERECTUS The first human ancestor that walked in a fully upright position and controlled his environment.

HOMO HABILIS (means "handy man") Believed to be the first human ancestor to produce and use stone tools.

HOMOLOGOUS ORGANS Organs derived from the same ancestor, these organs serve different functions in different organisms, such as the forelimb of humans and the wing of a bird.

HOMO SAPIENS (means "wise man") Genus and species named for the modern species of human.

HOMOZYGOUS The same alleles of a gene occur at one chromosomal site.

HOPEFUL MONSTER An organism with a drastic mutation in its genetic makeup.

HORNED DINOSAUR Bird-hipped dinosaur with a varying number (depending on the species) of sharp, bony points protruding from the head.

HUBBLE'S LAW Faster-moving galaxies are further away from their point of origin than slower galaxies.

HUMAN GENOME PROJECT Effort to map all 100,000 genes on the twenty-six human chromosomes.

HYDROSPHERE Liquid state of matter (especially water) comprising earth.

ICHTHYOSAUR Reptile that returned to the sea and bears a resemblance to dolphins (because of their similar occupation of an ecological niche).

IMPLODE When a star collapses on itself. Occurs when the star can no longer produce enough pressure in its core to resist the weight of its own gravity.

INCREASING DIVERSITY HYPOTHESIS The biosphere's diversity has increased through time.

INDICATOR SPECIES Species with known tolerances to specific pollutants or other types of environmental damage.

INHERITANCE OF ACQUIRED TRAITS The belief that traits acquired during an organism's life can be passed on to offspring, usually associated with Lamarck.

INNER PLANETS Rocky planets orbiting closest to the sun; consisting of Mercury, Venus, Earth, and Mars.

INSECTIVORE Generalized insect-eating mammal with long snout and ratlike appearance.

INSTINCT Genetically "programmed" behavior, usually a response to a specific stimulus.

INTRINSIC FACTORS Genes, behavior, and other factors that influence evolution from within the organism (compare **extrinsic factors**).

INVERTEBRATES Animals without backbones such as clams, snails, and worms.

IRIDIUM Element not common on the earth's surface, but abundant in meteors and space debris.

IRON AGE The period when man first started heating ores to great temperatures, allowing for the stronger components of the ores to be extracted, purified, and worked. Iron was among the first of these.

JAWLESS FISHES First fish to appear in the early Paleozoic. Some species still exist.

KAMIKAZE GENE Gene that destroys an organism when activated.

KEY INNOVATION A major trait that allows a group to adapt and exploit a new environment (for example, wings on birds).

KEYSTONE SPECIES A species that influences many other species in an ecosystem. Loss of a keystone species may cause **secondary extinctions.**

KILL CURVE Frequency curve that shows how often extinctions occur.

LABYRINTHODONT (Greek *labyrinth*, maze; and *dont*, teeth) General name for some Paleozoic amphibians.

LATITUDINAL DIVERSITY GRADIENT General increase in the number of species moving toward the equator.

LAW OF LATERAL CONTINUITY Horizontal strata are laterally continuous when deposited.

LAW OF ORIGINAL HORIZONTALITY Superposed sediments are deposited in horizontal layers.

LAW OF SUPERPOSITION Where sediments are layered, the top layers are younger than the ones on the bottom.

LEARNING An organism can modify its response to a stimulus by experience.

LIFEBOAT ETHICS Harsh rules for survival—for strong individuals to live, weaker ones must die.

LINEAGE TREND Trend seen in the evolution of a single lineage.

LITHOSPHERE (Greek *lithos*, rocks) The solid

part of a plant that includes rocks and their erosional products, sediments, and soils.

LIVING FOSSIL Group that has not changed significantly in many millions of years (for example, the horseshoe crab).

LIZARD-HIPPED DINOSAUR Dinosaurs with hip bones resembling those of modern day lizards. This group included sauropods and theropods.

LOBE-FINNED FISH Fish with fleshy lobes that reinforce the fin. Ancestors of these fish gave rise to amphibians.

LOCALIZATION The subdivision of the brain cortex into local areas of control and function.

LOWER PALEOLITHIC Period characterized by two stone industries, the **Pebble Culture** and the Acheulean.

LUNG Chamber containing internal air sacs that absorb air into the bloodstream.

LYCOPSID Plants identified by scaly bark; they were abundant in the Paleozoic era.

MACROEVOLUTION Long-term result of mutation, selection, and other processes that cause creation of new species or groups.

MAIN SEQUENCE The long-lasting stage in which a star uses hydrogen atoms for fuel by fusing them together to form helium atoms.

MAMMALLIKE REPTILE The predecessors of mammals.

MANTLE The thick middle layer of the earth beneath the crust that is composed of heavy elements.

MARSUPIAL Pouch-bearing mammals without a placenta, found mostly in Australia.

MASS EXTINCTION The disappearance of many types of species from many different groups within a relatively short period.

MATTER ERA The period of universe evolution when matter became more common than energy as temperatures and pressures dropped.

MEGAFAUNA Very large mammals which became common around two million years ago (for example, woolly mammoth).

MEIOSIS Process of generating haploid cells in which half the number of chromosomes is produced and **recombination** occurs.

MEMES Basic units of learned information (contrast with **genes**).

MESOLITHIC Middle stone age marks the short transition from.

MESOZOIC MARINE REVOLUTION The shift in abundance from sessile to modern mobile invertebrates in the Mesozoic.

METAL AGES The period that marked the first use of metals by man and the growth of modern technologies. This period is marked by the transition of hunting and gathering societies to agrarian-based economies.

METAMORPHOSIS A process of massive cellular rearrangement, usually occurring in insects and other organisms before becoming an adult.

METAPHYTE Multicellular plant.

METAZOAN Multicellular animals.

MICROEVOLUTION Changes in frequency of traits and genes in a population not giving rise to a new species.

MICROINJECTION Microscopic instruments are used to transfer **DNA.**

MICROPLATE The first small crustal plates that formed when the earth began to cool.

MICROSPHERES Crude cell-like structures containing amino acids, but no nucleus or organelles.

MILANKOVICH CYCLE A cyclical wobble that naturally occurs as the earth spins on its axis.

MILKY WAY GALAXY Spiral galaxy in which our solar system exists.

MILLER, STANLEY Discovered **amino acids** could be recreated by stimulating the early atmosphere of the earth with electrical energy (see also Urey, Harold).

MILLIPEDE A multi-legged arthropod, it was one of the first animals to migrate onto land.

MINIMUM VIABLE POPULATION Minimum population size necessary for the species or population to survive.

MITOSIS Cell multiplication in which all chromosomes of a cell are duplicated.

MODERN FAUNA Common after the Paleozoic era, composed of sea urchins, clams, and snails and other mobile invertebrates.

MODERN SYNTHESIS Synthesis of information gathered from the fields of genetics, paleontology, anatomy, and other fields supporting the theory of natural selection.

MOLECULAR RATES The biochemical rates of evolution.

MOLLUSK A phylum that contains many modern marine invertebrates, such as **clams, snails,** and **cephalopods.**

MOLTS Process in which the insect sheds the outer skeleton (exoskeleton).

MONGOLOIDS One of the four **Homo sapien** races. Traits include thick, straight black hair, sparse body and facial hair, broad face, and epicanthic fold.

MONKEY Arose from prosimians. The monkey has well-developed, grasping hands and relatively high intelligence.

MONOTREME Most primitive group of mammals in that they lay eggs and have poor temperature regulation (for example, the duck-billed platypus).

MOSAIC EVOLUTION Anatomical parts evolving at different rates.

MOSASAUR Giant prehistoric marine lizard with a similar appearance to alligators.

MULTIREGIONAL MODEL Modern human races evolved "in place" from separate populations of **Homo erectus.**

MUTATION An error in the DNA copying process resulting in genetic change.

MUTATIONIST SCHOOL School of thought which proposes that evolution occurs whenever mutations cause new forms to arise.

NATURALISTIC FALLACY Whatever is natural must therefore be good.

NATURAL SELECTION Environmental sorting of genetically produced variations.

NATURE-NURTURE DEBATE A controversial argument where "nature" refers to the idea that genes control most mental abilities and "nurture" holds that the environment is the major influence on an individual's mental capabilities.

NEANDERTHALS Ancient modern humans that retained some of the "brutish" traits of the *Homo erectus* such as protruding brow ridge and jaws and elongated skulls.

NEGATIVE ARTIFICIAL SELECTION Where humans actively try to eliminate all types of individuals in a species.

NEGROID (Latin, black) Prominent characteristics include dark skin, low body fat, small ears, and large broad noses.

NEOLITHIC New stone age that marks the beginning of plant and animal domestication.

NEURAL NETWORKS Process through which the human brain attempts to become mechanically stimulated through the use of highly interconnected processing elements.

NEURONS Nerve cells.

NEUTRAL THEORY Some traits are selectively neutral, meaning that they are neither useful nor harmful.

NEUTRON Neutrally charged subatomic particle located in the nucleus of an atom.

NEUTRON STAR Extremely dense leftover star material in which electrons and protons have been compressed to produce neutrons.

NICHE An organism's "occupation"—what it does in an ecosystem.

NICHE PACKING Process in which evolution invents "new ways of doing things" in a given ecosystem.

NONBRANCHING EVOLUTION The result of directional selection where an entire species is transformed into a new species.

NONGENETIC PLASTICITY Changes in an individual's development caused by environmental change (for example, stunted growth).

NOVA A star explosion in which large amounts of matter and energy are expelled into space at great speeds.

NUCLEAR FORCE Causes atomic nuclei to form by attraction of subatomic particles like protons and neutrons. There are two kinds of nuclear forces, strong and weak.

NUCLEAR FUSION When hydrogen atoms fuse and form helium atoms.

NUCLEAR WINTER Atmospheric phenomenon caused by the release of dust, soot, and other particles from nuclear explosion. The particle matter left in the atmosphere would block out the sun's rays and cause a global "wintering" reaction.

ODD-TOED UNGULATE A mammal that uses only one toe per foot to support its weight (for example, the horse).

OPEN SYSTEMS System that takes energy from the environment and uses it to construct complexity and organization.

OPEN UNIVERSE Theory that the universe will expand forever.

OPPORTUNISTIC SPECIES Species that evolved to take advantage of natural disturbances in the ecosystem.

ORIGINATION The evolution of a new group or the branching of new groups stemming off from an original species.

ORNITHOPODS "Duck-billed" dinosaur that is believed to be a social animal, it provided care for its young.

ORTHOGENETIC SCHOOL Belief that some

predetermined direction of evolution will bring a species to an inevitable goal.

ORTHOSELECTION Straight-line selection on a species by persistent environmental trends.

OUTER PLANETS The group of highly gaseous planets in the solar system, consisting of Jupiter, Saturn, Uranus, and Neptune.

OUTGASSING Volcanic release of gases into the atmosphere.

OUT OF AFRICA MODEL Modern humans evolved in Africa and migrated out of Africa about 100,000 years ago.

OZONE A rare gas in the upper atmosphere brought about when sunlight causes oxygen atoms to combine. It protects the earth from ultra-violet radiation.

PALEOLITHIC First of man's cultural stone ages. Lasted from 2.6 million years ago until 10,000 years ago.

PALEONTOLOGISTS People who study fossils.

PALEONTOLOGY Study of fossils.

PALEOSPECIES PROBLEM Difficulty in distinguishing between fossil species due to the inability to recognize interbreeding ability.

PALEOZOIC FAUNA Comprised mostly of stationary (**sessile**) filter feeders such as **corals** and **brachiopods.**

PANADAPTATIONISM A tendency to view any trait as adaptive.

PANGEA A supercontinent which formed when all the continents came together in the late Paleozoic.

PEBBLE CULTURE Also known as the Oldowan tradition. It was the first attempt by early humans to work with and use stone.

PERMINERALIZATION The filling of pores or cavities of a fossil with minerals, which infiltrate from the surrounding sediment.

PHENOTYPE Physical appearance as the result of gene expression for a particular trait (compare **genotype**).

PHENOTYPIC PLASTICITY Changes in an individual or group resulting from environmental conditions and not the gene pool.

PHOTOSYNTHESIS Process in which plants use light, carbon dioxide, and water to produce food and oxygen.

PHYLA Basic groups of life that exist or existed in the past (for example, echinoderms).

PHYSICAL EVOLUTION Long-term changes in nonliving matter caused by chemical and physical processes.

PISTIL Female organ of the flower.

PLACENTAL Mammal which evolved the placenta that protects the embryo by partially separating the mother's and child's circulatory systems.

PLACODERM Extinct primitive jawed fish with bony armor.

PLASMID DNA molecules that occur naturally within bacterial cells.

PLATE TECTONICS The migration of plates along the earth's surface; occurs because plates ride on underlying magma currents.

PLEIOTROPY A single gene controlling many traits.

PLESIOSAUR Slow reptile swimmer equipped with four peddle-like appendages and an elongated neck.

POLYGENETIC TRAIT A trait that is controlled by many different genes.

PORPOISE Highly specialized swimming mammal first appearing in the Cenozoic era.

POSITIVE ARTIFICIAL SELECTION Human encouragement of reproduction by purposely breeding certain individuals for scientific, aesthetic, or practical purposes.

POSITIVE FEEDBACK Occurs when a previous change causes an even greater amount of the same kind of change; a "snowball effect."

POSTMATING ISOLATING MECHANISM A reproductive isolating mechanism that allows mating to occur but prevents the production of healthy, fertile offspring.

PREADAPTATION A process in which an organ was previously used for one function and becomes used in another later function.

PREMATING ISOLATING MECHANISM A reproductive isolating mechanism which prevents mating from occurring.

PRESERVE DESIGN Shape and size considerations of wildlife preserves.

PRIMARY PRODUCERS Groups, especially plants that can synthesize their own food and are the foundation of any ecosystem, from which all consumers receive their nutrition; organisms capable of producing their own food source and located at the base of the food chain.

PRIMATE A group that can be recognized by such basic traits as grasping hands, stereovision, relatively large brains, and prolonged parental care of offspring.

PRIMITIVE TRAIT Traits shared by many organisms, such as hair in mammals.

PRIMORDIAL SOUP Description of earth's early ocean.

PRINCIPLE OF MEDIOCRITY The earth is not unique in its capability to support life.

PROKARYOTE Simple cells that lack an organized nucleus and specialized internal organs.

PROSIMIAN First primates to appear and the least "advanced" in primate characteristics.

PROTOCULTURE Rudimentary behavior shown in chimpanzees and apes that includes an ability to combine concepts.

PROTON Positively charged subatomic particle located in the nucleus of an atom.

PSEUDOEXTINCTION When an entire species evolves into another species by non-branching evolution.

PSILOPSIDS Spore-bearing plants that are the earliest known multicellular land-dwellers.

PTERODACTYLOID More specialized flyer of the two flying reptile groups. They were larger, lacked tails and teeth, had light skeletons, and a bony crest located on top of the head.

PTEROSAURS (means "wing lizard") Flying reptiles.

PULL OF RECENT The younger a group is the greater the opportunity for preservation. This is because a fossil has less chance of being destroyed or rendered inaccessible.

PUNCTUATED EQUILIBRIUM Theory that most of the species' evolutionary existence is relatively static and all or most change occurs during brief intervals.

PUNNETT SQUARE A genetic calculation in which each of the spaces are filled with possible combinations (**alleles**) for a particular gene.

RACE Populations of a species that are recognizably different, but still capable of interbreeding (see also **subspecies**).

RACIAL SENESCENCE A belief that a species dies of old age or develops such a highly specialized form of life that it becomes extinction prone.

RADIATION ERA The first 100 seconds of the universe when only energy (radiation) existed.

RADIOMETRIC DATING The dating of radioactive material in which the half-life of a parent material is known (for example, uranium).

RANDOM WALK Statistical trials carried out by using random processes such as coin flips that prove that random events can often cause patterns (trends) that seem to be nonrandom.

RATCHET EFFECT Allows increases in complexity to be added to previous gains and makes the process cumulative.

RATE The amount of change in some phenomenon over a period of time.

RAY-FINNED FISH One of the two types of bony fish; identified by small bones located in the fins.

REASONING Behavior in which associations and synthesis of learned behavior occur.

RECAPITULATION Organisms in their embryonic development tend to go through stages of similar anatomical make-up to their "simpler" ancestors.

RECESSIVE ALLELE Form of a gene that will not be expressed in the presence of a **dominant allele.**

RECOMBINANT DNA TECHNOLOGY Genes of one species are "spliced" and combined into the genes of another.

RECOMBINATION Genes in sex cells are shuffled about during the copying process, creating unique genetic combinations.

RED GIANT Stage in the death of a star in which nuclear fusion releases energy and the star expands, giving off a reddish light.

RED SHIFT A shift of radiation toward the red end of the light spectrum.

REDUCTIONISM Theory that emergent properties can be understood by studying the lower levels of the property's complexity.

REFLEX The most basic response to some stimulus.

REFUGIA Areas such as coral atolls that provide refuge for shallow water groups during sea level drops, preventing their extinction.

REGULATORY GENES Genes that control the timing of hormones and other biochemical mechanisms governing the growth of an organism.

RELATIVE RATE When differences in species' anatomical features can be observed, but time intervals cannot be accurately dated.

RELATIVE TIME Where sequential reconstruction of events are capable of being made, but no specific ages can be assigned.

RELAXED SELECTION Transmission of

genes that cause disease and various disabilities because of modern medicine's ability to help those individuals survive and reproduce.

REPRODUCTION Replication of an organism (see also **sexual** and **asexual reproduction**).

RESTRICTION ENZYME Enzyme that locates and cuts DNA strands at specific sites.

RETURN TIME Interval between environmental events of a specified size.

REVERSE RAREFACTION CURVE Graph that converts the number of species killed in a group to the number of families or genera killed.

RHAMPHORYNCOID One of the two groups of flying reptiles. Recognized by their tail, teeth, and heavier skeletons.

RISE OF MARINE REPTILES The stage in the evolution of life in the oceans occurring in the Mesozoic era.

RISE OF MOBILE INVERTEBRATES The second stage in the evolution of life in the oceans, occurring in the Mesozoic era.

RISE OF SESSILE INVERTEBRATES AND FISHES The first stage in the evolution of life in the earth's oceans, coming in the Paleozoic era.

RNA Ribonucleic acid; involved in **transcription** and **translation** processes.

ROBUST (means "large") An extinct species of the *Australopithicus* genus.

SALTATIONS (Latin, to jump) A drastic mutation that, if capable of surviving, could produce a new species (see **hopeful monster**).

SAUROPODS A herbivore, the largest known dinosaur and largest land life to have existed on earth.

SAVAGERY Proposed first stage of human cultural evolution by Lewis Henry Morgan (see also **barbarism** and **civilization**).

SCIENTIFIC CREATIONISM Religious concept that seeks to take the Bible literally, even though it may conflict with scientific facts.

SCIENTIFIC METHOD Generation of hypotheses based on observation and experimentation.

SCORPION One of the first land colonizers.

SEA URCHINS Starfish relatives that became common during the Mesozoic.

SECONDARY EXTINCTION A cascade through the food web as the extinction of one species causes the extinction of another.

SECOND GENERATION STAR A star that has formed from the remains of an extinct star.

SEDIMENTARY ROCKS Rocks formed by the deposition of erosional products in basins (for example, sandstone).

SEED Fertilized plant cell.

SEED FERN Plant that first appeared in the late Paleozoic era and probably evolved from seedless ferns.

SELECTIVE BREEDING Process by which humans actively control what organisms a plant or animal may breed.

SELF-FULFILLING PROPHECY A negative outlook can bring on the expected problem.

SELF-ORGANIZING SYSTEM Theory that a chemical system can be driven by an outside energy source so far out of its natural equilibrium it accidentally reaches an organized state.

SESSILE Immobility or staying in one place.

SEXUAL SELECTION Selection by mates such that some individuals have traits that allow them to have more offspring than others.

SHARK A fish with a skeleton made of cartilage.

SHOCKED MINERALS Deformed mineral grains that were created by the impact of a massive object at high speed such as a meteorite.

SLASH AND BURN AGRICULTURE A forest is cut down and subsequently burned to provide soil nutrients.

SNAIL Mobile invertebrate that became common in the Mesozoic era.

SOCIAL DARWINISM A scientific theory that became a justification for social and economic exploitation of poor and underprivileged classes. This is the idea of "survival of the fittest."

SOCIAL STRATIFICATION Differences in social status are important in governing social behavior.

SOFT EUGENICS The belief that the elimination of genetic human disabilities and diseases is ethical (compare **hard eugenics**).

SOLVENT A solution that can dissolve matter.

SPECIES AREA EFFECT The more area that is disturbed, the more species become extinct.

SPHENOPSID Plant first found in the Paleozoic era, that is recognizable by radiating whorls located at regular intervals along the trunk.

SPORE Unfertilized cell that has the capability to grow into a **gametophyte**.

SPOROPHYTE Spore producing plant.

STABILIZING SELECTION A type of selection that operates on both extremes of the trait distribution curve simultaneously.

STALKED ECHINODERM Sessile invertebrates that were prominent during the Paleozoic era.

STAMEN Male organ of the flower.

STASIS Periods where little or no evolutionary change has occurred.

STATE Highest level of organization in human social institutions. Characterized by people living in cities, maximum population density, full-time specialization on crafts, and inequitably distributed wealth (see also **band, tribe,** and **chiefdom**).

STEGOSAUR Bird-hipped dinosaur with a series of alternating plates running down the back. These plates were used for heat regulation.

STEREOVISION Occurs when the eyes face forward on the head; it causes overlapping fields of vision and allows for depth perception.

STONE INDUSTRY Archaeological term for a tradition of tool making.

STRATIGRAPHY Study and mapping of rock layers.

STROMATOLITE Large layered mounds consisting of colonies of algae.

SUBDUCTION BOUNDARY A boundary caused by a crustal plate being pushed underneath another plate.

SUBSPECIES Where populations of a species are recognizably different, but still capable of interbreeding (see also **race**).

SUPERNOVA An extremely spectacular **nova** event.

SYMBIOSIS HYPOTHESIS Idea that different kinds of single-celled organisms grouped together in a cooperative unit to form one multicellular organism.

SYMBIOSIS THEORY Theory that eukaryotic cells arose through the cooperative effort of different types of prokaryotic cells.

TAXONOMY Process of classifying organisms.

THECODONTS (means "socket tooth") Rep-

tile that flourished in the early Mesozoic and gave rise to dinosaurs.

THEISM Belief that God has knowledge and interest in all that occurs in the universe.

THEOCRACY A society governed by religious leaders, and generally associated with **chiefdoms.**

THEORY OF RECAPITULATION The theory that developmental sequences of an organism repeat the organism's evolutionary history.

THEROPODS Carnivorous relatives of the sauropods. This group includes the *Tyrannosaurus rex.*

THREATENED SPECIES Species likely to be endangered soon.

TINKERING Process by which natural selection uses available materials and adapts them to new environmental conditions.

TRACE FOSSIL The preservation of indirect evidence of past life. Examples would be burrows or trails.

TRANSGENESIS Transfer of genes between different types of organisms.

TRANSLATION Process whereby messenger **RNA** sequence is converted into proteins.

TRIBE Social system characterized by groups of people with a sense of private property and territoriality, and in which family descent is important for the passing of rank and possessions (see also **band, chiefdom,** and **state**).

TRILOBITE Paleozoic invertebrates, and the segmented ancestors to modern insects.

UMBRELLA SPECIES Species that have large area needs.

UNCOMMITTED CORTEX Areas of the brain that are not committed to any one function and contain neurons that store and synthesize incoming information.

UNIFORMITARIANISM The principle that present-day processes explain past events.

UPPER PALEOLITHIC A period that contains just one stone industry—the Mousterian.

URBAN STRESS SELECTION The selection of genes that help individuals cope with the mental and physical stresses of modern life.

UREY, HAROLD Scientist who recreated **amino acids** during an experiment about the early earth (see also **Miller, Stanley**).

VASCULAR SYSTEM (Latin *vascular*, small vessel) A plant's system of transporting water and nutrients from the roots to stems and leaves.

VECTOR Free-floating DNA molecule that can attach to cut DNA strands and make transfers to other cells.

VERTEBRATES Animals with backbones.

VESTIGIAL ORGANS Organs that are no longer useful for the survival of an organism, but have not been fully eliminated by natural selection.

VIRUS DNA molecules that force cells to reproduce their own DNA structure.

VITALISTS Group that has tried unsuccessfully to define some essence common to all life forms that separate life from nonlife.

VULNERABLE SPECIES Species that are likely to go extinct.

WANTON SELECTION Preferential survival of organisms during extinction events because they possess traits that just happen to be beneficial during the new demands of the catastrophe.

WHALE Large swimming mammal that first appeared in the early Cenozoic era.

WHITE DWARF The first stage in the death of a small star.

ZYGOTE Fertilized egg resulting from the union of gametes (sperm and egg).

Additional Credits

pp. 5, 6, 54, 58, 120, 145, Figures 1–2, 1–3, 2–19, 2–21, 3–25, 3–45: Reprinted by permission from pages 115, 118, 138, 137, 510, and 334 of *Historical Geology* by R. Wicander and J. Monroe. Copyright © 1989 by West Publishing Company. All rights reserved. **pp. 8, 49, 53, 60, 63, 104, 182, 198, 204,** Figures 1–4, 2–17, 2–18, 2–22, 2–25, 3–13, 3–66, 3–75, 3–80: Fred A. Racle, *Introduction to Evolution,* © 1979, pp. 14, 44, 55, 71, 72, 74, 78, 80, 126. Reprinted by permission of Prentice Hall, Englewood Cliffs, New Jersey. **p. 13,** Table 1–1: *The Gallup Poll Monthly,* November 1991 (#314), p. 34. **pp. 25, 100, 103, 105, 106, 127, 128, 140, 142, 164, 203,** Figures 2–3, 3–9, 3–11, 3–14, 3–15, 3–30, 3–31, 3–39, 3–41, 3–56, 3–79: Don L. Eicher/A. Lee McAlester, *History of the Earth,* © 1980, pp. 11, 37, 54, 115, 118, 125, 157, 237, 275, 347, 370. Reprinted by permission of Prentice Hall, Englewood Cliffs, New Jersey. **pp. 36, 328,** Figures 2–7 and 5–16: From *The New Biology,* 1987, by Robert Augros and George Stanciu, published by New Science Library. **pp. 44, 56, 75, 269, 270, 271,** Figures 2–13, 2–20, 2–30, 4–28, 4–29, 4–30: From Barrett/Abramoff/Kumaran/Millington, *Biology,* © 1986, pp. 414, 607, 755, 938, 939. Reprinted by permission of Prentice Hall, Englewood Cliffs, New Jersey. **p. 47,** Figure 2–16: Illustration by James Egleson from "The Genetic Code: II" by Marshall W. Nirenberg, *Scientific American.* Copyright © March 1963 by *Scientific American, Inc.* All rights reserved. **p. 64,** Figure 2–26: Figure 12, p. 106, in *Evolutionary Biology,* Douglas J. Futuyma, 2nd. edition, 1986, reprinted by permission of Sinauer Associates. **p. 65,** Figure 2–27: Reprinted by permission from page 145 of *Introduction to Physical Anthropology* 5th edition by Harry Nelson and Robert Jurmain. Copyright © 1991 by West Publishing Company. All rights reserved. **pp. 66 and 281,** Figures 2–28 and 4–37: *The Evolutionary Process,* Verne Grant, 1985, © Columbia University Press, New York. Reprinted by permission of the publishers. **p. 76,** Figure 2–31: From *A View of Life,* Luria/Gould/Singer, p. 558, Figure 24.4, © 1981 Benjamin/Cummings Publishing Company. **p. 80,** Figure 2–32: A. Jolly, *The Evolution of Primate Behavior,* 1972, Macmillan. **p. 93,** Figure 3–3: H. Hofman, "Attributes of Stromatolites," *Geological Survey of Canada Paper 69–39,* p. 4, 1969, Geological Survey of Canada, Department of Energy, Mines and Resources. Reproduced with the permission of the Minister of Supply and Services Canada, 1993. **p. 99,** Figure 3–7: NASA. **p. 106,** Table 3–2: From Carla W. Montgomery, *Fundamentals of Geology.* Copyright © 1989. Wm. C. Brown Communications, Inc., Dubuque, Iowa. All rights reserved. Reprinted by permission. **pp. 107, 113, 115,** Figures 3–16, 3–21, 3–22: M. Grant Gross, *Oceanography: A of the Earth,* 4th edition, © 1987, pp. 72, 147, 381. Reprinted by permission of Prentice Hall, Englewood Cliffs, New Jersey. **p. 108,** Figure 3–17: NASA. **p. 118,** Figure 3–24: R. Dott and R. Batten, *Evolution of the Earth,* 4th edition, © 1988, McGraw-Hill, Inc., reproduced with permission of McGraw-Hill. **p. 124,** Figure 3–27: From Morris S. Petersen and J. Keith Rigby, *Interpreting Earth History.* Copyright © 1978 Wm. C. Brown Communications, Inc., Dubuque Iowa. All rights reserved. Reprinted by permission. **p. 130,** Figure 3–33: Leo F. LaPorte, *Ancient Environments,* 2nd edition, © 1979, p. 32. Reprinted by permission of Prentice Hall, Englewood Cliffs, New Jersey. **pp. 136, 142, 143, 155, 156, 161, 162, 166, 180, 184,** Figures 3–37, 3–42, 3–43, 3–50, 3–51, 3–54, 3–55, 3–57, 3–65, 3–67: A. Lee McAlester, *The History of Life,* 2nd edition, © 1977, pp. 24, 80, 82, 92, 94, 101, 105, 132–33, 120. Reprinted by permission of Prentice Hall, Englewood Cliffs, New Jersey. **p. 147,** Figure 3–46: Reprinted with the permission of Merrill, an imprint of Macmillan Publishing Company from *Life of the Past,* 2nd edition by N. Gary Lane. Copyright © 1986 by Bell & Howell Company. **p. 159,** Figure 3–53: American Museum of Natural History. **p. 175,** Figure 3–62: Reprinted by permission of the artist, Donald Davis. **pp. 179, 199,** Figures 3–64, 3–76: D. Norman, *Dinosaur!,* 1991, Boxtree Limited, copyright 1991 by ML Design, London. **p. 196,** Figure 3–74: Carnegie Mellon. **p. 206,** Figure 3–81: From Hank Iken, "Hard Woods," by Phillip E. Ross. Copyright © April 1991 by Scientific American, Inc. All rights reserved. **p. 207,** Figure 3–82: H. Levin, *The Earth Through Time,* 1988, Saunders, p. 581. **p. 215,** Figure 3–85: Charles A. Weitz, *Introduction to Physical Anthropology and Archaeology,* © 1979, p. 235. Reprinted by permission of Prentice Hall, Englewood Cliffs, New Jersey. **p. 216,** Figure 3–86: *Annual Review of Ecology and Systematics,* Vol. 3, 1972, Annual Reviews, Inc. **p. 230,** Figure 4–4: Reprinted with the permission of Merrill Publishing Company, an imprint of Macmillan Publishing Company, Inc. from *Environmental Studies: Earth as a Living Planet,* 2nd edition by Daniel B. Botkin and Edward A. Keller. Copyright © 1987 by Merrill Publishing Company. **p. 235,** Figure 4–8: Reprinted from "Hydra as a Model for the Development of Biological Form" by P.D. Gingerich in *Nature* 284: 107–109, 1974. Copyright 1974 by Macmillan Magazines Limited. **p. 241,** Figure 4–12: From *Evolution* by Dobzhansky et al. Copyright © 1977 by W.H. Freeman and Company. Reprinted by permission. **p. 249,** Figure 4–14: James R. Beerbower, *Search for the Past: An Introduction to Paleontology,* 2nd edition, © 1968, p. 136. Reprinted by permission of Prentice Hall, Englewood Cliffs, New Jersey. **p. 255,** Figure 4–18: G. deBeer, *Atlas of Evolution,* p. 49. New York: Thomas Nelson, Inc., 1964. **p. 256,** Figure 4–19: From J. Bonner, "Size Change in Development and Evolution" from *J. Paleontology* 42: (Part II) 1–15. By permission of Paleontological Society. **p. 258,** Figure 4–20: Mound and Waloff, eds., Symposium Roy. Entolom. Soc. London, No. 9, pp. 188–204, Blackwell Scientific Publications, Ltd. **p. 294,** Figure 4–44: (top) R. Cowan, *History of Life,* 1990, Figure 17.13, p. 370, Blackwell Scientific Publications Ltd.; (bottom) modified from *Paleobiology,* 1991, p. 272, Fig. 2. **p. 296,** Figure 4–46: Reprinted from *The Ascent of Man* by David

Pilbeam. Copyright © 1972 by David R. Pilbeam. **p. 309,** Figure 5–3: *Theories of Everything*, John Barrow, 1991, by permission of the Oxford University Press. **p. 316,** Figure 5–7: J. Murray Mitchell, Jr., "Carbon Dioxide and Future Climate," *Environmental Data Service*, p. 8, March 1977. **p. 320,** Figure 5–10: "Rapid Increase in Human Population Is Paralleled by Rapid Increase in Extinction of Birds and Animals," as it appeared in Botkin and Keller, *Environmental Studies*, 2nd edition, Merrill, 1987, p. 26. Reprinted by permission of Springer Verlag. **pp. 321, 336,** Figures 5–11, 5–18: From *Atlas of the Environment* by Geoffrey Lean, Don Hinrichsen, and Adam Markham. © 1990 by Banson Marketing Ltd. **p. 335,** Figure 5–17: A Cockburn, *An Introduction to Evolutionary Ecology*, 1991, Figure 10.12, p. 306, Blackwell Scientific Publications Ltd. **p. 339,** Figure 5–19: From L. Durrell, *State of the Ark*, 1986, Anchor Books. Permission granted by DOUBLEDAY, a division of Bantam, Doubleday, Dell Publishing Group, Inc. **p. 341,** Figure 5–20: Copyright 1991, *USA Today*. Reprinted with permission. **pp. 350–51,** Figure 5–24: "Making Safer Pesticides" by Megan Jaegerman, July 17, 1991. Copyright © 1991 by The New York Times Company. Reprinted by permission. **p. 354,** Figure 5–25: Copyright © 1981 by Harrow House Editions. From *After Man* by Dougal Dixon. Reprinted by permission of St. Martin's Press, Incorporated. **p. 356,** Figure 5–26: C. Brace, *The Stages of Human Evolution*, © 1987, p. 141. Reprinted by permission of Prentice Hall, Englewood Cliffs, New Jersey. **p. 363,** Figure 5–28: Reprinted by permission of Mark S. Lesney and *Analog*. **p. 366,** Figure 5–30: L. Erlesmeyer-Kimling and L.F. Jarvik, SCIENCE, Vol. 142, 1963, p. 1478, copyright 1963 by the AAAS. **p. 367,** Figure 5–31: From *Interchange*, 1980. **pp. 369, 370,** Figures 5–32, 5–33: Judson/Kauffman/Leet, *Physical Geology*, 7th edition, © 1987, pp. 373, 394. **p. 379,** Figure 6–1: A. Iberall, *Toward a General Science of General Viable*, © 1972, McGraw-Hill. Reproduced with permission of McGraw-Hill.

Index

A

Absolute rates of evolution, 234
Absolute time, 104–6
Acid rain, 327
Acquired Immune Deficiency Syndrome
    (AIDS), 344
Acquired traits, inheritance of, 4
Adaptation, 34
    natural, to human-modified
        environments, 343–44
Adaptationism, 69
Adaptive radiation, 245
Additive nature of brains, 77
Adenine, 45
"Adulthood" phase of star's life, 94
Adult maintenance, genes and, 44–45
After Man: A Zoology of the Future (Dixon),
    353, 354
Age of the earth, 11–12
Aggressive behaviors, 18
Agnosticism, 13, 14
Agricultural societies, 215

Agriculture
    biotechnology uses in, 349–50
    global warming and, 315
    slash and burn, 323, 341
AIDS (Acquired Immune Deficiency
    Syndrome), 344
Air, evolution of life in, 188–202
    advanced insects and flying reptiles, 188,
        192–97
    birds and flying mammals, 197–202
    primitive insects, 189–92
    See also Atmosphere
Algae, green, 153
Alleles, 49–50, 52
Allmon, W., 326
Allometric plot, 254–56
Allometry, 254
Alvarez, Luis, 173
Amino acid chains, 46
Amino acids, 24
Amino acid sequencing, 243–44
Amphibians, 158–60
    observed declines in modern, 325

Analogous organs, 254
Anatomical intermediates, 59–60
Anatomical rates, 232–40
Anatomical similarities, 54–55
Angiosperms, 180–81, 264
Ankylosaurs, 168
Antarctica, 179–80
Anthropic principle, 13–14
Anthropologists, 10
Apes, 204, 205–6
Archaeology, 213–14
Archaeopteryx, 59, 60, 197, 198
Archaic reptiles, 162–63
Archaic ungulates, 183–84, 186
Arms race, 148, 272
Arthropods, 158, 189, 320
Artificial selection, 33, 34, 35, 61
Asexual reproduction, 50
Atavism, 54
Atheism, 13, 14
Atmosphere
    direction and rates of evolution, 228–29
    future of, 312–17
    history of, 117–19, 121, 122
    See also Air, evolution of life in
Atom, particles of, 23
Atom epoch, 91–92
Augros, R., 36, 328
Australia
    introduction of rabbits into, 329
    northward drift of, 330
Australoids, 211
Australopithecus, 206
    afarensis, 206, 207
    africanus, 206–8
    robustus, 206, 207, 208

B

Background extinctions, 277
    apparent decline in, 280–81
    difference between mass and, 279–80
    environmental change and, 283–84
    as selection by old rules, 292–93
Bacteria, genetic engineering of, 350–51
Baleen whales, 151
Band (social unit), 214–15, 216
Barbarism, 10

Barnes, R., 133, 191
Barrett, J., 44, 56, 269, 270, 271
Barrow, J.D., 309
Bats, 188, 200–202
Batten, R., 118
Beck, W., 186, 189
Beerbower, R., 249
Begon, M., 322
Berggren, W., 237
Bergmann's rule, 185, 250–51
Biased preservation of fossil record, 125
Big Bang, 23, 88–90, 93, 98, 310
    two eras after, 90–92
Big crunch, 310–11
Binet, Alfred, 365
Biochemicals, genetic engineering to
        produce, 350
Biochemical similarities, 54
Biodiversity, 320
    food and nonfood uses of, 334–37
Biological evolution
    cultural evolution compared with, 70–71
    discovery of, 4–10
    human. *See* Human evolution
    patterns of, 230–96
        extinction rates and directions, 275–96
        origination directions, 248–75
        origination rates, 232–48
    processes of, 32–69
        evidence for evolution by selection of
            variation, 53–66
        inherited variation (genes), 43–53
        misconceptions about, 66–69
        selection, 32–38
        speciation, 38–43
    products of. *See* Biosphere
    prospects for, 318–55
        future evolution without humans,
            353–55
        mass extinction, 319–42
        new diversity, 342–53
Biosphere, 109
    history of, 122–202
        fossil record. *See* Fossil record
        life in air, 188–202
        life in oceans, 137–52
        life on land, 152–88

multicellular life, evolution toward, 127–37
human decimation of, 319–42
Biosphere trends, 248–50, 258–75
diversity, 258, 261–68
ecological trends, 258, 268–75
Biostratigraphy, 103
Biotechnology, 346
ethical, legal, and environmental problems with, 351–53
potential uses of, 348–51
See also Genetic engineering
Bird-hipped dinosaurs, 165, 168, 169
Birds, 188, 197–200
diversification of, 198–200
flightless, 200
observed declines in modern, 325
origin of, 197–98
Black dwarf, 95, 96
Black holes, 96, 97
Blind Watchmaker, The (Dawkins), 12, 67
Body fossils, 122–23
Body size, trends in, 250–51
climate and, 185, 250–51
human, 297
increasing size, 257, 258
complexity and, 264–68
Body temperature in warm-blooded vs. cold-blooded animals, 181
Bonner, J., 235, 251, 256, 258, 265, 266, 267
Bony fish, 143, 159
Book gills, 158
Botkin, D., 230, 320
Bowler, Peter, 2, 10, 15, 16
Brace, C., 297, 356
Brachiopod, 139, 141, 243
anatomical rates in, 232–34
Brain
human, 71–81
as central organ of culture, 71–75
evolution of, 75–79
growing large complex, 79–81
organization of, 73–75
primate, 203
Brain size
in dinosaurs, 170, 171
human, 72–73

extrapolation of past, 355–56, 357
trends in, 296–97, 299
in mammals, evolution of, 272, 274
in porpoises, 151
relative, 72, 151
Branching evolution, 38–41
Breeding
in captivity, 340
selective, 342, 344–45, 346, 364, 365
Brennan, R., 46, 347
Briggs, D., 238, 286
Brontosaurus, 166
Bryozoans, 139, 141
Buffaloe, R., 23
Business, impact of evolutionary thought on, 16–18

C

Calow, P., 133, 191
Capron, Christiane, 367
Captivity, breeding in, 340
Carbon dioxide
excess of, in atmosphere, 312–13
greenhouse effects and, 117, 313–15
Cargo, D., 318
Carnivores, observed declines in modern, 325
Carrying capacity, 368
Cast, 123
Catastrophes, 287–89
defined, 307
global, 9–10
See also Mass extinctions
Catastrophism, 4
Caucasoids, 211, 212
Cenozoic era, 107, 117
birds and flying mammals, rise of, 188, 197–202
climate during, 119
flowering plants and mammals, rise of, 172, 179–88, 192, 264
marine mammals, rise of, 150–52
meaning of term, 139
primates in, 184–85, 203–5
Cephalopods, 147
Ceratopsians, 168

Cerebral cortex, 74
Cetus Corporation, 348
Chaisson, E., 29, 37, 72, 89, 91, 95, 208, 231, 312
Chaos epoch, 91
Chemistry and chemical processes, origin of life through, 23–30
  elsewhere in universe, 27–30
  experimental and theoretical evidence of, 24–26
  fossil evidence, 26
  frequency of, 26–27
Chiefdoms, 216–17
Chimpanzees, 204, 205
Chitin, 189
Chlorofluorocarbons (CFCs), 316
Chromosomes, 44
Civilization, 10
Clade, 248
Cladistics, 59
Cladogenesis, 38–41
Clams, 147
Classification of life, similarities and, 57–59
Climate, 119
  body size and, 185, 250–51
  extinction of dinosaurs and, 172
  global warming and, 314–15, 316
  mass extinction and change in, 285, 287
  megafauna extinction and warming, 186
  sea level and, 115, 117
Cline, 38
Closed universe, 308–11
Cloud, P., 102
Clouds, greenhouse effect and, 315
Coal, 158
Coccoliths, 150
Cockburn, A., 335
Coevolution, 192
Coevolutionary trends, 272
Colbert, E., 201
Cold-adapted organisms, 291
Cold-blooded animals, 181
Colonial hypothesis of origin of multicellular organisms, 133–34
Comets, 286, 287–88, 289
Competition, background extinctions due to, 292–93
Competitive replacement, 264

Complexity
  cultural evolution and increasing, 299
  evolutionary rate and, 245, 246
  extinction rate and, 295–96
  increasing body size and, 264–68
Condensing atmosphere, 117–18
Cone-bearing plants, 160–62
Conifers, 161–62
Conservation of momentum, law of, 92
Consistent directional selection, 250
Consumers, 130, 269, 271
Continental drift, 111–14, 115, 119, 312
Continents, creation of, 111–12
Contraction (birth) stage of stars, 93–94
Convection, 111–12
Convention on International Trade in Endangered Species (CITES), 339
Convergence, correlated traits and, 257–58, 260
Convergent evolution, 148
Convergent trends, 252–54
Cooper, J., 125
Cooperation, 273
Copernicus, 1
Cope's rule, 257
Corals, 139, 141, 146
Core of earth, 109
Correlated traits, 239, 250
  convergence and, 257–58, 260
  trends in, 254–57
Correlation, 102, 104
  spurious, 314
Cosmic background radiation, 89
Cosmologists, 308
Cosmos (Sagan), 29
Counseling, genetic, 362
Cowan, R., 294
Created from Animals (Rachels), 377
Creationism, 12, 13
Crocodiles, 165
Crosby, Alfred, 329
Crowther, P., 238, 286
Crustaceans, 147, 148
Crust of earth, 109–10
Cultural evolution, 22, 69–81, 213–17, 298, 299–300
  brain and, 71–81
    as central organ of culture, 71–75

evolution of, 75–79
  growing large complex brain, 79–81
comparison of biological processes with,
    70–71
developmental mechanisms in, 79–81
directional patterns in, 299–300
discovery of, 10
rate patterns in, 300
social evolution, 214–17
technological evolution, 214, 215, 298,
    299–300
Culture
  defined, 69
  primate preadaptations for, 78–79
Curtis, H., 108
Cuvier, Georges, 4, 10
Cycads, 161
"Cycles" of extinction, 281–83
Cytosine, 45

D

Dark matter, 310
Darwin, Charles, 32, 236, 237
  decline of ideas of, 7–9
  impact of, 1–2
  natural selection, theory of, 5–7, 9,
      11–14
  renewal and refinement of ideas of,
      9–10
Darwinian Revolution, 1–2
Dawkins, Richard, 12, 48, 67
Debt-for-nature swap, 342
De Chardin, Teilhard, 15
Decomposers, 269
Deforestation, tropical, 323–25
Deism, 12–13
Delayed development, 79–81
Derived traits, 59
Development, genes and, 43–44
Developmental similarities, 55–57, 268
Devonian period, 108
Diatoms, 150
Differentiation, 109
Diminishing resources, social evolution
    and, 368–70
Dinosaurs, 165–79
  bird-hipped, 165, 168, 169

extinction of, 171–79
lizard-hipped, 165–68
misconceptions about, 168–70
Diploid state, 44
Direction
  defined, 226
  of evolution of atmosphere, 228–29
  extinction directions, 280–83
  extinctions and general evolutionary,
      289–93
  statistical nature of directionality, 248,
      250–52
  trends in human evolution, 296–97
    biological evolution, 296–97
    cultural evolution, 299–300
  See also Extinction directions; Origination
      directions; Trends
Directional selection, 36–38, 39, 61, 62
  consistent, 250
Disabilities, relaxed selection toward, 358
Disease
  extinction of dinosaurs and, 172
  introduction into ecosystem, 329
  relaxed selection toward, 358
Divergent boundaries of plates, 112
Divergent trends, 252, 253
Diversity
  biosphere, 258, 261–68
  new, in future, 342–53
Dixon, D., 354
DNA (deoxyribonucleic acid), 26, 45
  code, 54
  Eve hypothesis and, 213
  genetic engineering methods and,
      346–48
  in human and chimpanzee genes, 205
  silent, 47, 51
  structure of, 45, 46
DNA hybridization, 243
DNA-ligase enzymes, 346
Dobzhansky, T., 233, 241
Dodo bird, extinction of, 327, 328
Dolphin deaths, reduction of, 341
Domestication, adoption of, 214
Dominant gene, 49
Doomsday equation, 368
Dott, R., 118
Double fertilization, 181

Drug industry, genetic engineering used in, 348–49
Durrell, L., 321, 339
Duyme, Michel, 367

E

Early-Mesozoic mass extinction, 278
Early-Paleozoic mass extinction, 277–78
Earth
    age of the, 11–12
    cyclical nature of, 229–30
    direction and rates of evolution, 227–30
    future of, 311–18
    history of, 101–22
        atmosphere, 117–19, 121, 122
        geologic time scale, 101–9, 110
        hydrosphere, 114–17, 121
        lithosphere, 109–14, 119, 121
Echinoderms, 139, 145–46
Echolocation, 200–201
Ecological economics, 341
Ecological Imperialism (Crosby), 329
Ecological reason for saving species, 337
Ecological release, 172, 245, 264
Ecological trends, 258, 268–75
Ecology, 268
Ecosystems
    disruption of, 327–30
    first, 130–31
    functions of, 337
    predation in, 269–73
Eddy, F., 296
Ediacara Formation, 136
Egg, hard-shelled, 162
Ehrlich, Paul, 307
Eicher, D., 25, 101, 105, 127, 128, 140, 142, 164, 203
Eldredge, Niles, 285
Electromagnetic force, 22, 23
Electrons, 23
Elements
    of life, six most important, 24
    star death and creation of, 97–98
    See also Chemistry and chemical processes, origin of life through
Embryological development, evolution and, 55–57, 268

Emergent properties, concept of, 30
Encephalization quotient (EQ), 171
Endangered species, 339–40
Endangered Species Act (1973), 340
End-Mesozoic mass extinction, 150, 173–79, 277, 283, 287, 291
Endoskeletons, 194
End-Paleozoic mass extinction, 146, 160, 277
Energy
    cultural evolution and use of, 299, 300
    diminishing resources of, 368–69
    leakages in food pyramid, 270–72
Entropy, law of, 31
Environmental change
    evolutionary rates and, 245
    extinctions and, 283–87, 318–19
        background extinctions, 283–84
        categories of changes causing, 326–32
        mass extinctions, 284–87
    origination caused by, 318–19
Environmental disturbance by genetically engineered organisms, 352
Environments, natural adaptation to human-modified, 343–44
Epicontinental seas, 117
Equilibrium, punctuated, 236–39, 247–48, 249
Equilibrium diversity hypothesis, 262
Eras, 107–9
    See also Cenozoic era; Mesozoic era; Paleozoic era; Precambrian era
Esthetic reasons to save species, 334
Ethical reasons to save species, 334
Ethics
    genetic engineering and, 351, 352
    impact of evolutionary thought on, 14–16
    lifeboat, 378
Eugenics, 360–64
    hard, 362–64
    soft, 361–62
Eugenics Education Society, 360
Eukaryotes, 128, 131–32
European Renaissance, 3
Eve hypothesis, 213
Even-toed ungulates, 184

Evolution
  basic facts and interpretations of, 375–77
  defined, 2
  embryological development and, 55–57,
    268
  rates and directions of, 226–27
  social and personal implications of,
    377–79
  See also Biological evolution; Cultural
    evolution; Human evolution;
    Physical evolution
Evolution: The History of an Idea (Bowler),
  10
Evolutionary humanism, 15
Evolutionary mysticism, 15
Evolutionary Progress (Hull), 15
Evolutionary thought
  birth of modern, 3
  history of, 2–10
  social impact of, 10–18
    in business and politics, 16–18
    in philosophy and ethics, 14–16, 378
    in religion, 10–14, 377–78
Evolution in Action (Huxley), 15
Exoskeleton, 158, 193–94
Expanding universe, 88
Extinction, 230–32, 275–96
  causes of past, 283–89
  classifying, 276–77
  defining, 230, 276–77
  of dinosaurs, 171–79
  direction of evolution and, 289–93
  group extinction patterns, 293–96
  modern, 325–42
    causes of, 325–32
    methods of stopping, 337–42
    reasons to stop, 333–37
  secondary, 327, 328
  See also Background extinctions; Mass
    extinctions
Extinction: Bad Genes Or Bad Luck? (Raup),
  285, 290
Extinction directions, 280–83
Extinction rates
  in fossils, 277–80
  modern, 323–25
Extinction selectivity, 176–78

Extinction (Stanley), 285
Extrapolation, 305–8, 368
  of human brain size, 355–56, 357
Extra-terrestrial impacts, mass extinctions
    caused by, 286, 287. See also
    Meteorite bombardment
Extra-terrestrial theories of cycles of
    extinction, 281–82
Extrinsic factors, 245

F

Fallacy, naturalistic, 15–16, 18
Fauna
  modern, 262–64
  Paleozoic, 262
Feedback, positive, 71
Ferns, 156, 158
  seed, 161
Fertilization, double, 181. See also
    Reproduction
Feynman, Richard P., 375
Field of bullets selection, 290–92
Fish, 141–46
  bony, 143, 159
  jawless, 141, 142
  lobe-finned, 143, 158–59
  observed declines in modern, 325
  origin of, 143–46
  ray-finned, 143
Fitness, defined, 34
Flagship species, 338
Flightless birds, 200
Flowering plants, 172, 180–81, 192, 264
Flying mammals, 200–202
Flying reptiles, 188, 194–97
Food chain, 269
Food pyramid, 270–72, 273, 333
Food uses of biodiversity, 334–37
Food web, 269–70, 274
  habitat disturbance effect on, 327
Forces, four basic, 22–23
Forest-sustainable resources, 337, 341
Fossilization, categories of, 123, 124
Fossil record, 26, 122–26
  completeness (incompleteness) of,
    123–26, 237–38

Fossil record (*cont'd*)
  eukaryotes in, 132
  evidence for evolution in, 59–60
  of flying groups, 189
  human, 206–10
  of multicellular evolution, 134–37
  prokaryotes in, 128–31
Fossils, 4
  body, 122–23
  formation of, 122–23
  living, 243, 246
  measuring extinction rates in, 277–80
  measuring origination rates from, 232–41
  trace, 123, 135
  use of, for stratigraphy, 102–4, 105
Founder effect, 41
Fox, Sidney, 24
Frontal lobes, 74–75
Fuel exhaustion (death) of star, 94–97
Fundamentalism, 12
Futuyma, D., 64, 211, 236

G

Gaia hypothesis, 274–75
Galaxies, speed and direction of motion of, 309
Galaxy epoch, 92
Galton, Francis, 360
Gametes, 43, 44
Gametophytes, 155
Gene banks, 336
Gene flow, 50–51
Genentech, 348
Gene pool, 48–53
  expanding, 50–52
  gradualist evolution vs. punctuated equilibrium of, 247–48
  shuffling the, 48–50, 52
Generation length, evolutionary rate and, 246–47
Genes, 21, 43–53
  adult maintenance and, 44–45
  development and, 43–44
  in evolution, 48–53
  functioning of, 45–48

human mind and, 364–67
  regulatory, 80–81
  See also DNA (deoxyribonucleic acid)
Gene splicing, 346
Gene therapy, 362, 363
Genetic analysis of races, 210
Genetic counseling, 362
Genetic drift, 51, 212–13
Genetic engineering, 335–36, 342, 345–53, 357
  ethical, legal, and environmental problems with, 351–53
  methods of, 346–48
  removal of debilitating genes from gene pool by, 362
  uses of, 348–51
Genetics, population, 52–53
Genome map, 349, 361
Genotypes, 50
Geologic time scale, 101–9, 110
Germ cells, 4–5
Gibbons, 205
Gill arches, 146
Gills, book, 158
Ginkoes, 161
Glaciation, 119, 120, 317, 318
Global warming, 314–15, 316
Goldilocks's paradox, 28
Gorillas, 205
Gould, Stephen Jay, 76, 137
Gradualistic evolution, 236–39, 247–48, 249
Grant, V., 66, 244, 281
Gravitational force, 22, 88
Great American Interchange, 292–93, 294
Great chain of being, 2, 3, 11
Green algae, 153
Greenhouse effect, 117, 313–15, 316, 317
Greenhouse gases, 314–15
Gribbin, John and Mary, 81
Gross, M., 113, 115
Ground upwards theory, 197, 199
Group extinction patterns, 293–96
Group longevity, 293, 295
Group origination rates, 232, 240–41
Guanine, 45
Gymnosperms, 160–62, 264

H

Habitable planet, criteria for, 28–29
Habitat disturbance, 327
Hadron epoch, 91
Half-life, 105
Hands, primate, 202
Haploid state of gametes, 44–45
Hard eugenics, 362–64
Hardin, Garrett, 378
Hard-shelled egg, 162
Hardy-Weinberg equation, 52–53
Helium atoms, 91–92
Heritability of trait, 365
Herrnstein, Richard, 358
Heterochrony, 144–45
Heterozygotes, 50, 52
History of evolutionary thought, 2–10
Hitler, Adolf, 17, 360, 364
Hollow curve, 283, 284, 288
Homo, 206, 208–10
  erectus, 209
  habilis, 208–9
Homologous organs, 54, 253
  embryonic precursors of, 55–57
Homo sapiens, 209
Homozygous, 50
Hoofed mammals, 183–84
Horse, evolution of, 254, 255, 257, 259
Hubble, Edwin, 88
Hubble's law, 88–89
Hubble Telescope, 309
Hull, David, 15
Human evolution
  biological evolution, 70–71, 202–13,
    296–99
    directional patterns, 296–97
    fossil record, 206–10
    future, 355–67
    human races, 210–13
    planned selection in, 360–67
    primate evolution, 202–6
    rate patterns, 297–99, 300
    unplanned selection in, 357–60
  cultural evolution, 213–17
    biological vs. cultural processes, 70–71
    brain and, 71–81
    developmental mechanisms in, 79–81

discovery of, 10
patterns of, 298, 299–300
social evolution, 214–17
technological evolution, 214, 215, 298,
  299–300
future evolution of society, 367–70
patterns of, 296–300
processes of, 69–81
prospects for, 355–70
Human genome project, 361
Humanism, evolutionary, 15
Human overkill, theory of, 188
Humans
  decimation of biosphere by, 319–42
  future evolution without, 353–55
  population growth, 340–41, 367–68
Hunting, extinctions caused by human,
  330–32
Hunting and gathering, 214, 215
Hutton, James, 3
Huxley, Julian, 15
Huxley, T.H., 10, 11, 14
Hybridization, DNA, 243
Hydrogen, formation of star from, 93–94
Hydrogen atoms, 91–92
Hydrosphere
  direction and rates of evolution, 228–29
  evolution of, 114–17, 121
  evolution of life in, 127–52
    marine mammals, 150–52
    mobile invertebrates and marine
      reptiles, 146–50
    multicellular life, evolution toward,
      127–37
    sessile invertebrates and fishes, 139–46
  future of, 317–18

I

I.Q.
  in identical twins, 366–67
  number of offspring and, 359
I.Q. test, 364
Iberall, A., 379
Ice Age, 119, 120, 279
  megafauna, 185–88
Ichthyosaurs, 148, 149, 152
Identical twins studies, 365–67

Impersonal God, 12–13
Inbreeding, genetic bottleneck of, 326
Increasing diversity hypothesis, 262
Indicator species, 338
Industrialization, 313, 320
Inertial body heat, 171
Information storage in brain, 73
Inheritance of acquired traits, 4
Inherited variation, 43–53
Inner planets, 98
Innovation, key, 245
Insectivores, 183
Insects, 158
    advanced, 192–94
    fossil record of, 125–26
    pollination of flowering plants by,
        180–81
    primitive, 189–92
    in tropics, 320, 321, 325
    wings of, 188–89
Instinct, 75
Integration in multicellular organisms, 134
Intelligence
    in dinosaurs, 170, 171
    intermediate heritability of, 366
    number of offspring and, 358–59
    social, 79
    See also Brain; Brain size
Interbreeding, criterion of, 41–43
Interconnections in biosphere, 273–75
Interconnectivity of brain's neurons, 73–74
Interglacials, 186
Internalist theories, 7–9
Intrinsic factors, 245–47
Invertebrates, 153
    defined, 139
    first land, 158
    mobile, 146–48
    observed declines in modern, 325
    sessile, 139–40
Iridium, 173–74

J

Jawless fish, 141, 142
Jaws
    origin of, 145, 146
    single-boned, in mammals, 181

Jefferson, Thomas, 4
Judeo-Christian groups, reaction to
        evolution, 11–12
Judson, S., 369, 370
Jurmain, R., 65, 208

K

Kamikaze genes, 352
Keller, E., 230, 320
Key innovation, 245
Keystone species, 327, 330, 338
Kill curve, 288

L

Labyrinthodonts, 160
Lamarck, Jean Baptiste, 4, 5, 6
Lambert, D., 167, 168, 171
Land, evolution of life on, 152–88
    flowering plants and mammals, 172,
        179–88, 192, 264
    major problems for, 152, 153–54
    seedless plants and amphibians, rise of,
        153–60
    seed plants and reptiles, rise of, 160–79
    See also Lithosphere
Lane, N.G., 141, 147
Laporte, L., 130
Larvae, insect, 192
Latin America, deforestation in, 323, 324
Latitudinal diversity gradient, 322–23
Law(s)
    endangered species protection, 339–40
    genetic engineering and problems of,
        351–52
Leakey, Louis, 208
Learning, 75
Leaves, 156
Lepton epoch, 91
Levin, H., 149, 157, 207
Levinton, J., 282
Lewontin, Richard, 210
Life
    classification of, similarities and, 57–59
    key features of, 24–26
    as open system, 31–32
    origin through chemical processes,
        24–30
    elsewhere in universe, 27–30

experimental and theoretical evidence of, 24–26
fossil evidence, 26
frequency of, 26–27
physical laws and, 30–32
six most important elements of, 24
species as fundamental unit of, 43
See also Biological evolution; Earth; Human evolution
Lifeboat ethics, 378
Limbic system, 77
Lineage trends, 248, 250–58
Linnaeus, Carolus, 57
Lithosphere
evolution of, 109–14, 119, 121
direction and rate of, 227–28
future of, 312
See also Land, evolution of life on
Living fossils, 243, 246
Lizard-hipped dinosaurs, 165–68
Lizards, 165
Lobe-finned fish, 143, 158–59
Lobotomies, 74
Localization of brain, 74
Longevity, group, 293, 295
Lovelock, James, 274
Lungs, 194
Luria, S., 76, 259, 261
Lycopsids, 156–58
Lyell, Charles, 3, 102

M

McAlester, A., 25, 101, 105, 127, 128, 135, 140, 142, 155, 156, 159, 161, 162, 164, 166, 180, 185, 203
McMahon, T., 256
McNamara, K., 274
Macroevolution, 66
Main sequence of star's life, 94, 95
Mallory, B., 318
Mammal-like reptiles, 163, 181
Mammals
brain size in, evolution of, 272, 274
flying, 200–202
land, 181–88
Ice Age megafauna, 185–88
radiation of, 182–85
traits of, 181–82

marine, rise of, 150–52
observed declines in modern, 325
speculative, of the future, 353, 354
Mammary gland, 181
Mantle of earth, 109, 110–14
Marine mammals, rise of, 150–52
Marine reptiles, 148–50
Mars, 28
Marsupials, 182, 183
Mass extinctions, 276–77, 279
as bottleneck, 290–92
difference between background and, 279–80
early-Mesozoic, 278
early-Paleozoic, 277–78
end-Mesozoic, 150, 173–79, 277, 283, 287, 291
end-Paleozoic, 146, 160, 277
environment changes and, 284–87
future prospects for, 319–42
causes of, 325–32
methods of stopping extinctions, 337–42
rates of modern extinction, 323–25
reasons to stop extinctions, 333–37
susceptibility of species to, 332–33
tropical deforestation, 323–25
tropical extinctions, 320–23
middle-Paleozoic, 278
as selection by new set of rules, 290–92
timing of, 281
Matter
dark, 310
rapid depletion of, 369–70
in universe, amount of, 309–10
Matter era, 90, 91–92
Medicine, 337
relaxed selection allowed by, 358
Mediocrity, principle of, 27
Megafauna, 185–88
Megatrends. *See* Biosphere trends
Megolithic Age (middle Stone Age), 214
Meiosis, 45, 48
Memes, 69–70
Mencken, H.L., 1
Mendel, Gregor, 8, 43
Mental diseases and disabilities, relaxed selection toward, 358

Mental traits, measuring, 364–67
Mesozoic era, 107, 108, 117, 119
  advanced insects in, 192–94
  mass extinctions in
    early-Mesozoic, 278
    end-Mesozoic, 150, 173–79, 277, 283, 287, 291
  meaning of term, 139
  mobile invertebrates, rise of, 146–48
  reptiles, rise of, 162–79
    dinosaurs, 165–79
    flying, 188, 194–97
    marine, 148–50
  seed plants, rise of, 160–62
Mesozoic marine revolution, 147–48
Metal Ages, 214
Metamorphosis, 192
Metaphytes, 134–37
Metazoans, 134–37
Meteorite bombardment, 27, 281–82
  craters from, 287–88, 289
  mass extinctions caused by, 173–79, 286, 287
  nuclear winter and, 317
Meteorite hypothesis of dinosaur extinction, 173–79
Methane, 314
Microbes, single-celled, 128–32
  complex, 128, 131–32
  simple, 128–31
Microevolution, 63–66
Microinjection, 348
Microspheres, 24–25
Mid-Atlantic ridge, 114
Middle-Paleozoic mass extinction, 278
Milankovich cycle, 119
Milky Way, 92, 98, 99, 101
Miller, R., 125
Miller, Stanley, 24, 25
Millipedes, 158
Mind, genes and human, 364–67
Minerals, shocked, 174, 179
Miner's Canary, The (Eldredge), 285
Minimum viable population (MVP), 326
Mitosis, 44, 45
Mobile invertebrates, 146–48
Mobility, evolutionary rate and, 245–46
Modern synthesis, 9–10

Mold, 123
Molecular evolution, 243–44
Molecular rates, 244
Molecules, 23–24
Mollusks, 147, 158
Molnar, S., 47, 366, 367
Molts, 193
Momentum, law of conservation of, 92
Mongoloids, 210, 211, 212
Monkeys, 204–5
Monotremes, 182–83
Monroe, S., 5, 6, 54, 58, 104, 120, 124, 129, 145, 173, 200, 253
Montgomery, C., 106
Montreal Accord (1987), 316
Morgan, Lewis Henry, 10
Mosaic evolution, 239, 240
Mosasaurs, 149, 152, 165
Multicellular life, evolution toward, 127–37
  complex single-celled microbes, 128, 131–32
  multicellular organisms, 132–37
  simple single-celled microbes, 128–31
Multiregional model of human evolution, 213
Mutation
  defined, 51
  DNA, 213
  evolutionary rate and tendency to undergo, 247
  expanding gene pool by, 51–52
Mutationist school, 8–9
Mutualism, 273
Mysticism, evolutionary, 15

N

Natural adaptation to human-modified environments, 343–44
Naturalistic fallacy, 15–16, 18
Natural selection, 1, 33, 34
  creation of new opportunities for, 344, 345
  misconceptions about, 67–69
  ongoing, 62–63
  theory of, 5–7, 9
    reaction among religious groups to, 11–14

Nature
  religious view of man apart from, 11
  search for guidance from, 14–16
Nature-nurture debate, 77, 365–67
Nazi Germany, 360
Neanderthals, 209
Nebel, B., 35, 40, 255
Negative artificial selection, 61
Negroids, 211
Nelson, H., 65, 208
Nemesis, 281
Nemesis Affair, The (Raup), 283
Neolithic Age (new Stone Age), 214
Neurons, 73–74
Neutral theory of molecular rates, 244
Neutrons, 23
Neutron stars, 96, 97
Newton, Isaac, 3
New World, New Mind (Ornstein and
     Ehrlich), 307
New World monkeys, 204
New Zealand, introduction of new species
     into, 328–29
Niche, 262
Niche packing, 262, 264, 323
Nonbranching evolution, 38, 39, 234
Nondinosaurs, 163–65
Nonfood uses of biodiversity, 337
Norman, D., 179, 199
North American and South America,
     linkage of, 330
Nova, 96, 98
Nuclear forces, 22, 88
Nuclear fusion, 94
Nuclear weapons, threat of, 316
Nuclear winter, 176, 316–17

O

Oceans
  formation of, 114–17, 121
  primordial soup of, 26, 27
  See also Hydrosphere
Odd-toed ungulates, 184
Offspring care
  among dinosaurs, 168
  primate, 203

Oil, diminishing reserves of, 368–69
Old World monkeys, 204
Olive, P., 133, 191
O'Neill, R., 228
One Percent Advantage, The (Gribbin and
     Gribbin), 81
Ongoing evolution, 60–66
*On the Origin of Species by Means of Natural
     Selection: Or the Preservation of
     Favoured Races in the Struggle for Life*
     (Darwin), 1, 5
Open systems, 31–32
Open universe, 308–10
Opportunistic evolution, 253
Opportunistic species, 343, 344
Organelles, 13, 128, 131
Organization of atoms in living matter,
     30–31
Organs
  analogous, 254
  homologous, 54, 55–57, 253
  vestigial, 54
Original horizontality, law of, 102
Original lateral continuity, law of, 102, 103
Origination, 230–32
Origination directions, 248–75
  biosphere trends, 248–50, 258–75
  lineage trends, 248, 250–58
Origination rates, 232–48
  anatomical, 232–40
  causes of, 245–48
  correlation of group extinction rates
       and, 293–96
  group, 232, 240–41
  measurement from fossils, 232–41
  measurement using living organisms,
       241–44
Ornithopods, 168, 170
Ornstein, Robert, 307
Orthogenetic school, 7–8
Orthoselection, 251
Outer planets, 98
Outgassing, 114, 117, 118
Out of Africa model of human evolution,
     213
Overintelligence problem, 79
Oxygen accumulation in atmosphere, 118
Ozone, 117, 315–16

P

Paleolithic Age (old Stone Age), 214
Paleontologists, 122
Paleontology, 213
Paleospecies problem, 42–43
Paleozoic era, 107, 108, 112, 138–46
  changing land-sea patterns in, 116–17
  climate during, 119
  "explosion" of life in early, 137
  mass extinctions in
    early-Paleozoic, 277–78
    end-Paleozoic, 146, 160, 277
    middle-Paleozoic, 278
  meaning of term, 138–39
  primitive insects, rise of, 189–92
  seedless plants and amphibians, rise of,
    153–60
  sessile invertebrates and fishes, rise of,
    139–46
Paleozoic fauna, 262
Pangea, 112–14, 116, 119, 146, 150, 153,
  179
Pantheism, 13
Parasites, 269, 329
Patents for genetically engineered species,
  351–52
Patterson, J., 125
Periods, 108
Personalized God, 12, 377
Personal philosophy, evolution and, 378
Phenomenon of Man (de Chardin), 15
Phenotype, 50
Phenotypic plasticity, 66
Philosophy, evolution and personal, 14–16,
  378
Photosynthesis, 27, 118, 154
Photosynthetic atmosphere, 118
Phyla, 137
Phyletic evolution and anagenesis, 38, 39,
  234
Physical evolution
  discovery of, 3
  patterns of, 227–30
    directions and rates in earth's
      evolution, 227–30
    sun and universe, 230, 231
  processes of, 22–32

    chemical processes, origin of life
      through, 24–30
    chemistry of life, 23–24
    four basic forces, 22–23
    life vs. physical laws, 30–32
  products of, 88–122
    earth, history of, 101–22
    solar system, history of, 98–101
    stars, history of, 93–98, 311
    universe, origin and history of, 88–92
  prospects for, 308–18
    future of earth, 311–18
    future of sun, 311
    future of universe, 308–11
Phytoplankton, 150
Pistil, 180
Placental mammals, 182, 183
Placoderms, 143
Planets
  inner, 98
  outer, 98
  relative sizes in relation to sun, 100
Planned selection in future human
  evolution, 360–67
Plant-animal relationship as cycle, 130–31
Plants, land
  appearance and diversification of, 264,
    265
  extinction selectivity and, 176–78
  flowering, 172, 180–81, 192, 264
  origin of, 153–55
  seed, 160–62
  seedless, 155–58
Plasmids, 346
Plasticity, phenotypic, 66
Plate tectonics, 111–14, 115, 119, 312
Plato, 360
Pleiotropy, 50
Pleistocene epoch, 185–88
Plesiosaurs, 148–49, 152
Poaching, 339
Politics, impact of evolutionary thought on,
  16–18
Pollen, 160–61
Pollination, 180
Pollution, 327, 333, 344, 345
Polygenetic trait, 50

Polypeptide chain, assembly of, 47
Population, minimum viable (MVP), 326
Population genetics, 52–53
Population growth, human, 340–41, 367–68
Porpoises, 150, 151–52
Positive feedback, 71
Postmating isolating mechanisms, 41
Potato famine, 334
Preadaptation, 68
   to humans, 343–44
   primate, for evolution of culture, 78–79
Precambrian era, 107, 108, 114
   evolution toward multicellular life in, 127–37
Predation
   background extinctions due to, 292–93
   coevolutionary trends in, 272
   in ecosystem, 269–73
Prediction
   extrapolation and, 305–8
   spatial scale and, 306–8
Prehoda, R., 300
Premating isolating mechanisms, 41
Preserves and preserve design, 338–39
Primary producers, 130
Primate preadaptations for culture, 78–79
Primates, 184–85
   evolution of, 202–6
   kinds of, 203–5
   observed declines in modern, 325
Primate traits, 202–3
Primitive insects, 189–92
   advanced insects compared to, 192–93
Primitive traits, 59
Primordial soup, 26, 27
Principles of Geology (Lyell), 102
Processes of evolution. *See under* Biological evolution; Human evolution; Physical evolution
Producers, 269, 271
   primary, 130
Products of evolution. *See under* Biological evolution; Human evolution; Physical evolution
Progress, philosophical beliefs and ethical systems based on, 14–15

Prokaryotes, 128–31
Prosimians, 203–4
Proteins, 24
Protoculture, 78
Protons, 23
Protozoans, 134
Pseudoextinction, 276, 295
Psilopsids, 156
Pteranodon, 196
Pterodactyloids, 195
Pterosaurs, 195–97
Pull of the recent, 126
Punctuated equilibrium, 236–39, 247–48, 249
Punnett square, 49, 50

R

Races, 42, 62–63, 64, 210–13
Rachels, James, 377
Racial senescence, 67–68
Racie, F., 8, 49, 53, 63, 104, 182, 204
Radiation
   adaptive, 245
   cosmic background, 89
Radiation era, 90–91
Radioactive decay, 104–5
Radiometric dating, 105–6
Rain forests, destruction of, 320–23
Randomness of evolution, 66–67
Random walk, 252
Rapoport's rule, 292, 323
Rate
   absolute, 234
   complexity and evolutionary, 245, 246
   correlated traits versus mosaic evolution, 239–40
   defined, 226
   of evolution of atmosphere, 228–29
   patterns of human evolution, 297–99, 300
   punctuated or gradual, 236–39, 247–48, 249
   relative, 234–36
   See also Extinction rates; Origination rates
Raup, David, 126, 279, 280, 283, 285, 286, 288, 289, 290

Ray-finned fish, 143
Reasoning, 75–77
Recapitulation, 57, 268
Recessive gene, 49
Recombinant DNA technology, 346
Recombination, 45
Recrystallization, 123
Red giant stage, 94–95, 96, 311
Red shift, 88
Reflex, 75
Refugia, 285
Regulatory genes, 80–81
Relative brain size, 72, 151
Relative rates of evolution, 234–36
Relative time, 104–6
Relaxed selection, 358–59
Religion, evolution and, 10–14, 377–78
Replacement, 123
    competitive, 264
Reproduction, 26, 34, 44
    asexual, 50
    of flowering plants, 180–81
    in seedless plants, 155
    sexual, 48–50, 132
Reproductive isolating mechanisms, 41–43
Reptiles, 162–79
    archaic, 162–63
    dinosaurs, 165–79
    flying, 188, 194–97
    mammal-like, 163, 181
    marine, 148–50
    observed declines in modern, 325
    origin of birds in, 197–98
Restriction enzymes, 346
Return time, 288
Reverse rarefaction curve, 278–79
Rhamphorynchus, 195, 196
Rhamphoryncoids, 195
Rhinos, killing of, 330, 331
RNA (ribonucleic acid), 45
Robber barons, 17
Robust australopithecine species, 206, 207, 208
Rodents, 185
Romer, A., 144, 151
Roots, 154
Ross, R., 326

S

Sagan, Carl, 29
Sauropods, 166–68, 170
Savagery, 10
Scale dependence of trends, 227, 228
Scientific method, 3
Scorpions, 158
Sea level, changes in, 115–17
    glaciation and, 317, 318
    mass extinctions and, 285, 286, 287
Sea urchins, 147, 148, 249
Secondary extinction, 327, 328
Sedimentary rocks, 122
Seed, 161
    of flowering plant, 180–81
Seed banks, 336
Seed ferns, 161
Seedless plants, 155–58
Seed plants, 160–62
Seismosaurus, 166
Selection, 32–38
    artificial, 33, 34, 35, 61
    basic types of, 33–34
    consistent directional, 250
    directional, 36–38, 39, 61, 62
        consistent, 250
    in future human evolution
        planned, 360–67
        unplanned, 357–60
    by new rules vs. no rules, 290–92
    relaxed, 358–59
    sexual, 33–34
    stabilizing, 38
    urban stress, 359–60
    of variation, evidence for evolution by, 53–66
        fossil record, 59–60
        ongoing evolution, 60–66
        similarities among living organisms, 53–59
    See also Natural selection
Selective breeding, 342, 344–45, 346, 364, 365
Selfish Gene, The (Dawkins), 48
Self-organizing chemical and physical systems, 25
Self-reliance, philosophy of, 16, 378

Self-understanding, evolution and, 378
Senescence, racial, 67–68
Sessile invertebrates, 139–40
SETI (Search for Extraterrestrial
    Intelligence), 30
Sexual reproduction
    among eukaryotes, 132
    variation produced by, 48–50
Sexual selection, 33–34
Sharks, 143
Sharpton, V., 176
Shaw, George Bernard, 360
Shocked minerals, 174, 179
Silent DNA, 47, 51
Similarities among living organisms, 53–59
Simpson, G., 186, 189
Singer, S., 76
Size criteria for habitable planet, 28–29
Slash and burn agriculture, 323, 341
Slugs, 158
Snails, 147, 158
Snakes, 165
Social Darwinism, 16–18
Social evolution, 214–17
Social intelligence, 79
Society, future evolution of, 367–70
Socioeconomic causes of extinction,
    reducing, 340–42
Soft eugenics, 361–62
Solar system, history of, 98–101. *See also*
    Sun
Spatial scale, 306–8
Specialization in multicellular organisms,
    134
Speciation, 38–43
Species
    defining, 41
    endangered, 339–40
    extinctions of modern, 332–42
        methods of stopping, 337–42
        reasons to stop, 333–37
        susceptibility to, 332–33
    flagship, 338
    as fundamental unit of life, 43
    human-created, 342–53
        direct creation of life forms, 344–51
        ethical, legal and environmental
            problems in, 351–53

natural adaptation to human-modified
    environments, 343–44
indicator, 338
introduction into ecosystem of new,
    327–30
keystone, 327, 330, 338
opportunistic, 343, 344
research and description of, 337–38
threatened, 340
umbrella, 338
vulnerable, 338
Species area effect, 284, 285
Sphenopsids, 156–58
Spores, 155
Sporophyte, 155
Spurious correlations, 314
Stabilizing selection, 38
Staked echinoderms, 139
Stamen, 180
Stanciu, G., 36, 328
Stanley, Steven, 126, 131, 242, 285, 295
Stars
    history of, 93–98, 311
    neutron, 96, 97
Stasis, 234, 247
States, 216, 217
Static universe, 11
Statistical nature of directionality, 248,
    250–52
Stegosaurs, 168, 170
Stellar epoch, 91, 92
Stereovision, 202–3
Stokes, W., 99, 112, 135, 187, 196, 260
Stone Age, 214
Stratigraphy, 102–4, 105
Stromatolites, 130
Subduction, 112
Subspecies, 42
Sun, 230
    future of, 311
    size of planets in relation to, 100
    solar output, 314–15
Supernova, 96, 98
Superposition, law of, 102, 103
Survival of the fittest, 16–18, 34
Survival traits during wanton selection, 291
Susceptibility, extinction, 332–33
Swanson, C., 42, 298, 299

Symbiosis theory, 131–32
Symbiotic hypothesis of origin of
    multicellular organisms, 132–33

T

Taxonomy, 57–59
Technological evolution, 214, 215, 298,
    299–300
Teeth
    human tooth size, trends in, 297, 299
    rates of change in, 234, 235
    specialized, in mammals, 181
Tektites, 179
Temperature
    body, in warm-blooded vs. cold-blooded
        animals, 181
    throughout evolution, 119
Temperature criteria for habitable planet,
    28–29
Thecodonts, 165
Theism, 12, 13
Theocracies, 216–17
Thermodynamics, second law of, 31
Theropods, 166, 170
Threatened species, 340
Throneberry, J., 23
Thymine, 45
Time
    geologic time scale, 101–9, 110
    relative versus absolute, 104–6
Titanotheres, 254–57
Toothed whales, 151
Trace fossils, 123, 135
Traits
    correlated, 239, 250
        convergence and, 257–58, 260
        trends in, 254–57
    derived, 59
    heritability of, 365
    inheritance of acquired, 4
    of mammals, 181–82
    mental, measuring, 364–67
    primate, 202–3
    primitive, 59
    to promote survival during "wanton
        selection," 291
    trends in single, 250–52

Transcription, 45–46
Transgenesis, 346
Translation, 46–47
Transportation speeds, exponential
    increase in, 300
Trees downwards theory, 197, 199
Trends, 226, 228
    biosphere, 248–50, 258–75
    coevolutionary, 272
    divergent and convergent, 252–54
    lineage, 248, 250–58
    scale dependence of, 227, 228
    statistical tendencies of, 248, 250–52
    See also Biosphere trends; Direction
Tribes, 215–16
Triceratops, 168
Trilobites, 139, 141, 240, 241
Tropical deforestation, 323–25
Tropical extinctions, 320–23
Tsunamis, 176, 179
Turk, J., 331, 332
Turtles, 150, 163–65
Twins studies, identical, 365–67
Tyrannosaurus rex, 166, 175

U

U.S. Department of Agriculture, 349
U.S. Fish and Wildlife Service, 340
Ultrasaurus, 166
Umbrella species, 338
Uncommitted cortex, 74–75
Ungulates, 183–84
Uniformitarianism, 3, 11, 101
Universe
    amount of matter in, 309–10
    closed, 308–11
    conscious design of, 11, 12
    expanding, 88
    future of, 308–11
    open, 308–10
    origin and history of, 88–92
    patterns of evolution, 230, 231
    size of, 308–9
    static, 11
Urban stress selection, 359–60
Urey, Harold, 24, 25
Ussher, Archbishop, 3, 101, 102

V

Van Andel, T., 229
Van Couvering, J., 237
Variation
  inherited, 43–53
  limits of genetic, 65
  selection by, evidence for evolution by,
    53–66
    fossil record, 59–60
    ongoing evolution, 60–66
    similarities among living organisms,
      53–59
Vascular system, 154
Vectors, 346
Venus, 28
Vertebrates
  comparison of development in major,
    56–57
  fish, rise of, 141–46
  flying, 188, 194–202
  land
    appearance and diversification of, 264,
      265
    first, 158–60
Vestigial organs, 54
Viruses, 346
Vitalists, 30
Volcanoes, 110, 174, 281, 288
Voltaire, 69
Vulnerable species, 338

W

Wallace, A.R., 5
Wanton selection, 290–92
Ward, P., 176
Warm-bloodedness of mammals, 181
Warming, global, 314–15, 316
Water. *See* Hydrosphere
Weapons, evolution of, 298, 299
Weiss, C., 215, 216
Weitz, S., 80
Wells, H.G., 360
Whales, 150–52
White, Leslie, 300
White dwarf, 95, 96, 311
Wicander, R., 5, 6, 54, 58, 104, 120, 124,
    129, 145, 173, 200, 253
Wildfires, global, 176
Wilson, E.O., 337
Wings
  evolution of, 188–89, 191
  of primitive vs. advanced insects, 192,
    193
  See also Air, evolution of life in
Wonderlife Life (Gould), 137

Z

Zygote, 44

57530
8